Haemophilus, *Actinobacillus*, and *Pasteurella*

Haemophilus, *Actinobacillus*, and *Pasteurella*

Edited by

W. Donachie, F. A. Lainson, and
J. C. Hodgson
Moredun Research Institute
Edinburgh, Scotland

Springer Science+Business Media, LLC

Library of Congress Cataloging-in-Publication Data

International Conference on Haemophilus, Actinobacillus, and Pasteurella (3rd : 1994: Edinburgh, Scotland)
Haemophilus, actinobacillus, and pasteurella / edited by W. Donachie, F.A. Lainson, and J.C. Hodgson.
p. cm.
Includes bibliographical references and index.

DOI 10.1007/978-1-4899-0978-7
1. Hemophilus--Congresses. 2. Actinobacillus--Congresses. 3. Pasteurella--Congresses. I. Donachie, W. II. Lainson, F. A. III. Hodgson, J. C. IV. Title.
QR82.B95I57 1994
616'.0145--dc20 95-35588
CIP

Proceedings of the Third International Conference on *Haemophilus*, *Actinobacillus*, and *Pasteurella*, held July 31–August 4, 1994, in Edinburgh, Scotland, United Kingdom

Originally published by Plenum Press, New York in 1995
MyCopy version of the original edition 1995

10 9 8 7 6 5 4 3 2 1

PREFACE

The Third International Conference on *Haemophilus, Actinobacillus*, and *Pasteurella* (HAP94) was held in July and August at the Edinburgh Conference Centre, Heriot-Watt University, Riccarton Campus, Edinburgh, Scotland, UK. Previous conferences in 1981, Copenhagen, and 1989, Guelph, had indicated widespread interest in this group of pathogenic bacteria and the timing of the Edinburgh conference was prompted by the major advances in our knowledge of the HAP group that had occurred in the five years since the Guelph meeting. These organisms are considered as a group because of their close relationship in an evolutionary sense and because of the similarities in the types of diseases that they preduce. The main objectives of the meeting were to review and discuss current knowledge and present experimental findings relating to the fundamental, applied, clinical, and therapeutic aspects of disease research involving the HAP group of organisms.

HAP 94 was attended by 160 delegates from around the world and included many of the foremost researchers studying HAP organisms. The conference was structured around 16 talks from invited speakers who covered key areas of HAP research including taxonomy, mechanisms of pathogenesis, animal disease models, virulence factors, molecular biology, immunolgy, antigen analyses, and experimental and commercial vaccine development. The talks provided a review of the current state of research in each field and allowed each speaker to focus on his or her personal research interests.

We are now pleased, in association with Plenum Press, to publish the conference proceedings as a comprehensive, authoritative, and up-to-date account of many aspects of research into the HAP group of organisms.

One area of major interest to researchers is that of plasmids and potential vectoring systems. As a result of the great interest shown in a workshop at the conference on this topic and the quality of the contributions presented, we have included a detailed report which we hope will be welcomed by researchers in this field.

In addition to invited talks, the full abstracts of offered talks and posters are presented in this book.

W. Donachie
F.A. Lainson
J.C. Hodgson

ACKNOWLEDGMENTS

We thank the main sponsors of the meeting, the Moredun Foundation, Lothian and Edinburgh Enterprise Ltd., and Hoechst Animal Health Ltd., who provided the necessary initial funding and administrative assistance essential to the setting up of the conference. The financial help received from the United States Department of Agriculture, SmithKline Beecham, and Solvay Duphar is gratefully acknowledged.

The cooperation of the many Moredun Research Institute staff members who were involved in the conference and the advice and encouragement of the Director, Professor Ian Aitken, was crucial to the success of the conference and is gratefully acknowledged. We thank particularly the members of the Pasteurella Project Group and Mrs. Sandra McGill for their help in organizing the conference and Mrs. Christine Curran for her advice and patience in typing and formatting the proceedings for publication.

Lastly, we would like to record our appreciation of all the time and effort the contributors have made to this book.

CONTENTS

INVITED PAPERS

WORKSHOP SUMMARY

ABSTRACTS

TAXONOMY OF THE FAMILY *PASTEURELLACEAE* POHL 1981

M. Bisgaard

Department of Veterinary Microbiology
The Royal Veterinary and Agricultural University
DK-1870 Frederiksberg C
Copenhagen
Denmark

INTRODUCTION

The family *Pasteurellaceae* Pohl 1981 was conceived to accomodate a large group of Gram-negative chemoorganotrophic, facultatively anaerobic, and fermentative bacteria including the genera *Pasteurella* (Trevisan, 1987), *Actinobacillus* (Brumpt, 1910), *Haemophilus* (Winslow *et al.*, 1917), and several other groups of organisms that exhibit complex phenotypic and genotypic relationships with the aforementioned genera (Pohl, 1979; Pohl, 1981). Most members of the family may cause disease in mammals including humans, birds, and/or reptiles (Bisgaard, 1993; Frederiksen, 1993).

Several methods have been used to examine the taxonomy of *Pasteurellaceae*. Evidence supporting that these organisms constitute a family came from phenetic numerical analysis (Sneath and Johnson, 1973; Broom and Sneath, 1981). Subsequent studies of strains of *Actinobacillus*, *Pasteurella* and *Yersinia*, with some allied bacteria, showed 23 reasonable distinct groups (Sneath and Stevens, 1985), and confirmed that *Actinobacillus* and *Pasteurella* are hard to differentiate. The existence of both V-factor dependent and independent strains within the same species/taxon of all 3 genera has been reported for *[H.] aphrophilus* (Pohl, 1979), *A. pleuropneumoniae* (Pohl *et al.*, 1983), *P. avium* (Mutters *et al.*, 1985a), *P. multocida ssp. multocida* (Krause *et al.*, 1987), *H. parainfluenzae* (Gromkova and Koornhof, 1990), a nonclassified haemolytic *Actinobacillus*-like organism (Eckart *et al.*, 1991), Kilian's taxon B/Bisgaard's taxon 22 (Ryll *et al.*, 1991), and *[H.] paragallinarum* (Mouahid *et al.*, 1992). Studies on phenotypic traits of these organisms thus indicate an overlapping interrelationship between genera, and attemps to create genera and higher taxa based upon phenotypic characterization have resulted in highly artificial groupings as discussed by De Ley *et al.* (1990).

THE FAMILY *PASTEURELLACEAE* POHL 1981

The genetic relationships of these organisms and their relationship with other bacteria have been elucidated through DNA:DNA hybridization (Pohl, 1979; Pohl, 1981; Mannheim, 1983; Mutters *et al.*, 1989), rRNA-DNA hybridization (De Ley *et al.*, 1990), and 16S rRNA sequencing (Dewhirst *et al.*, 1992; Dewhirst *et al.*, 1993).

rRNA-DNA hybridizations showed beyond doubt that the family *Pasteurellaceae* forms one cluster which is linked with both branches of the genus *Alteromonas*, the family *Enterobacteriaceae*, the family *Vibrionaceae*, and the family *Aeromonadaceae*. All of these clusters make up superfamily I *sensu* De Ley which corresponds to part of the gamma subclass of the *Proteobacteria*. The results also showed that the family *Pasteurellaceae* is as heterogeneous as the *Enterobacteriaceae* and *Vibrionaceae* and that, in addition to the three accepted genera, there is room for at least four other new genera around each of the following rRNA branches: "*Histophilus ovis*", *[H.] ducreyi*, *[H.] aprhophilus*, and *[A.] actinomycetemcomitans*. Some of these genera were quite remote from the three authentic genera in the family.

Virtually complete 16S rRNA sequencing of representative *Pasteurellaceae* and beta and gamma *Proteobacteria* have confirmed the location of the family in the gamma division of the *Proteobacteria* (Dewhirst *et al.*, 1992). Subsequent investigations have shown the existence of seven major clusters, four of which contained two or more subclusters (Dewhirst *et al.*, 1993). The seven rRNA branches of De Ley et al. (*P. multocida* NCTC 10322^T, *A. lignieresii* NCTC 4189^T, *H. influenzae* NCTC 8143^T, *[H.] aprhophilus* NCTC 5906^T, *[A.] actinomycetemcomitans* NCTC 9710^T, *H. ducreyi* CIP 542^T and "*Histophilus ovis*" HIM 896-7) were recognized as subclusters 3B, 4A, 1C, 1B, 1A, 4B and 2B, respectively, by Dewhirst *et al.* (1993).

The general complexity of the branching within *Pasteurellaceae* indicates that it will be difficult to divide the family into genera that are phenotypically coherent.

Cellular lipids and carbohydrates of the *Pasteurellaceae* have been reported by Mutters *et al.* (1993). The cellular fatty acid patterns proved to be uniform with minor variations, but the separation from *Neisseriaceae* and *Moraxella* was possible. Also the distribution of phospholipids was uniform within the family. The lipoquinone contents were useful for the descrimination of groups within the family, not necessarily reflecting the degree of genomic relatedness. Finally, analysis of cellular carbohydrates showed a common sugar pattern with all members of the family in addition to characteristic profiles discriminating groups, often at the species level.

Polyamine patterns have been studied by Bunka *et al.* (1994), who detected several different polyamine patterns among strains phylogenetically allocated the family as well as not yet described species classified with the family, indicating that analysis of polyamine patterns might represent useful markers for classification within the family.

As a consequence of the above mentioned studies several species previously classified with the *Pasteurellaceae* have been eliminated from the family i.e. *[H.] equigenitalis*, *[H.] piscium*, *[H.] vaginalis*, *[P.] anatipestifer*, *[P.] piscicida*, *Pasteurella*-like Bovine-lymphangitidis group and *Pasteurella*-like CDC group EF-4 (De Ley *et al.*, 1990; Mutters *et al.*, 1989).

GENUS *PASTEURELLA* TREVISAN 1887

DNA:DNA hybridizations have shown that this genus consists of at least 11 species some of which have not yet been named (Mutters *et al.*, 1989; Mutters *et al.*, 1985b). The *Pasteurella sensu stricto* includes *P. multocida* with its 3 subspecies (*multocida*, *septica*, and *gallicida*), *P. dagmatis*, *P. gallinarum*, *P. canis*, *P. stomatis*, *P. avium*, *P. volantium*, *P. anatis*, *P. langaa* and *Pasteurella* species A and B. Species and taxa excluded from the genus include *[P.] aerogenes*, *[P.] haemolytica* biovars A and T, *P. multocida* biovar 1, *[P.] pneumotropica* biovars Heyl and Jawetz, *[P.] testudinis*, *[P.] ureae*, and the SP-group.

According to De Ley *et al.* (1990) strains belonging to a well defined genus are all on the same rRNA branch. However, the *P. multocida* NCTC 10322^T rRNA branch only contained 3 out of 5 accepted species investigated of genus *Pasteurella sensu stricto* (De

Ley *et al.*, 1990). *P. multocida* ssp. *gallicida* SSI P426^T and *P.* sp. B SSI P683^T were both located outside the *Pasteurella* branch, underlining the heterogeneity in this genus. The phylogenetic studies of Dewhirst *et al.* (1992) clearly separated the genus *Pasteurella sensu stricto* into two subclusters. Cluster 3A included *P. gallinarum*, *P. volantium*, *P. avium*, *P.* sp. A, *P. anatis*, and *P. langaa* in addtion to *[H.] paragallinarum* and *[H.]* taxon C. Cluster 3B contained *P. multocida* ssp. *multocida*, *P. dagmatis*, *P. stomatis*, *P. canis* and *P. sp. B*. Members of cluster 3B are indole positive or indole variable and have mammalian hosts, while members of cluster 3A are indole negative and have avian hosts. Extended studies including 16 new species/taxa further separated clusters 3A and 3B (Dewhirst *et al.*, 1993). A new subcluster 3D included *P. langaa*, Bisgaard's taxon 2 and taxon 3, *P. anatis* and *[A.] salpingitidis*.

According to Dewhirst *et al.* (1993) Bisgaard's taxon 13 strain CCUG 16497 (K117) clusters with *P. multocida* ssp. *multocida* NCTC 10322^T. DNA:DNA hybridizations have previously classified this strain with *P. avium* biovar 2 (Mutters *et al.*, 1985a; Madsen *et al.*, 1985). Considering that DNA:DNA hybridizations are presently recommended for outlining of species among strains showing more than 98% sequence similarity of their 16S or 23S rRNA this finding should be reinvestigated.

Six species have been described as belonging to *Pasteurella sensu stricto* since the genus was redefined, i.e. *[P.] granulomatis* (Ribeiro *et al.*, 1989), *[P.] caballi* (Schlater *et al.*, 1989), *[P.] bettii*, *[P.] lymphangitidis*, *[P.] mairi* and *[P.] trehalosi* (Sneath and Stevens, 1990). With the exception of *[P.] granulomatis* these species have been validly published. However, *[P.] bettyae*, *[P.] lymphangitidis*, *[P.] mairi*, and *[P.] trehalosi* do not belong to *Pasteurella sensu stricto* as previously pointed out. Whether *P. caballi* represents a true member of the genus could be questioned. The type strain of *P. caballi* showed only 25-29 RBR with *Pasteurella sensu stricto* as outlined by Dewhirst *et al.* (1993). Fifty-three per cent binding was observed between the type strain and *P. langaa* which has been excluded from the other avian pasteurellas in subcluster 3A (Dewhirst *et al.*, 1992) and transferred to a new subcluster 3D (Dewhirst *et al.*, 1993), which in addition to *P. langaa* contains *P. anatis*, Bisgaard's taxon 2 and taxon 3, and *[A.] salpingitidis*.

In general, concordance was observed between rRNA:DNA hybridizations and 16S rRNA sequencing. Both methods indicate that the genus *Pasteurella sensu stricto* is heterogenous. According to Dewhirst *et al.* (1993) the named *Pasteurella sensu stricto* species fell into three of four disconnected subclusters. Both investigations, however, have the weakness that only one strain of a species/taxon has been included. Several strains covering the heterogeneity of each species/taxon needs to be investigated before safe taxonomic conclusions can be drawn.

GENUS *ACTINOBACILLUS* BRUMPT 1910

The relative closeness of the genera *Pasteurella* and *Actinobacillus* has previously been mentioned. Based upon DNA homology studies the genus *Actinobacillus* should be limited to the following species: *A. lignieresii*, *A. equuli*, *A. capsulatus*, *A. suis*, *A. ureae*, *A. hominis*, *A. pleuropneumoniae*, and Bisgaard's taxa 5, 9 and 11 (Escande *et al.*, 1984; Mutters *et al.*, 1989). True actinobacilli were included in cluster 4A of Dewhirst *et al.* (1992) which in addition to *A. lignieresii*, *A. pleuropneumoniae*, Bisgaard's taxon 11, *A. ureae*, *A. equuli*, and *A. suis* also included the type strain of *[H.] parahaemolyticus*. Additional investigations added *A. hominis* (Dewhirst *et al.*, 1993). Cluster 4B organisms which branched more deeply without actually forming a separate cluster included *[H.] paraphrohaemolyticus* NCTC 10670^T, *[H.] ducreyi* CPI 542^T, *[P.] haemolytica* NCTC 9380^T, *[H.] parainfluenzae* strains ATCC 33392^T and ATCC 29242, and *[H.]* minor group strain 202. Sequencing of additional strains added Bisgaard's taxa 8, 9 and 5 and *[P.]*

bettyae to cluster 4 (Dewhirst *et al.*, 1993). Bisgaard's taxa 8 and 9 were closely related to the *Actinobacillus sensu stricto*.

[H.] ducreyi CIP 542^T, *[P.] bettyae* CCUG 2042^T, *[P.]haemolytica* NCTC 9380^T and Bisgaard's taxon 5 formed a subcluster within cluster 4 confirming previous results obtained by rRNA:DNA hybridizations (De Ley *et al.*, 1990).

A. capsulatus did not cluster with the true actinobacilli (Dewhirst *et al.*, 1992; Dewhirst *et al.*, 1993) and Dewhirst *et al.* (1992) discussed whether NCTC 11408^T and CCUG 12396^T, used in their studies, were the same as SSI strain P243^T and related to strains P244, P564, P572 and CIP 704.80 as shown by Escande *et al.* (1984). *[A.] actinomycetemcomitans*, *[A.] salpingitidis*, *[A.] muris* and *[A.] seminis* should be excluded from *Actinobacillus sensu stricto* based upon rRNA:DNA (De Ley *et al.* (1990) and sequencing results (Dewhirst *et al.*, 1992; Dewhirst *et al.*, 1993). *[A.] rossii* was excluded from the genus by Escande *et al.* (1984).

GENUS *HAEMOPHILUS* WINSLOW *ET AL.*, 1917, 561^{AL}

The genus *Haemophilus* also appears to be very heterogenous. In Bergeys Manual of Systematic Bacteriology (Kilian and Biberstein, 1984) the genus contains 16 species. Based upon DNA:DNA hybridizations the genus has been restricted to *H. influenzae* ssp. *influenzae* (Kilian's biotypes II, III and VII), *H. influenzae* ssp. *meningitidis* (Kilian's biotypes I, IV, V and VI), *H. aegyptius*, *H. haemolyticus*, *H. intermedius* ssp. *intermedius* (*H. parainfluenzae* biovar IV), *H. intermedius* ssp. *gazogenes* (*H. influenzae* biotype VIII) and the *H. parainfluenzae* complex (Kilian's biotypes I, II, III and IV in addition to *H. taxon* A) which contains several species (Mutters *et al.*, 1989). A considerable lower genomic relationship was observed between members of the redefined genus and other species of the former genus.

Three *Haemophilus* rRNA branches were found among selected strains by De Ley *et al.* (1990). Neither the type strain of *H. parainfluenzae* nor unlocalized *Haemophilus* strains were, however, located on these branches.

16S rRNA sequencing (Dewhirst *et al.*, 1992; Dewhirst *et al.*, 1993) confirmed the existence of the rRNA branches outlined by De Ley *et al.* (1990). *H. influenzae*, *H. aegyptius* and *H. haemolyticus* made up cluster 1C which represents *Haemophilus sensu stricto*. Strains classified as *H. intermedius* were not investigated by Dewhirst *et al.* (1992 and 1993). The type strain of *H. parainfluenzae* was only related to an other strain of *H. parainfluenzae*. These strains branched alone in a deep position. This observation is unexpected and difficult to explain considering the DNA:DNA results.

The *H. aphrophilus/paraphrophilus* and *H. ducreyi* rRNA branches were located in clusters 1B and 4B of Dewhirst *et al.* (1992 and 1993) respectively, underlining the heterogeneity of the genus *Haemophilus* as defined in the latest edition of Bergeys Manual of Determinative Bacteriology (Kilian and Biberstein, 1984).

NEW CANDIDATES FOR GENERA WITHIN THE FAMILY *PASTEURELLACEAE*

More than 80 species or taxa from human, mammalian, avian and reptilian sources were listed and discussed in a review by Mutters *et al.* (1989) who also designated reference strains for each of the described, but not yet formally recognized taxa. Updates have been published by Bisgaard (1993) and Frederiksen (1993).

Subsequent publications have supported the view that *[H.] aphrophilus* and *[H.] paraphrophilus* should be regarded as one species separated from *[A.]*

actinomycetemcomitans (Sedlacek *et al.*, 1993), that *[P.] aerogenes* is heterogenous (Lester *et al.*, 1993), and that *[H.]* sp. taxon minor group can be divided into at least two major genetic groups (Hampson *et al.*, 1993). In addition, further characterization has been published on *[P.] haemolytica*-like bacteria from swine (Oberst *et al.*, 1993), and a new porcine V-factor dependent organism provisionally designated *Haemophilus* sp. sorbitol$^+$ has been described by Guttierrez *et al.* (1993). During the present conference DNA binding studies have been presented showing that *[H.]* sp. taxon minor group (three homology groups) and the taxa previously designated C, D/E (two homology groups), and F (two homology groups) form distinct groups only distantly related to each other and to other selected members of the family (K. Møller, M. Kilian, F. Grimont and P. A. D. Grimont). Additional taxa exist in the author's collection.

Although not included in the three authentic genera most of these species/taxa have now been shown to be members of the family *Pasteurellaceae* (De Ley *et al.*, 1990; Dewhirst *et al.*, 1993).

The rRNA branches *[H.] aphrophilus* NCTC 5906^T, *[A.] actinomycetemcomitans* NCTC 9710^T, *[H.] ducreyi* CIP 542^T, and "*Histophilus ovis*" HIM 896-7 represent candidates for new genera within the family (De Ley *et al.*, 1990). The 16S rRNA sequences for serovars a, b and c of *[A.] actinomycetemcomitans* differed from one another by more than many species, and Dewhirst *et al.* (1992). discussed the possible change of serovars to subspecies. Additional strains of the three serovars should, however, be sequenced before such conclusions can be drawn. *[H.] segnis* clustered with the *[H.] aphrophilus* branch. Genera represented by *[H.] aprophilus* and *[A.] actinomycetemcomitans* were related to the genus *Haemophilus sensu stricto* while the type strain of *[H.] parainfluenzae* clustered separately indicating the existence of a new genus unrelated to *Haemophilus sensu stricto*. "*Histophilus ovis*" (*[H.] somnus*) also represents a candidate for a new genus. Bisgaard's taxon 6 clusters with *[H.] somnus* according to Dewhirst *et al.* (1993). This is surprising as it has recently been shown that taxon 6 is related to the *[P.] aerogenes* complex (Lester *et al.*, 1993). Another genus-like structure is formed by *[A.] seminis*, *[P.] mairi* and *[P.] aerogenes* (Dewhirst *et al.*, 1993). *[P.] pneumotropica* biovar Jawetz NCTC 8141^T which showed 42% DNA binding with the type strain of *[A.]* muris (Ryll *et al*, 1991) formed a separate cluster according to Dewhirst *et al.* (1993) and candidates for a new genus. Strains classified with Kilian's taxon B/Bisgaard's taxon 22 might also belong to this genus just like other related isolates from rodents.

[A.] capsulatus and *[H.] paracuniculus* might also represent a new genus (Dewhirst *et al.*, 1993). Bisgaard's taxon 2 and taxon 3 might also represent a new genus-like structure as already suggested by Piechulla *et al.* (1985). The same can be maintained for the avian *[P.] haemolytica*-complex (Piechulla *et al.*, 1985). Both groups cluster with *P. anatis* and *P. langaa* according to Dewhirst *et al.* (1993). Both genus-like clusters, however, contain several species, and additional strains covering the heterogeneity of each taxon should be investigated before final conclusions can be made.

Taxon 14 of Bisgaard which was shown to belong to the family by De Ley *et al.* (1990) clusters with *[P.] testudinis* (Dewhirst *et al.*, 1993). This cluster branches substantially deeper than any other clusters of the *Pasteurellaceae*. These species which lack base signatures found in all other species of the family finally also represent a new genus-like structure.

The phylogenetic trees presented so far have been modified as new sequence data have been added. 16S rRNA sequences for 16 additional strains have been presented at this conference (F. E. Dewhirst, B. C. Paster, G. J. Fraser, and I. Olsen), and according to the new tree it would seem that the family needs to be divided into over 20 genera, or reduced to a single genus. The "final" tree will depend on sequence data from classified, but not yet sequenced species/taxa.

At its latest meeting in Prague in 1994 the ICSM Subcommittee on *Pasteurellaceae* and Related Organisms decided to set up a subgroup for outlining new genera within the family, underlining the significance of a polyphasic approach. Researchers within this area were invited to contact the Subcommittee for co-operation.

In conclusion, the family *Pasteurellaceae* is presently as heterogenous as the family *Enterobacteriaceae*, and due to a high degree of host specificity within the HAP group the existence of many more species can be foreseen, as more animal species are examined.

REFERENCES

Bisgaard, M., 1993, Ecology and significance of *Pasteurellaceae* in animals. *Zbl. Bakt.* 279:7-26.

Broom, A.K. and Sneath, P.H.A., 1981, Numerical taxonomy of *Haemophilus. J. Gen. Microbiol.* 126:123-149.

Bunka, S., Petzold, K., Lubitz, W. and Busse, H.-J., 1994, Distribution of polyamine patterns within members of the family *Pasteurellaceae. In:* Abstracts of the 7th International Congress of Bacteriology and Applied Microbiology Division, Prague, Czech Republic, July 3rd-8th, 1994, pp. 226, *Czechoslovak Society for Microbiology.*

De Ley, J., Mannheim, W., Mutters, R., Piechulla, K., Tytgat, R., Segers, P., Bisgaard, M., Frederiksen, W., Hinz, K.-H. and Vanhoucke, M., 1990, Inter- and intrafamilial similarities of rRNA cistrons of the *Pasteurellaceae. Int. J. Syst. Bact.* 40:126-137.

Dewhirst, F.E., Paster, B.J., Olsen, I. and Fraser, G.J., 1992, Phylogeny of 54 representative strains of species in the family *Pasteurellaceae* as determined by comparison of 16S rRNA sequences. *J. Bact.* 174:2002-2013.

Dewhirst, F.E., Paster, B.J., Olsen, I. and Fraser, G.J., 1993, Phylogeny of the *Pasteurellaceae* determined by comparison of 16S ribosomal ribonucleic acid sequences. *Zbl. Bakt.* 279:35-44.

Eckart, F., Stenzel, A., Mutters, R., Frederiksen, W. and Mannheim, W., 1991, Some unusual members of the family *Pasteurellaceae* isolated from human sources - phenotypic features and genomic relationships. *Int. J. Med. Microbiol.* 275:143-155.

Escande, F., Grimont, F., Grimont, P.A.D. and Bercovier, H., 1984, Deoxyribonucleic acid relatedness among strains of *Actinobacillus* spp. and *Pasteurella ureae. Int. J. Syst. Bact.* 34:309-315.

Frederiksen, W., 1993, Ecology and significance of *Pasteurellaceae* in man - an update. *Zbl. Bakt.* 279:27-34.

Gromkova, R. and Koornhof, H.J., 1990, Naturally occurring NAD-independent *Haemophilus parainfluenzae. J. Gen. Microbiol.* 136:1031-1035.

Guttierrez, C.B., Tascon, R.I., Rodriguez Barbosa, J.I., Gonzalez, O.R., Vazquez, J.A. and Rodriguez Ferri, E.F., 1993, Characterization of V factor-dependent organisms of the family *Pasteurellaceae* isolated from porcine pneumonic lungs in Spain. *Comp. Immun. Microbiol. infect. Dis* 16:123-130.

Hampson, D.J., Blackall, P.J. Woodward, J.M. and Lymbery, A.J., 1993, Genetic analysis of *Actinobacillus pleuropneumoniae*, and comparison with *Haemophilus* spp. taxon "minor group" and taxon C. *Zbl. Bakt.* 279:83-91.

Kilian, M. and Biberstein, E.L. *Haemophilus* Winslow, Broadhurst, Buchanan, Krumwiede, Rogers and Smith 1917, 1984, 561[AL]. *In:* N. R. Krieg and J. G. Holt, eds., Bergeys Manual of Systematic Bacteriology, vol. 1, pp. 558-569. The Williams & Wilkins Co., Baltimore.

Krause, T., Bertschinger, H.U., Corboz, L. and Mutters, L.R., 1987, V-factor dependent strains of *Pasteurella multocida* subsp. *multocida. Zbl. Bakt.* A266:255-260.

Lester, A., Gerner-Smidt, P., Gahrn-Hansen, B., Søgaard, P., Schmidt, J. and Frederiksen, W., 1993, Phenotypical characters and ribotyping of *Pasteurella aerogenes* from different sources. *Zbl. Bakt.* 279:75-82.

Madsen, E.B., Bisgaard, M., Mutters, R. and Pedersen, K.B., 1985, Characterization of *Pasteurella* species isolated from lungs of calves with pneumonia. *Can. J. Comp. Med.* 49:63-67.

Mannheim, W., 1983, Taxonomy of the family *Pasteurellaceae* Pohl 1981 as revealed by DNA:DNA hybridization. *INSERM* 114:211-226.

Mouahid, M., Bisgaard, M., Morley, A.J., Mutters, R. and Mannheim W., 1992, Occurrence of V-factor (NAD) independent strains of *Haemophilus paragallinarum. Vet. Microbiol.* 31:363-368.

Mutters, R., Piechulla, K., Hinz, K.-H. and Mannheim, W., 1985a, *Pasteurella avium* (Hinz and Kunjara 1977) comb. nov. and *Pasteurella volantium* sp. nov. *Int. J. Syst. Bact.* 35:5-9.

Mutters, R., Ihm, P., Pohl, S., Frederiksen, W. and Mannheim, W., 1985b, Reclassification of the genus *Pasteurella* Trevisan 1887 on the basis of deoxyribonucleic acid homology, with proposals for the new species *Pasteurella dagmatis*, *Pasteurella canis*, *Pasteurella stomatis*, *Pasteurella anatis*, and *Pasteurella langaa*. *Int. J. Syst. Bact.* 35:309-322.

Mutters, R., Mannheim, W. and Bisgaard, M., 1989, *In:* Taxonomy of the group. *Pasteurella* and pasteurelloses, C. Adlam and J. M. Rutter, eds., pp. 3-34. Academic Press, London.

Mutters, R., Mouahid, M., Engelhard, E. and Mannheim, W., 1993, Characterization of the family *Pasteurellaceae* on the basis of cellular lipids and carbohydrates. *Zbl. Bakt.* 279:104-113.

Oberst, R.D., Chengappa, M.M., Arndt, T., Staats, J., Hays, M.P., Reed, A. and Chang, Y.F., 1993, Further characterization of *Pasteurella haemolytica*-like bacteria isolated from swine enteritis. *Vet. Microbiol.* 34:287-302.

Piechulla, K., Bisgaard, M., Gerlach, H. and Mannheim, W., 1985, Taxonomy of some recently described avian *Pasteurella/Actinobacillus*-like organisms as indicated by deoxyribonucleic acid relatedness. *Avian Path.* 14:281-311.

Pohl, S., 1979, Reklassifizierung der Gattungen *Actinobacillus* Brumpt 1910, *Haemophilus* Winslow et al. 1917 und *Pasteurella* Trevisan 1887 anhand phänotypischer und molekularer Daten, insbesondere der DNS-Verwandtschaften bei DNS:DNS - Hybridisierung in vitro und Vorschlag einer neuen Familie, *Pasteurellaceae*. Thesis, Fachbereich Biologie der Philipps-Universität Marburg/Lahn.

Pohl, S., 1981, DNA relatedness among members of *Haemophilus*, *Pasteurella* and *Actinobacillus*. *In: Haemophilus, Pasteurella* and *Actinobacillus*, M. Kilian, W. Frederiksen, and E. L. Biberstein, eds., pp. 245-253. Academic Press, London.

Pohl, S., Bertschinger, H.U., Frederiksen, W. and Mannheim, W., 1983, Transfer of *Haemophilus pleuropneumoniae* and the *Pasteurella haemolytica*-like organism causing porcine necrotic pleuropneumonia to the genus *Actinobacillus* (*Actinobacillus pleuropneumoniae* comb. nov.) on the basis of phenotypic and deoxyribonucleic acid relatedness. *Int. J. Syst. Bact.* 33:510-514 .

Ribeiro, G.A., Carter, G.R., Frederiksen, W. and Riet-Correa, F., 1989, *Pasteurella haemolytica*-like bacterium from progressive granuloma of cattle in Brazil. *J. Clin. Microbiol.* 27:1401-1402.

Ryll, M., Mutters, R. and Mannheim, W., 1991, Untersuchungen zur genetischen Klassifikation des *Pasteurella - pneumotropica*-Komplexes. *Berl. Münch. Tierärztl. Wschr.* 104:243-245.

Sedlácek, I., Gerner-Smidt, P., Schmidt, J. and Frederiksen, W., 1993, Genetic relationship of strains of *Haemophilus aphrophilus, H. paraphrophilus*, and *Actinobacillus actinomycetemcomitans* studied by ribotyping. *Zbl. Bakt.* 279:51-59.

Schlater, L.K., Brenner, D.J., Steigerwalt, A.G., Wayne Moss, C., Lambert, M.A. and Packer, R.A., 1989, *Pasteurella caballi*, a new species from equine clinical specimens. *J. Clin. Microbiol.* 27:2169-2174.

Sneath, P.H.A. and Johnson, R., 1973, Numerical taxonomy of *Haemophilus* and related bacteria. *Int. J. Syst. Bact.* 23:405-418.

Sneath, P.H.A. and Stevens, M., 1985, A numerical taxonomic study of *Actinobacillus, Pasteurella* and *Yersinia*. *J. Gen. Microbiol.* 131:2711-2738.

Sneath, P.H.A. and Stevens, M., 1990, *Actinobacillus rossii* sp. nov., *Actinobacillus seminis* sp. nov., nom. rev., *Pasteurella bettii* sp. nov., *Pasteurella lymphangitidis* sp. nov., *Pasteurella mairi* sp. nov., and *Pasteurella trehalosi* sp. nov. *Int. J. Syst. Bact.* 40:148-153.

HAEMORRHAGIC SEPTICAEMIA (*PASTEURELLA MULTOCIDA* SEROTYPE B:2 AND E:2 INFECTION) IN CATTLE AND BUFFALOES

M.C.L. de Alwis

Veterinary Research Institute
PO Box 28, Peradeniya
Sri Lanka

INTRODUCTION

The genus Pasteurella consists of a large number of species with a worldwide distribution and causing a wide range of diseases in animals and man. Diseases of economic importance to agricultural animals is associated with two species *Pasteurella multocida* and *Pasteurella haemolytica*.

The earliest records of disease in deer, cattle and swine caused by Pasteurella dates back to Bollinger (1878). The disease called 'Barbone' described in Italy by Oreste and Armani (1887) is believed to have been haemorrhagic septicaemia (HS).

Some common diseases caused by Pasteurella sp. in agricultural animals is shown in table 1.

Table 1. Some common diseases caused by Pasteurella species in production animals.

Species	Disease
Cattle	Haemorrhagic septicaemia
	Bovine pneumonic pasteurellosis
Buffalo	Haemorrhagic septicaemia
Sheep and goats	Pneumonic pasteurellosis
	Septicaemic pasteurellosis
Pigs	Atrophic rhinitis
	Pneumonia
	Septicaemia (rare)
Poultry/turkeys	Fowl cholera
Rabbits	Snuffles

Pasteurella have also been associated with disease in a variety of other host species. Records of these diseases are summarised in table 2.

Haemophilus, Actinobacillus, and Pasteurella
Edited by W. Donachie *et al.*, Plenum Press, New York, 1995

Table 2. Pasteurella infections in other species of animals.

Animal	Country	Serotypes	References
Bison	USA	B	Carter 1957,59
Yak	China	B?	Carter 1957,59
Deer	England, Australia	B, A	Jones and Hussaini 1982; Carrigan etal 1991
Elephants	Sri Lanka	B	De Alwis and Thambithurai 1965; Wickramasuriya and Kendaragama 1982
Camels	Sudan		
Horses	India, Egypt, USA	B, D	Pavri and Apte 1967; Carter 1957,59
Snow leopard	India	F	Chaudhuri etal 1992
Cat	USA,	France	Carter 1957,59
Mink	USA	A, D	Carter 1957,59
Monkey	USA	A	Carter 1957,59

HAEMORRHAGIC SEPTICAEMIA

The recognition of haemorrhagic septicaemia (HS) as a specific disease entity became apparent with the advent of the methods of serotyping of *Pasteurella multocida* in the 1950's. Haemorrhagic septicaemia was recognised as an acute septicaemic disease principally affecting cattle and buffaloes. It causes significant economic losses in Asia and Africa, and is associated with two specific serotypes. The disease can be reproduced in susceptible animals using pure cultures alone. If treated in time, chemotherapy using appropriate antibacterial drugs is effective, and vaccines using the specific serotypes can prevent the disease. HS is therefore recognised as a primary pasteurellosis.

OCCURRENCE AND DISTRIBUTION

HS occurs in southern and southeast Asia including Indonesia, Malaysia, Thailand and Philippines, in the near and middle east and in the northern, eastern, central and southern Africa. Some southern European states and USSR record a low sporadic occurrence whilst low sporadic or endemic status is reported in many south American states. (FAO 1991a). The disease was recognised in Japan in 1923 (Carter 1982), but there are no records of occurrance thereafter. The disease was reported among bison in the National Parks of the USA in the years 1912, 1922 and 1965, among dairy cattle in 1969 (Carter 1982), and in beef calves as recently as 1993 (Rimler and Wilson 1994).

The disease has never been reported from Western Europe, Australia, Oceania and Canada. (FAO, 1991a). Reports very suggestive of HS emerge from Italy and some southern and central American states where susceptible species exist and the husbandry practices are similar to those of HS-endemic areas. Yet no laboratory confirmation based on serotype identification is available.

Serotype Distribution

Asian countries have recorded only the Asian serotype, and the African countries only the African serotype. A few countries such as Egypt and Sudan have recorded both serotypes (Shigidi and Mustafa, 1979, Farid *et al.*, 1980). There is some evidence that type B may be present in some African states, particularly eastern (De Alwis, 1984).

The USA strain was originally identified as the Asian type B:2. More recent investigations by Rimler and Wilson (1994), using the Carter: Heddleston system, and

DNA fingerprinting, indicated the presence in USA of more than one serotype within the B group Viz B:2 and B:3,4.

ECONOMIC LOSSES

There is more speculation than accurate estimates of the economic losses due to HS. This situation is to be expected since the disease occurs mainly in regions where husbandry practices are primitive and consequently disease surveillance systems are poorly developed. Reported losses may therefore, reflect the trends, the actual losses being considerably higher. Few countries have attempted to quantify losses. Indonesia recently reported an annual loss of US $ 4000-6000, Laos US $ 1.4 million and Malaysia approximately US $ 1 million (FAO, 1991b). The criteria used in arriving at these estimates vary, and are therefore not comparable.

Massive epidemics, naturally draw the attention of the authorities. In endemic countries or regions, such epidemics no longer occur. Adult animals in such areas have developed what is described as 'naturally acquired immunity' and the disease continues to smoulder among young animals only. Such losses are insidious in nature and difficult to estimate, but may be economically significant. This pattern is well established in Sri Lanka. (De Alwis, 1981; De Alwis and Vipulasiri, 1980).

Economics of disease control is itself a matter of controversy. The cost of regular disease control measures, mainly vaccination, has to be weighed against actual economic losses that would occur in the absence of vaccination, which is a difficult parameter to assess. A study in Sri Lanka, carried out in an HS endemic area in the 1970's when alum precipitated vaccine was used with a coverage of 35%, indicated that approximately 14% of buffaloes and 8% of cattle died of HS (De Alwis and Vipulasiri, 1980). Compared with a 21% buffalo mortality in the first year of life (Kumaratileke and Buvanendran, 1979) it indicated that HS accounted for 2 out of every 3 buffaloe deaths. A similar study conducted in the 1990's after implementing improved vaccination programs from 1983 onwards showed that deaths in both the HS vulnerable species i.e. buffalo, and the age group (under 2 years) had diminished (Wijewardana- unpublished data).

Most HS vaccines are produced in the respective countries and are relatively cheap. The vaccine produced in Sri Lanka for instance in 1993 cost approximately 4 US cents per dose (De Alwis, unpublished data). The justification for regular mass vaccination is often questioned but most countries accept this as an insurance against epidemics.

THE CLINICAL SYNDROME

Following exposure, clinical signs appear after a short incubation period. In indigenous buffaloes in Sri Lanka following experimental infection by aerosol or oral dosing with live cultures, clinical signs appeared after an average incubation period of approximately 30 hrs. When naturally exposed by tying them in close contact with clinically diseased ones, clinical signs appeared in 46-80 hours of initial contact. (De Alwis *et al.*, unpublished data). In these transmission experiments, the clinical syndrome displayed three phases, an initial phase of temperature elevation followed by a phase of respiratory distress leading to a terminal phase of recumbency and death. Submandibular oedema may be observed from the initial stages. The course of the disease ranged from 2-5 days, and there were varying degrees of overlap between the three phases depending on the duration. (De Alwis *et al.*, - unpublished data). Among free range animals, the first reports are that of animals being found dead without any clinical signs been observed. The course of the disease recorded in natural outbreaks is shorter than in experimental transmission suggesting that in natural field outbreaks the initial phase may escape unnoticed.

Occasionally atypical syndromes caused by the HS serotypes in the natural host species has been recorded. A protracted pneumonic form was reported among buffaloe calves in Sri Lanka, terminating in septicaemia. (De Alwis *et al.,* 1975). Syndromes described in the literature are the septicaemic form, respiratory form, cutaneous, pectoral or oedematous forms. Presumably these clinical syndromes described are based on dominant clinical signs.

Bacteriological Findings

Septicaemia in HS is essentially terminal. At death counts of pasteurellae in the blood are in the region of 10^5 - 10^6 colony forming units (cfu)/ml. Considerable multiplication occurs after death and blood drawn from the jugular vein around 20 hr. after death had counts of 10^{12} cfu/ml. (De Alwis *et al.,* unpublished data). Similar multiplication is found to occur in muscle tissue after death. The presence of the organism in the saliva and nasal secretions during the clinical disease was found to be inconsistent.

Pathology

On autopsy, the first observed lesion is sub-cutaneous oedema with sero-sanguinous fluid particularly in the submandibular, throat pharyngeal and brisket region. The subcutaneous connective tissue may be studded with petechial haemorrhages. Varying degrees of lung involvement are seen ranging from generalised congestion to patchy or extensive consolidation. Thickening of the interlobular septa is evident giving rise to a lobulated appearance. There is often pleurisy and pericarditis with thickening of the pericardium and accumulation of sero-sanguinous fluid in the pericardial sac. Petechial haemorrhages on the myocardium is a consistent feature, particularlly if the base of the ventricles. Lymph nodes in the thoracic region may be swollen and haemorrhagic. Changes in the gastro-intestinal tract range from varying degrees of congestion to petechial or echchymotic haemorrhages.

Rhoades *et al.,* (1967) found that in experimentally produced HS, the nature of the lesions depended on the route of infection. Pneumonic lesions dominated when the route of infection was intranasal. De Alwis *et al.,* (1975) in experimental transmission of HS, showed that the extent of lesions depended on the duration of the disease. In peracute cases where death occurred in 24-36 hours no more than a few scattered petechial haemorrhages were seen. When the course of the disease was more than 72 hours, there was extensive pneumonia, pleurisy and pericarditis with marked adhesions.

EPIDEMIOLOGY

Seasonal Incidence

It is generally believed that HS is associated with wet humid weather, and most outbreaks occur during the wet seasons. The massive outbreaks in Sri Lanka in the mid fifties (Perumalpillai and Thambiayah, 1957; Dassanayake, 1957), and in Zambia (Francis *et al.,* 1980) were associated with the rainy seasons. In Sudan outbreaks are reported to occur during rainy seasons (Mustafa *et al.,* 1978). Practically all Asian countries associate HS outbreaks with the rain. It seems likely that losses occur mostly in countries with well-defined wet and dry seasons.

Few countries have attempted to quantify the seasonal losses. One study in Sri Lanka (De Alwis and Vipulasiri, 1980) indicated that outbreaks occurred throughout the year and that whilst dry season outbreaks were contained, wet season outbreaks tended to spread, presumably due to the enhanced survival of the causative bacterium. Other wet season-associated factors may also contribute towards the spread of disease. In Asia where buffaloes are used for draught power in the rice fields, there is a considerable movement of animals with the onset of the rains.

Host Susceptibility

Conventionally the term HS is used to describe disease in cattle and buffaloes. There is general agreement that buffaloes are more susceptible than cattle although quantitative evidence is scarce. In Malaysia, 73% of all recorded losses due to HS during 1970-79 and 90% during 1980-89 were among buffaloes despite the fact that its buffalo population is half that of cattle (FAO 1979, 1991). That the losses in buffaloes are higher than in cattle has been shown in epidemiological studies in Sri Lanka. (De Alwis and Vipulasiri, 1980; De Alwis, 1981). One of these studies further showed that whilst the herd infection rate was not significantly different in cattle and buffaloes, once infected morbidity within the herd was greater in buffalo herds.

Sporadic outbreaks of HS have been reported among pigs in Sri Lanka (De Alwis, unpublished data), Thailand (FAO, 1979), Malaysia (FAO, 1979, 1991) and India (Dhanda, 1959a; Murthy and Kaushik, 1965). Although reports are available associating the HS serotypes with disease in goats in India and Malaysia (FAO, 1979), experiments carried out in Sri Lanka showed that no natural transmission occurred from diseased cattle to in-contact goats, nor did the goat, succumb to experimental infection using challenge doses ranging from 10^{10} - 10^{12} viable organisms. In the latter instance only about 10% animals died of HS whilst the rest merely showed a transient temperature elevation and swelling at site of inoculation. (Wijewardana *et al.*, 1986b).

Morbidity and Mortality

Once clinical disease is established death usually results. Recoveries are rare and may occur in well managed herds where early detection is done by regular checking of the rectal temperatures of in-contact animals. Case fatality is thus near 100%. Information emerging from disease reporting systems, however, indicate highly variable figures for 'number affected' and 'number dead'. India, Nepal and Philippines for instance report case fatality rates ranging from 5% to over 90%. (FAO, 1991b). It is likely that in the absence of accurate information on individual animals, all animals in affected herds are reported as 'affected animals'.

Morbidity and mortality are affected by a variety of factors and their interactions.

Host Species. The higher susceptibility of buffaloes has already been dealt with.

Age. Most countries recognise higher morbidity among young animals but quantitative estimates are scarce. Surveys in Sri Lanka have shown that of all losses due to HS, 65% in buffaloes and 77% in cattle were among animals under two years. (De Alwis and Vipulasiri, 1980). Another study of two outbreaks of HS in neighbouring farms showed that the most vulnerable age was six months to two years (De Alwis *et al.*, 1976). The disease rarely occurred among under six month old calves. These studies were carried out in endemic areas where adult populations had high levels of immunity and the abscence of disease in very young animals may be due to maternal immunity. One case of HS was reported in Sri Lanka in a 2-3 month old calf born to an animal imported from a HS free country. (De Alwis, unpublished data).

Immunological Factors. Morbidity in a herd where the disease breaks out will evidently depend on the immune status of the herd, either acquired naturally or induced by vaccination. The greater the percentage of immune to non-immuned animals the lower will be the morbidity.

Endemic and Non - endemic Areas. Studies in Sri Lanka have shown that when sporadic outbreaks occur in non-endemic areas, morbidity and mortality is heavy and occur in animals of all ages. In contrast, when periodical, seasonal outbreaks occur in endemic areas, morbidity is low and only a few young animals die each year. This pattern presumably has an immunological basis. Through repeated exposure adult animals in endemic areas have acquired natural immunity. Alternatively vaccination which is carried out mainly in endemic areas also would induce immunity. Thus, only young hitherto unexposed and perhaps yet unvaccinated animals will succumb to HS periodically, and such animals will naturally be restricted to a specific age group.

Husbandry Methods. One epidemiological study in Sri Lanka (De Alwis and Vipulasiri, 1980), showed the disease occurred more frequently among large, nomadic herds as compared with small, confined herds. The same study showed, however, that in large herds with frequent occurrence, the morbidity in each outbreak was low, whereas when the disease occurred less frequently among small herds, morbidity was greater.

Naturally Acquired Immunity

Bain (1954) was the first to observe this phenomenon among buffaloes in Thailand. He concluded that 10% of all buffaloes in Asia were naturally immune. The significance of this phenomenon in the epidemiology of HS was not realised for two decades until epidemiological studies in Sri Lanka in the 1970's revealed that the proportion of such immune animals varied from herd to herd, and within the same herd from time to time. In general, the percentage of naturally immuned animals was highest in endemic areas and least in non endemic areas. (De Alwis, 1982b; De Alwis and Sumanadasa, 1982). Following exposure, immunity developed within a few weeks and persisted in some animals for over a year. The phenomenon of naturally acquired immunity was largely responsible for the different morbidity and mortality patterns in endemic and non-endemic areas.

A few instances of natural immunity in cattle in the absence of exposure to HS has been reported in Australia (Bain 1954), USA (Bain *et al.*, 1982; Sawada *et al.*, 1985), and in Chad (Bain *et al.*, 1982). In these instances, immunity may have been induced by exposure to related serotypes of Pasteurella.

The Carrier Status

The presence of HS - causing pasturellae in the nasopharynx of healthy cattle and buffaloes is well documented, the percentage of such animals detected varying considerably and ranging from less than 1% to over 40%. (Singh, 1948; Mohan *et al.*, 1968; Wijewantha and Karunaratne, 1968; Mustafa *et al.*, 1978; Hiramune and De Alwis, 1982). With further analysis of these findings, it became apparent that the percentage carriers detected in each instance was related to recent outbreaks of HS in the herd. Gupta (1962) found that during an outbreak of HS, 7.5% healthy animals were carriers. Forty days later, none were detected in the same herd. The studies of De Alwis *et al.*, (1986) in Sri Lanka and Saharee *et al.*, (1992) in Malaysia showed that pasteurellae were found not only in nasopharynx but also in the associated lymph nodes. All of these studies were confined to a single swabbing of the nasopharynx before and/or after slaughter or culture from lymph nodes after slaughter.

More light was cast on the carrier status when De Alwis *et al.,* (1986) swabbed the nasopharynx of cattle and buffaloes repeatedly at intervals in herds where outbreaks had occurred and subsided. They found that different animals appeared as carriers on different days. Transient or intermittent appearance of the organism in the nasopharynx became evident. In a subsequent study designed to confirm the above findings, De Alwis *et al.,* (1990) produced carrier animals by experimental exposure and observed them for up to one year. Some animals were slaughtered at varying intervals after exposure. Attempts were made to isolate the pasteurellae used for exposure from 14 different sites using a marker. It was revealed that in a high percentage of exposed animals, pasteurellae were lodged in their tonsils for long periods. Intermittently, it appeared in the nasopharynx. It became apparent that there were 'Latent Carriers' which harboured the organism in the tonsils only and that intermittently, the organism presumably multiplied and spilled over into the nasopharyngeal secretions. Thus, in all previous investigations with single swabbing, only the 'Active Carrier' or 'Shedder' status may have been detected. All carriers had high antibody levels and the number of animals showing high antibody may reflect the total carrier status.

It was also found that when latent carrier animals were treated with antibiotics to which the infecting organisms were sensitive *in vitro* the organisms persisted in the tonsils (De Alwis *et al.,* 1990). Further investigations on experimentally produced carriers by Horadagoda and Belak (1990) using immunochemical techniques revealed that the site of localisation of the organism was the crypts of the tonsils and not the tonsillar tissue itself, thus explaining the persistence despite antibiotic therapy.

The factors that caused the latent carriers to become active intermittently are yet obscure. Experimental application of well known stressors such as excessive ploughing excercise, long distance walking with inadequate food and water and immunosupression induced by steroid treatment to buffaloes failed to convert a latent HS carriers into active form, whereas other latent infections such as trypanosomiasis precipitated into clinical disease (Abeynayake *et al.,* personal communication).

The Epidemiological Cycle

Summing up all available epidemiological information, the epidemiological, cycle in haemorrhagic septicaemia may be reconstructed as follows (see figure 1).

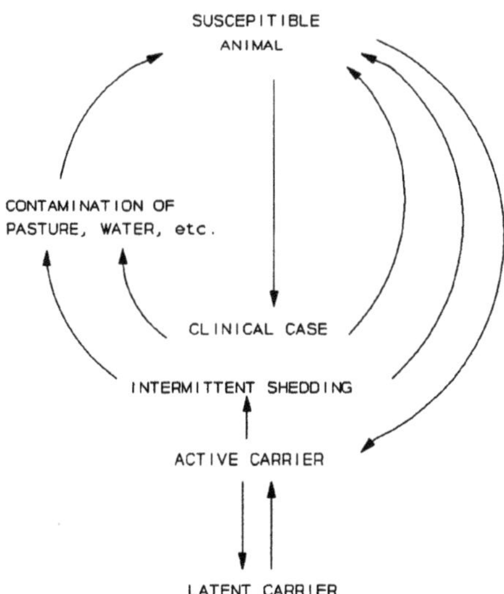

Figure 1. Showing the presumptive epidemiological cycle in haemorrhagic septicaemia.

The old theory that under 'Stress' carrier animals break down into clinical cases does not appear to hold, since all carrier animals also have high antibody levels believed to be protective. More investigation is evidently necessary to determine the factors which activate latent carriers. Experimental infection experiments have also shown that large numbers of organisms are required to set up an infection by the intra-nasal or oral route. (De Alwis, unpublished data). Whether an activated carrier could shed sufficient organisms or whether an excessive multiplication occurs within the susceptible animal after entry, under certain circumstances, is yet uncertain.

PATHOGENESIS

The successful transmission of disease by the intranasal and oral routes, producing a syndrome with clinical signs and lesions resembling natural disease (De Alwis, unpublished data) indicates that these may be the natural routes of infection. The virulence of the organisms is believed to be associated with the presence of surface, 'capsular' material, which is easily lost on sub - culture with associated loss of virulance. The organisms isolated from clinically normal animals even in the latent carrier state, have been found to be as virulent as those from clinical cases (Wijewardana *et al.,* 1986). These workers also produced evidence that carrier animals possessed high levels of humeral immunity thus maintaining a host - pathogen balance.

It appears that upon entry into an unexposed susceptible animal, the initial site of multiplication is the tonsillar tissue. Thus on the one hand the pathogen will multiply *in vivo* and on the other the host defence mechanisms both specific and non specific will interact and depending on which process dominates will lead either to clinical disease or 'arrested infection'. The dose of infecting organisms, amongst other factors, will be an important determinant of the outcome. The role of bacterial adherence in the multiplication of pasteurellae in the respiratory tract in HS is an area which merits investigation. This phemomenon could be an important determinant of the outcome of infection.

There is as yet no evidence of an exotoxin in HS. The role of hyaluronidase produced by type B strains (Carter and Chengappa, 1980) in the pathogenesis is yet uncertain, particularly in view of the fact that it has not been demonstrated in type E strains, which also cause HS. Endotoxin is likely to play an important role in HS as was evidenced by the work of Rhoades *et al.,* (1967) who found similarities in the pathological lesions in natural HS and endotoxic shock in calves.

DIAGNOSIS

Two aspects of diagnosis, field and laboratory, are considered as important.

Field Diagnosis

This is important since the immediate instigation of control measures to prevent spread of disease is based on a field diagnosis. A field diagnosis can be made on the basis of characteristic clinical signs, gross pathological lesions, herd history (endemicity of the area, vaccination history), morbidity and mortality patterns, species susceptibility, age group affected etc. Since morbidity and mortality are highly variable and are dependent on a number of factors and their interactions these parameters must be viewed in the background of surrounding circumstances.

Laboratory Diagnosis

Laboratory confirmation of diagnosis is by isolation and identification of the agent. This is done by cultural and biochemical as well as serological methods. Direct microscopic examination of smears from material is not conclusive.

Collection of Material. The septicaemia in HS is essentially terminal. Pasteurellae are not consistently present in nasal secretions. Thus, blood or nasal secretions from animals during the clinical phase does not constitute reliable material, and negative results using these is not conclusive. From a fresh carcass, a blood sample or swab collected within a few hours of death, from the heart or from the jugular vein is satisfactory. From older carcasses, a long bone is collected. Blood can be packed in ice, or placed in a transport medium during despatch. The long bone is cleaned of muscle tissue and despatched. Upon reaching the laboratory it is immersed in alcohol, sterilised by flaming, split open and the marrow scooped off.

Cultural and Biochemical Methods. Direct culture yields results only with fresh material. Contaminants and *postmortem* invaders when present overgrow the pasteurellae. A small volume of the eluted blood in 2-3 ml of physiological saline, or bone marrow suspended in saline is injected to a mouse. If pasteurellae are present, even in small numbers, the mouse would die in 24-36 hours. Pure cultures can be obtained by culturing the blood of the mouse, which serves as a 'biological screen' for extraneous organisms. Smears made from the mouse blood or cultures of mouse blood will show Gram negative, short, cocobacillary bipolar staining forms. A degree of pleomorphism may be noted particularly in old cultures with longer rods of varying length. Short, cocobacillary forms are typically observed from fresh material and the organism tends to appear larger in tissue smears. Leishman's or methylene blue stain would make the bipolar staining more evident.

HS organisms produce oxidase, catalase and indole and will reduce nitrates. They do not grow on McConkey agar, produce hydrogen sulphide or urease, and fail to liquify gelatin or utilise citrate. Glucose and sucrose are always fermented while other sugars are fermented by some strains with acid only.

Serological Methods

The serological classifications of pasteurella that were developed over the years, the number of serotypes identified by each method and the position of the HS serotypes is shown in table 3.

Table 3. Methods adopted for serotyping of *Pasteurella multocida* and the position of the HS Serotypes.

Authors	Technique	No. of Types	HS Types
Little and Lyon (1943)	Slide Agglutination and passive mouse protection tests.	1,2 and 3	2
Roberts (1947)	Passive mouse protection test.	I -V	I
Carter (1955,1961) Rimler and Rhoades (1987)	'Capsular' typing IHA test using heat labile (56°C /30min.) antigen and rabbit antisera.*	A B D E and F	B and E
Namioka and Murata (1961a)	Simplified capsular typing using fresh cultures and rabbit antisera.	A B D and E	B and E
Namioka and Murata (1961b,c) Namioka and Bruner (1963) Namioka and Murata (1964)	'Somatic' typing by Agglutination test using HCL treated cells and rabbit antisera.	1 - 11	6
Heddleston *et al.*, (1972)	'Somatic' typing by AGPT using heat stable antigen and chicken antisera.	1 - 16	2 and (5)

* Originally used fresh human 'O' erythrocytes. Subsequently modified by Carter and Rappay (1962) using formalinised cells; by Sawada *et al.*, (1972) using glutaraldehyde fixed sheep erythrocytes and by Wijewardana *et al.*, (1986a) using fresh sheep erythrocytes.

Brogden and Packer (1979) found that serotypes determined by one system did not correlate with those of another. This is not surprising because the antigen preparations used in the different systems were widely different. The designations of 'Capsular' and 'Somatic' antigens were purely empirical, and would very well have shared common components. For instance De Alwis (1987) reported that the results of the 'Capsular' typing by the indirect haemagglutination test described by Carter (1955) using 56°C /30 min supernate was the same when Heddlestons 'Somatic' antigen (100°C/1 hr) supernate was substituted. Furthermore, the same results were obtained when the Carters supernate was subsequently heated to 100°C for 1 hour. What Namioka and Murata (1961b) called the 'Somatic' antigen, on the other hand, was a different entity altogether and was the residue left after a more drastic HCL treatment of the cells. Phenol - water extracts (Westphal type) of this residue when used to coat RBC in IHA tests, gave results that parallelled the Namioka's typing. The complex nature of the serum absorptions required to prepare the type specific sera for Namioka's typing has resulted in this technique becoming virtually obsolete. Within the 'Capsular' type B, however, since only two Namioka's 'Somatic' types have been recorded, straight, unabsorbed sera can be used and if antigen preparations from reference Asian (6:B) , African (6:E) and Australian non HS (11:B) strains are tested simultaneously with the antigen of the test strain, interpretation is facilitated.

The present serotyping systems may therefore be looked upon purely as a convenient diagnostic tool, rather than a classification on a rational basis.

Serotype Designation. Two methods of serotype designation are used. The more popular method at present is to use the Carter: Heddleston system, evidently for the ease of carrying out the Heddlestons typing. Occasionally, particularly with regard to HS serotypes, the Namioka: Carter system is also used. The Asian and African HS serotypes are designated B:2 and E:2 respectively by the former method and 6:B and 6:E by the latter system.

Routine Laboratory Diagnosis. For routine diagnosis using material from HS suspected animals, the procedure outlined is figure 2, is of practical value.

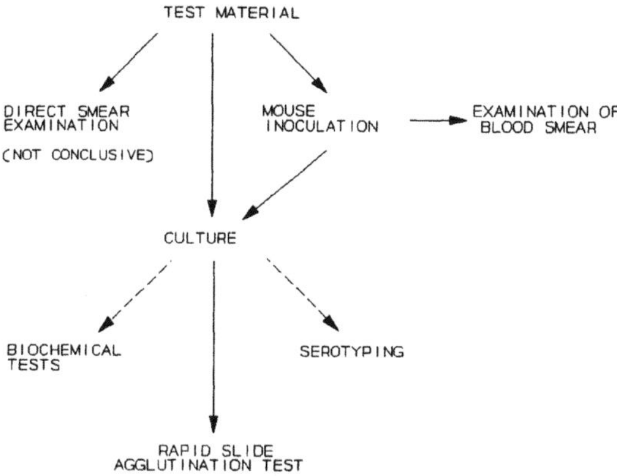

Figure 2. Scheme for routine laboratory diagnosis of haemorrhagic septicaemia.

The tests indicated by the bold line can be accomplished within 48 hours, when a confirmatory report can be issued.

Detection of Antibodies

Antibody detection is not used for routine diagnosis. The epidemiological finding that most exposed but clinically unaffected animals develop high antibody titres, unattainable by vaccination, is of diagnostic value in situations where deaths have occurred in a flash and no carcasses are available for collection of material for laboratory tests. High levels of antibody among surviving animals is indicative of recent exposure. The IHA and passive mouse protection tests have been used to detect antibodies and more recently an ELISA technique has been developed (Dawkins *et al.*,1992)

TREATMENT

The rapid onset of disease, its short course and the fact that the disease is most prevalent among animals which are least likely to be observed leaves little room for treatment, which is effective only if carried out in the early stages. The only practical approach in the face of an outbreak is to check the rectal temperatures of all in-contact animals in diseased herds and carry out antibacterial therapy immediately. Traditionally intravenous administration of sulphadimidine 33 1/3% solution has been recommended. The large volume required and the consequences of leakage while injecting weighs against its use.Tetracyclines, chloramphenicol, and penicillin + streptomycin combinations have been found to be useful (De Alwis, unpublished data). A collection of HS strains from Malaysia, Indonesia, Thailand, Myanmar, India and Sri Lanka when tested were found to be sensitive to most of the commonly used antibiotics, (De Alwis 1984). Abeynayake *et al.*, (1993) examined 24 Sri Lankan isolates from haemorrhagic septicaemia outbreaks and found all to be highly sensitive to penicillin, ampicillin, enrofloxacin, oxytetracycline nitrofurantoin cefalothin and chloramphenicol whilst 11 of them were resistant and 8 showed partial sensitivity to sulphonamides.

Treatment with hyperimmume serum has been attempted experimentally, but is of no practical value. (Keng and Phay, 1963).

CONTROL

Vaccination is generally accepted as a means of disease control.

Vaccines

Several types of vaccines have been developed. These include plain bacterins, alum precipitated vaccine, aluminium hydroxide gel vaccine and the oil adjuvant vaccine (OAV). Most countries use local strains as seed. It was believed that the Indian 'P52' strain and the Burmese 'Katha' strain possessed special immunogenic merit but this fact could not be established experimentally (De Alwis, 1984). Presently fermentors are used in most countries. The seed is periodically passaged in a natural host in order to retain its full complement of antigens. It has been estimated that at least 1.5 - 2.0 mg of dry whole bacteria, are required to immunise an animal. Thus vaccines require the production of dense cultures. Agar-wash cultures harvested from Roux flasks meet this requirement but involves testing individual flasks, a laborious process. Enriched media are now used to produce dense broth cultures. Growth promoting additives such as yeast extract and autodigest of pancreas are incorporated. Growth is enhanced by aeration. Various vortex and sparger aeration systems are used. Formalin at a final concentration of 0.3 - 0.5 percent is used to inactivate the cultures.

Efficacy and Duration of Immunity

Bacterins confer only a few weeks immunity. Dense bacterins may cause shock reactions. The alum precipitated and aluminium hydroxide gel vaccines are believed to confer 4-5 months immunity and need be given twice yearly. The longest duration of immunity is conferred by the OAV. (Bain *et al.*, 1982). This vaccine is traditionally prepared by emulsifying bacterin and mineral oil with lanoline to produce a pure white, thick emulsion. Two initial doses followed by annual revaccination is reckoned to provide adequate immunity. This vaccine is recommended for prophylactic use. The optimum age for primary vaccinationis 4-6 months as it has been found that the response of younger animals is poor (De Alwis *et al.*, 1978).

Recent work has shown that the OAV is stable at 4°C for around 6 months (Vipulasiri *et al.*, 1982, Chandrasekeran *et al.*, 1987).Stability is increased by increasing the lanoline content, but this increases the viscosity and makes it difficult to inject (Gomis *et al.*, 1989). Other emulsifying agents such as 'Arlacel A' or modern oil adjuvants will help to reduce viscosity, but will increase the cost of production.

Quality Testing

A rapidly growing large colony sized seed, its good agglutinability, purity and agglutinability of harvest and a minimum turbidity standard equivalent to 1.5 mg dry weight/volume has to be maintained in order to ensure a good vaccine. Potency has been tested in natural hosts as well as in laboratory animals (Nagarajan *et al.*, 1972; Gupta and Sareen 1976; Chandrasekeran and Yeap, 1978; Vipulasiri *et al.*, 1982; Gomis *et al.*, 1989). Tests in natural host is not feasible with every batch of vaccine. The active mouse protection test (AMPT) using the method of Ose and Muenster (1968) is widely used.

Gomis *et al.*, (1989) found that a good vaccine gave above 4 logarithmic units protection in AMPT.

These workers further found that a minimum antigen content of 1.5 mg per ml. dry weight/volume was also required to ensure good protection in cattle, with a 3 ml dose containing 1.5 ml of bacterin.

Vaccination Programmes

Most countries still use the alum precipitated vaccine, annually, before the rainy season, so that peak immunity is provided when most needed. Thailand uses an alhydrogel vaccine twice yearly. The OAV is the principal vaccine used in Sri Lanka, Malaysia, Iraq, Egypt and Indonesia, (De Alwis, 1984; FAO, 1991 b).The coverage attained is often low and may vary from 20-50%, but may be high for specific areas. A near 100% is achieved for instance in the Lombok island of Indonesia (FAO, 1991b).

Experimental Vaccines

Although several new vaccines have been experimented with, none have become established. Bhatty (1973), produced a vaccine using sodium alginate as adjuvant. Oil adjuvant vaccines using purified cell extracts have been produced but were no better than the whole cell OAV which is easier to produce. (Bain 1955; Dhanda 1959; Nagy and Penn 1976). A more easily injectable double emulsion vaccine (DEV) was produced (Gupta *et al.*, 1979; Yadev and Ahooja 1983; Chandrasekeran *et al.*, 1991). Although the earlier DEV was less potent than the OAV, Chandrasekeran *et al.*, demonstrated immunity at 52 weeks after use of their vaccine.

Based on the observation that the naturally acquired immunity resulting from 'arrested infections' was highly protective, De Alwis proposed two lines of investigation for development of better vaccines. Assuming that the better immunity in natural infection was the result of specific antigens produced *in vivo*, an attempt to identify them and cause their expression *in vitro* was considered worthwhile. This aspect is presently being pursued by Wijewardena *et al.*, in Sri Lanka and Mukkur *et al.*, in Australia and Malaysia. The other area proposed for investigation was the simulation of natural infection by live vaccines containing avirulent or low virulent antigenically identical or related strains. De Alwis and Carter (1980) successfully immunised calves using a live streptomycin dependent mutant of the HS organism. Its poor multiplication *in vivo* and the large dose consequently required excluded the possibility of developing a practical vaccine. A serologically related strain from deer has also been used as a live vaccine. (Myint *et al.*, 1987; Myint and Carter, 1989, 1990). This vaccine when administered by the sub -cutaneous route was lethal to a small proportion of young buffalo calves, hence was considered unsuitable for primary vaccination of young calves. When used intra nasally it was safe, and 50% protection was recorded against sub-cutaneous challenge. It has been used extensively by the intranasal route in several provinces in Myanmar, and absence of outbreaks in these provinces is reported, suggesting its efficacy against infection by natural routes. This vaccine deserves a more comprehensive trial. De Alwis, Carter and Chengappa (1980) developed streptomycin independent revertants of dependent mutants. Most of these revertants had their virulence restored. A few displayed low virulence. One of these has given immunity in calves, both by the intranasal and intra-muscular routes and has proved to be safe, even in doses of 10^{10} organisms intramuscularly (De Alwis *et al.*, unpublished data). This vaccine needs to be further tested. When using mutants which are not genetically defined, their stability is a matter for concern.

REFERENCES

Abenayake, P., Wijewardana, T.G., and Thalagoda, S.A., 1993, Sulphonamide resistance in *Pasteurella multocida* (Serotype 6:B) in Sri Lanka, Presentation at the 46th Annual Sessions of the Sri Lanka Veterinary Association December 1993.

Bain, R.V.S., 1954, Studies on haemorrhagic septicaemia in cattle. I naturally acquired immunity in Siamese buffaloes, *Br.Vet.J.* 110:481.

Bain, R.V.S., 1955, A preliminary examination of antigens of *Pasteurella multocida* type I, *Br.Vet.J.* 111:492.

Bain, R.V.S., De Alwis, M.C.L., Carter, G.R. and Gupta, B.K., 1982, Haemorrhagic Septicaemia, Animal Production and Health Paper No. 33, FAO, Rome.

Bhatty, M.A., 1973, A preliminary comparison of sodium alginate and oil adjuvant haemorrhagic septicaemia vaccines in cattle, *Bull. epizoot. Dis. Afr.* 21:171.

Bollinger 1878, Cited by Hudson, J.R., `Infectious Diseases of Animals: Diseases due to Bacteria, Vol.2. Pasteurellosis, A.W. Stableforth and I.A. Galloway, P.421, Butterworth Scientific Publications, London.

Brogden,K.A. and Packer, R.A., 1979, Comparison of *Pasteurella multocida* serotyping systems, *Am. J. Vet.Res.* Res. 40:1332.

Carrigan, M.J., Dawkins, H.J.S., Cockram, F.A. and Hausen, A.T. 1991, *Pasteurella multocida* septicaemia in fallow deer (*Dama dama*), *Aust. Vet. J.* 68:201.

Carter, G.R., 1957, Studies on *Pasteurella multocida*. III A serological survey of bovine and porcine strains from various parts of the world, *Am.J.vet.Res.*18:437.

Carter, G.R., 1959, Studies on *Pasteurella multocida* IV. Serological types from species other than cattle and swine, *Am. J.vet.Res.* 21:173.

Carter, G.R., 1961, A new serological type of *Pasteurella multocida*, *Vet.Rec.* 73:1052.

Carter, G.R., 1982, Whatever happened to haemorrhagic septicaemia? *J.Am.vet. med. Ass.* 180:1176.

Carter, G.R., and Chengappa, M.M., 1980, Hyaluronidase production by type B *Pasteurella multocida* from cases of haemorrhagic septicaemia, *J. clin. Microbiol.* 11:94

Carter, G.R. and Rappay, D.E., 1962, Formalinised erythrocytes in the haemagglutination test for typing *Pasteurella multocida*, *Br. Vet.J.*118:289.

Chandrasekeran, S., Muniandy, N., Kennet,I. and Mukkur, T.K.S., 1991, Protection, antibody response and isotype specificity in buffaloes immunised with conventional haemorrhagic septicaemia vaccines, Proc. FAO/APHCA workshop on haemorrhagic septicaemia, Feb. 1991, Kandy, Sri Lanka.

Chandrasekeran, S. and Yeap, P.C., 1978, Safety and potency testing of haemorrhagic septicaemia oil adjuvant vaccine by mouse protection test, *Kajian Veterinar*, 10:28.

Chandrasekeran, S. Yeap,P.C. and Saad, R., 1987, Effect of storage conditions on the stability and potency of haemorrhagic septicaemia oil adjuvant vaccine, *Kajian Veterinar*, 19:71.

Chaudhuri, S., Mukherjee, S.K., Chatterjee, A. and Ganguli, J.L., 1992, Isolation of *Pasteurella multocida* serotype F:3,4 from a stillborn snow leopard, *Vet. Rec.* 130:36.

Dassanayake, L., 1957, The haemorrhagic septicaemic outbreak of 1955-56, *Cey. Vet.J.* 5:56.

Dawkins, H.J.S., Johnson, R.B., Spencer, T.L. and Patten, B.E., 1990, Rapid identification of *Pasteurella multocida* organisms responsible for haemorrhagic septicaemia using an enzyme linked immunosorbent assay (ELISA), *Res. Vet. Sci.* 49:261.

De Alwis, M.C.L., 1981, Mortality among cattle and buffaloes in Sri Lanka due to haemorrhagic septicaemia, *Trop. anim. hlth Prod.* 13:195.

De Alwis, M.C.L., 1982a, *Pasteurella multocida* serotype 6:B from an elephant, *Sri Lanka Vet. J.*30:28.

De Alwis, M.C.L., 1982b, The immune response of buffalo calves exposed to natural infection with haemorrhagic septicaemia, *Trop. anim. hlth Prod.* 14:29.

De Alwis, M.C.L, 1984, Haemorrhagic septicaemia in cattle and buffaloes, *Rev. Sci. Tech. Off. Int. Epiz.* 3:707.

De Alwis, M.C.L., 1987, Serological classification of *Pasteurella multocida*, *Vet. Rec.* 121:44.

De Alwis, M.C.L. and Carter, G.R., 1980, Preliminary field trials with a streptomycin dependent live vaccine against haemorrhagic septicaemia, *Vet. Rec.* 106:435.

De Alwis, M.C.L., Carter, G.R. and Chengappa, M.M., 1980, Production and charecterisation of streptomycin dependent mutants of *Pasteurella multocida* from bovine haemorrhagic septicaemia, *Can.J. comp. Med.* 44:418.

De Alwis, M.C.L., Gunatileke, A.A.P. and Wickramasinghe, W.A.T. 1978, Duration of immunity to haemorrhagic septicaemia in cattle following immunisation with alum preciptated and oil adjuvant vaccines, *Ceylon vet. J.* J. 26:35.

De Alwis, M.C.L., Jayasekera, M.U. and Balasunderam, P., 1975, Pneumonic pasteurellosis in buffalo calves associated with *Pasteurella multocida* serotype 6:B, *ceylon vet. J.*23:58.

De Alwis, M.C.L., Kodituwakku, A.O. and Kodituwakku,S., 1976, Haemorrhagic septicaemia: an analysis of two outbreaks of disease among buffaloes, *Ceylon vet. J.* 24:18.

De. Alwis, M.C.L., Sumanadasa, M.A., 1982, Naturally acquired immunity to haemorrhagic septicaemia among cattle and buffaloes in Sri Lanka, *Trop. anim. hlth Prod.* 14:27.

De Alwis, M.C.L. and Thambithurai, V., 1965, A case of haemorrhagic septicaemia in a wild elephant in Sri Lanka, *Ceylon vet. J.* 13:17.

De Alwis, M.C.L. and Vipulasiri, A.A., 1980, Anepizootiological study of haemorrhagic septicaemia in Sri Lanka, *Ceylon vet. J.* 28:24.

De Alwis, M.C.L., Wijewardana, T.G., Gomis, A.I.U. and Vipulasiri, A.A.,1990, Persistence of the carrier status in haemorrhagic septicaemia (*Pasteurella multocida* serotype 6:B infection) in buffaloes, *Trop. anim. hlth Prod.*22:185.

De Alwis, M.C.L., Wijewardana, T.G., Sivaram, A. and Vipulasiri, A.A. 1986, The carrier and antibody status of cattle and buffaloes exposed to haemorrhagic septicaemia: investigations on survivors following natural outbreaks, *Sri Lanka vet. J.* 34:33.

De Alwis, M.C.L., Wijewardana, T.G. , Athureliya, D.S. and Vipulasiri,A.A., 1986, Prevalence of haemorrhagic septicaemia carriers among cattle and goats in endemic areas of Sri Lanka , *Sri Lanka vet. J.*34:16.

Dhanda, M.R., 1959, Immunisation of cattle against haemorrhagic septicaemia with purfied capsular antigens (a preliminary communication), *Indian vet. J.* 36:6.

FAO, 1979, Proc. FAO/APHCA Workshop on Haemorrhagic Septicaemia, Dec. 1979, Colombo, Sri Lanka.

FAO, 1991a , FAO- WHO-OIE Animal Health Yearbook 1991, FAO, Rome.

FAO, 1991b, Proc. FAO/APHCA Workshop on haemorrhagic Septicaemia, Feb.1991., Colombo, Sri Lanka.

Farid, A., El Ghani, M.A., Khalil, A., El Ghawas, A. and Kamel, A., 1980, Identification of serological types of *Pasteurella multocida* isolated from apparently healthy buffaloes, *Agric. Res. Rev.*, 58:107.

Francis, B.K.T., Schels, H.F. and Carter, G.R., 1980, Type E *Pasteurella multocida* associated with haemorrhagic septicaemia in Zambia, *Vet. Rec.* 107:135.

Gomis , A.I.U., De Alwis, M.C.L., Radhakrishnan, S. ,and Thalagoda, S.A., 1989, Further studies on the stability and potency of haemorrhagic septicaemia oil adjuvant vaccine, *Sri Lanka vet. J.* 36:9.

Gupta, B.K., 1962, Studies on the carrier problem in haemorrhagic septicaemia, Thesis, Punjab University, Chandigar, India.

Gupta, B.K., Mittal, K.R. and Jaiswal, T.N., 1979, Multi-emulsion haemorrhagic septicaemia oil adjuvant vaccine, Proc. FAO/APHCA Workshop on Haemorrhagic Septicaemia, Dec. 1979, Colombo , Sri Lanka.

Gupta, M.L. and Sareen, R.L., 1976, Evaluation of haemorrhagic septicaemia oil adjuvant vaccine by mouse protection test, *Indian vet. J.* 53:489.

Hiramune, T. and De Alwis, M.C.L., 1982, Haemorrhagic septicaemia carrier status of cattle and buffaloes in Sri Lanka, *Trop. anim. hlth Prod.*14:92.

Heddleston, K.L., Gallergher, J.E. and Rebers, P.A., 1972, Fowl Cholera: Gel diffusion precipitin test for serotyping *Pasteurella multocida* from avian species, *Avian Dis.*16:925.

Horadagoda, N. and Belak, K., 1990, Demonstration of *Pasteurella multocida* type 6:B (B:2) in formalin fixed paraffin embedded tissues of buffaloes by the peroxidase antiperoxidase (PAP) technique, *Acta vet. scand* 31:493.

Kumaratileke, W.L.J.S. and Buvanendran, V., 1979, A survey of production charecteristics of indigenous buffaloes in Sri Lanka, *Ceylon vet. J.* 27:10.

Little, P.A., and Lyon, B.M.,1943, Demonstration of serological types within non-haemolytic pasteurellae, *Am.J.vet. Res.* 4:110.

Mohan, K., Singha, M.N., Singh, R.P. and Gupta, G.M., 1968, A study of immunity against *Pasteurella multocida* in buffalo calves, and their carrier status, *Vet. Rec.* 83:155.

Mustafa, A.A., Ghalib, H.W. and Shigidi, M.T., 1978, Carrier rate of *Pasteurella multocida* in cattle associated with an outbreak of haemorrhagic septicaemia in Sudan, *Br. vet. J.* 134:375.

Myint, A., and Carter, G.R., 1989, Prevention of haemorrhagic septicaemia with a live vaccine, *Vet. Rec.* 124:508.

Myint, A.and Carter, G.R., 1990, Field use of haemorrhagic septicaemia live vaccine, *Vet. Rec.* 126:648.

Myint, A., Carter, G.R. and Jones, T.O., 1987, Prevention of experimental haemorrhagic septicaemia with a live vaccine, *Vet. Rec.* 120:500.

Nagarajan, V., Sundaram, S. and Ramani, K., 1972, Evaluation of the potency of haemorrhagic septicaemia alum precipitated vaccine by mouse protection test, *Indian vet. J.* 49:1080.

Nagy, C.K. and Penn, C.W., 1976, Protection of cattle against haemorrhagic septicaemia by capsular antigens of *Pasteurella multocida* serotypes B and E, *Res. vet. Sci.* 20:249.

Namioka, S. and Bruner, D.W., 1963, Serological studies on *Pasteurella multocida* IV. Type distribution of organisms on the basis of their capsule and O groups, *Cornell Vet.* 53:41.

Namioka, S. and Murata,M. , 1961a, Serological studies on *Pasteurella multocida* I.A. simplified method for capsular typing of the organisms, *Cornell Vet.* 51:498.

Namioka, S. and Murata, M., 1961b, Serological studies on *Pasteurella multocida* II Charecteristics of the somatic 'O' antigen of the organism, *Cornell Vet.* 51:507.

Namioka, S. and Murata, M., 1961c, Serological studies on *Pasteurella multocida* III 'O' antigen analysis of cultures isolated from various animals, *Cornell Vet.* 51:522.

Namioka, S. and Murata, M., 1964, Serological studies on *Pasteurella multocida* IV. Charecterisitcs of the somatic `O' antigen, *Cornell vet.* 54.520.

Oreste and Armani, 1887, Cited by Hudson, J.R. in `Infections Diseases of Animals: Diseases due to Bacteria, Vol.2,Pasteurellosis,A.W.,Stableforth and I.A. Galloway, P. 421, Butterworths Scientific Publications, London.

Ose, E.E. and Muenster, O.A., 1968, A method of evaluating vaccines against *Pasteurella multocida*, *Am.J.vet. Res.* 29:1863.

Pavri, K.M. and Apte, V.H., 1967, Isolation of *Pasteurella multocida* from a fatal disease of horses and donkeys in India, *Vet. Rec.* 80:437.

Perumalpillai, C. and Thambiayah, V.S., 1957, Outbreaks of haemorrhagic septicaemia in an epizootic form in Ceylon, *Ceylon vet. J.* 5:24.

Rhoades, K.R., Heddleston, K.L. and Rebers, P.A., 1967, *Can. J. comp. Med.*, 31:226.

Rimler, R.B. and Rhoades, K.R., 1987, Serogroup F, a new capsule serogroup of *Pasteurella multocida*, *J. clin. Microbiol.* 25:615.

Rimler, R.B. and Wilson, K.R., 1994, Re-examination of *Pasteurella multocida* serotypes that caused haemorrhagic septicaemia in North America, *Vet. Rec.* 134:256.

Roberts, R.S., 1947, An immunological study of *Pasteurella septica*. *J. comp. Path.* 57.261.

Saharee, A.A., Salin, N.B., Rasedee, A. and Jainudeen, M.R., 1962, Haemorrhagic septicaemia carriers among cattle and buffaloes in Malaysia, Proceedings of the International Workshop on Pasteurellosis in Production animals, Aug. 1992, Indonesia ACIAR No. 43:89.

Sawada, T., Rimler, R.B. and Rhoades, K.R. , 1982, Indirect haemagglutination test that uses glutaraldehyde fixed sheep erythrocytes sensitised with extract antigens for detection of pasteurella antibody, *J. clin. Microbiol.* 15:752.

Shigidi, M.T.A., and Mustafa, A.A., 1979, Biochemical and serological studies on *Pasteurella multocida* isolated from cattle in Sudan, *Cornell vet* 69:77.

Singh, N., 1948, Nasal carriers in bovine pasteurellosis, *Indian J. vet. Sci.anim. Husb.* 18:261.

Vipulasiri, A.A., Wijewardena, T.G. and De Alwis,M.C.L., 1982, Effect of storage temperature and time on the potency of haemorrhagic septicaemia oil adjuvant vaccine, *Sri Lanka vet. J.* 30:19.

Wickremasuriya, U.G.J.S. and Kendaragama, K.M.T., 1982, A case report of haemorrhagic septicaemia in a wild elephant, *Sri Lanka vet. J.* 30:34.

Wijewantha, E.A. and Karunaratne, T.G., 1968, Studies on the occurence of *Pasteurella multocida* in the nasopharynx of healthy cattle, *Cornell Vet.* 58:462.

Wijewardana, T.G., De Alwis, M.C.L. and Bastianz, H.L.G., 1986a, Cultural biochemical and pathogenicity studies on strains of *Pasteurella multocida* isolated from carrier animals and outbreaks of haemorrhagic septicaemia, *Sri Lanka vet. J.* 34:43.

Wijewardana , T.G., De Alwis, M.C.L. and Vipulasiri, A.A., 1986b, An investigation into the possible role of the goat as a reservoir host in haemorrhagic septicaemia, *Sri Lanka vet. J.* 34:24.

Yadev, M.S. and Ahooja, M.L., 1983, Immunity trials in mice rabbits and calves with oil adjuvant vaccine against haemorrhagic septicaemia. *Indian J. anim. Sci.* 53.705.

VACCINE DEVELOPMENT AGAINST *PASTEURELLA HAEMOLYTICA* INFECTIONS IN SHEEP

W. Donachie

Moredun Research Institute
408 Gilmerton Road
Edinburgh EH17 7JH, Scotland

INTRODUCTION

Pasteurellosis caused by *P. haemolytica* is one of the most common bacterial infections of sheep, and by far the most important respiratory one, with a widespread distribution, occurring in temperate, subtropical and tropical climates (Gilmour and Gilmour, 1989). Pneumonic pasteurellosis in sheep was first described in Iceland by Dungal (1931) and the work of Biberstein (1960) and Smith (1961) in the 1960's on serotyping and biotyping was important in defining the epidemiological aspects of the disease. In sheep 2 biotypes comprising a total of 16 serotypes are recognised, with approximately 90% of all isolates serotypable and the remaining 10% currently untypable (Fraser *et al.*, 1982, Ball *et al.*, 1993). The biotypes, which were differentiated on their ability to ferment either arabinose (A) or trehalose (T), are each associated with distinct clinical syndromes (Smith, 1961).The A biotypes are responsible for pneumonic pasteurellosis in sheep of all ages while the T biotypes cause a systemic disease in 6 - 10 month old lambs. Recent taxonomical studies based on DNA homologies and 16s RNA work have suggested that the T biotype strains should be placed in a new genus, *P. trehalosi*, (Sneath and Stevens, 1990) but for the purposes of this chapter they will still be referred to as T biotypes of *P. haemolytica*. A number of virulence factors which enhance the bacterium's ability to cause disease have been associated with *P.haemolytica*. These include iron-regulated proteins (IRPs) (Donachie and Gilmour, 1988; Deneer and Potter,1989), leukotoxin (Kaehler *et al.*, 1980), lipopolysaccharide, (Rimsay *et al.*, 1981) and polysaccharide capsule (Adlam, 1989).

THE DISEASE

The biotype A serotypes of *P. haemolytica* are associated with pneumonic disease in all ages of sheep although a more severe form of the disease involving septicaemia, pleurisy and pericarditis also occurs in young lambs. The natural disease is sporadic, erupts suddenly and affects up to 10% of the flock with the peak incidence of disease in the spring and summer (Gilmour and Gilmour, 1989).

In the UK biotype T strains give rise to a distinct syndrome, known as systemic pasteurellosis, in young sheep. It occurs characteristically in the period from September to December (Stamp *et al.*, 1955) and mortality is generally low, around 3%.

Although *P. haemolytica* can act very effectively as a primary pathogen it is more usually involved with a predisposing factor. Environmental and management stressors such as dipping for external parasites, castration of male lambs, dosing with anthelmintics, gathering in warm weather and with particular reference to biotype T disease, moving to improved nutrition, are associated with outbreaks of pasteurellosis. However, definition and measurement of these stressors in controlled experiments is extremely difficult and thus far we can only speculate as to how these influence *Pasteurella* pathogenesis. There is both field and experimental evidence to confirm that infections with other micro-organisms are by far the most important predisposing factors. While most of these predisposing agents can give rise to clinical disease acting by themselves their combination with *P. haemolytica* results in severe disease. Some viruses such as parainfluenza type III (PI_3) , adenovirus and sheep pulmonary adenomatosis have been shown to have a temporal relationship with *P. haemolytica* in ovine pneumonia (Hore *et al.*, 1968, Davies *et al.*, 1980, LeaMaster *et al.*, 1987; Sharp, 1991a) while others such as reovirus and respiratory syncytial virus have a questionable role (Sharp, 1991b). Prior infection with *Mycoplasma ovipneumoniae* followed by *P. haemolytica* A biotype strains results in the development of atypical pneumonia of sheep, a proliferative exudative pneumonia affecting sheep from 2-12 months of age (Jones *et al*, 1986).

The involvement of other bacterial species in pasteurellosis has rarely been reported and when they have, they are often assumed to be opportunistic co-infections. However, *Bordetella parapertussis,* previously regarded as an obligate human pathogen had recently been isolated from pneumonic and apparently normal lungs, both in New Zealand and in Scotland (Cullinane *et al.*, 1987; Porter *et al.*, 1994). In the latter report B. parapertussis was recovered from 33 out of 52 lungs examined. Combined infection with both organisms in mice results in a more severe disease than when either organism is administered alone (Jian *et al.*, 1991) and more recently this combination effect has also been demonstrated in SPF lambs at Moredun Research Institute (MRI) (J. Porter, unpublished observations).

Control

Isolates of *P. haemolytica* from sheep are generally more susceptible to antibiotics than those recovered from cattle (Gilmour and Gilmour, 1989). Treatment of ovine pasteurellosis can be very effective and the use of oxytetracycline in prophylaxis, metaphylaxis and therapy is well documented (Gilmour *et al.*, 1982; Gilmour *et al.*, 1988). Although antibiotic treatment is of great value during an outbreak it is not a realistic way of controlling a disease which is sporadic and unpredictable in nature and which has such a short clinical course. It is also an inappropriate use of anti-microbials which encourages the development of antibiotic resistance. Although safe management practices are still likely to be the most economic way of controlling pasteurellosis it is generally believed that vaccination offers the best practical means of control.

GENERAL CONSIDERATIONS IN THE DEVELOPMENT OF SHEEP VACCINES

The development of vaccines against ovine pasteurellosis has to take into account a number of factors including.

1. The variety of strain types involved in disease. An effective vaccine should offer protection against all serotypes in the targeted species but a more practical approach is to protect against the most prevalent serotypes. Although there is variation in the prevalence of individual serotypes associated with sheep disease throughout the world, serotype A2 is usually the most common (Frank, 1982; Prince *et al.*, 1985). Figure 1 shows the prevalence of serotypes isolated in the UK and referred to MRI between 1982 and 1993. The demonstration

by Gilmour *et al.*, (1983) that cell-free extracts of *P. haemolytica* elicited serotype-specific protection in SPF lambs indicated the need to incorporate representative strains of the most commonly occurring serotypes in sheep vaccines. The traditional approach to vaccine formulation has been to incorporate all of the most commonly isolated serotypes but this leads to complex formulations which imposes a very heavy antigenic load on vaccinated animals. The identification of shared protective antigens is therefore important in the development and improvement of this type of vaccine as their inclusion will reduce the overall quantity of antigen required.

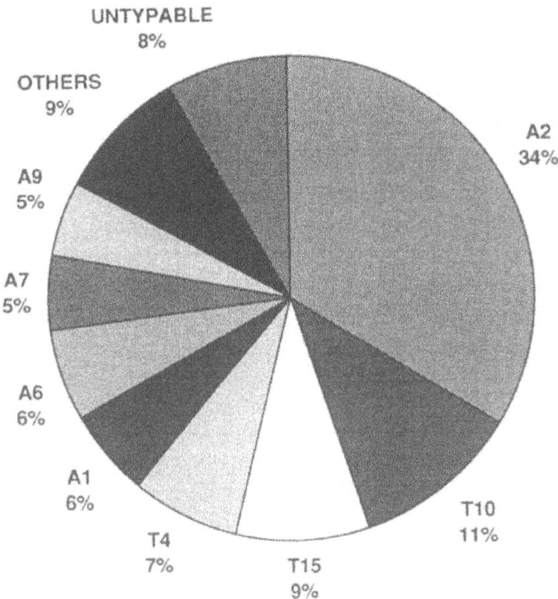

Figure 1. Relative prevalence of serotypes in ovine isolates of *Pasteurella haemolytica* referred to MRI for serotyping 1982-1993.

2. The specificity of the bacterium for host species excludes to a large extent the use of laboratory animals. The specific association of *P. haemolytica* with pathogenesis in ruminants is well documented and recent findings have suggested that this bacterium is well adapted to the ruminant host. Leukotoxin was shown to be highly specific for ruminant leukocytes (Kaehler *et al.*, 1980; Sutherland *et al.*, 1983) with little observable activity against the cells of any other species while transferrin binding activity was also shown to be restricted to ruminant transferrin (Schryvers and Ogunnariwo, 1990). A consequence of this specificity is that laboratory animal models cannot reflect the full pathogenic potential of *P. haemolytica*. A model such as the mouse septicaemic one (Evans and Wells, 1979) is useful as an indicator for vaccine efficacy and passive protection studies (Evans and Wells, 1979; Donachie *et al.*, 1986) but it should be borne in mind that that results obtained can be misleading when extrapolated into sheep.

3. The wide distribution of the pathogen in apparently normally healthy animals confounds experimental work in conventionally reared animals. With the limitations of laboratory animal models the development of sheep disease models has been essential. Early studies concentrated on the delivery of high numbers of bacteria directly into the lung or intravenously but such high numbers of bacteria gave rise to effects which were not related to

clinical development of pasteurellosis (Smith, 1960a,b; Smith, 1964). A major factor in this difficulty is the colonisation and the carriage of *P. haemolytica* by apparently healthy sheep. *P. haemolytica* is recoverable from young lambs hours after birth (Al-Sultan and Aitken, 1982) and is carried by a large proportion of healthy sheep in their naso-pharynx or in the tonsils (Gilmour and Gilmour, 1989). These animals are therefore compromised when used to study most aspects of pasteurellosis. These sheep also have a natural resistance to experimental infection. These problems were overcome at MRI by the use of specific pathogen free (SPF) lambs which are derived by hysterectomy and raised in microbiologically secure facilities (Hart *et al*, 1971). Biotype A disease is produced by infecting SPF lambs with parainfluenza virus type 3 (PI_3) prior to a *P. haemolytica* aerosol infection. The combination of PI_3 as a predisposing infectious agent and *P. haemolytica* administered in an aerosol produces a disease that is clinically and pathologically identical to that found in a naturally infected animal. The experimental production of biotype T type disease has always been problematical but recent experiments at MRI have demonstrated that this disease can be reproduced injecting SPF lambs subcutaneously with a log phase growth culture of T10 or T15. Studies into the pathogenesis of T type disease using this system suggest that endotoxin is a major virulence factor and that endotoxic shock is the main clinical manifestation of the disease (Hodgson *et al.*, 1993).

BACKGROUND TO MOREDUN WORK

The SPF lamb models described above have enabled us to assess the efficacy of potential individual vaccine components and proposed complete vaccines. Work by Gilmour *et al.*, (1983) with sodium salicylate extract (SSE) vaccines had shown that serotype specific immunity against *P. haemolytica* serotypes A1, A6 and A9 could be achieved. However for the control of pasteurellosis the most important objective was the development of effective vaccines against the A2 serotype. As well as being the most prevalent serotype it is also the least immunogenic in sheep and mice (Evans and Wells, 1979; Gilmour *et al.*, 1983). Experiments in sheep had indicated that protection against A2 was difficult to generate using conventional cell-free extract or whole cell vaccines (Gilmour *et al.*, 1983) and a different strategy for the development of effective vaccines against A2 was required. The basis of this new strategy was to show that immunity to A2 disease could be produced, to recover and examine A2 cells grown *in vivo*, to identify possible immunogens and to assess the protective potential of any immunogens.

Convalescent immunity

A major aim of the strategy was achieved when SPF lambs, which were infected with serotype A2 cells by aerosol, and allowed to recover were shown to be fully protected against subsequent infection with PI3 virus and *P. haemolytica A2* (Donachie *et al.*, 1986). These lambs exhibited no clinical signs of disease during the second infection, did not have lung lesions or evidence of bacterial infection/colonisation while control, naive lambs developed severe pasteurellosis. When sera and lung washings from convalescent lambs were measured by ELISA and immunoblotting with *P. haemolytica* cell walls high titres of IgG and IgA were present against cell surface components of *P. haemolytica*. In contrast there was little evidence of a cell-mediated response. Control, uninfected lambs did not show any specific immune responses to *P. haemolytica*. The studies with SPF convalescent lambs recovered from *P. haemolytica* pneumonia indicated that a strong local response was being mounted and probably played an important role in immunity. This posed a question about the relevance of humoral immunity in A2 pasteurellosis and prompted further work to determine whether or not it played a role in protection.

Passive immunity

The importance of circulating antibodies in immunity was demonstrated by Jones *et al.*, (1989) who hyperimmunised conventionally-reared sheep with *P. haemolytica* SSE and bacterin vaccines and then used the sera to passively immunise SPF lambs. In addition to showing that good protective immunity was achievable by this method the results proved that protection was antibody mediated and that circulating antibody was delivered to the lungs when the lamb was challenged by aerosol. While local pulmonary mucosal immunity is important, and the ideal objective for potential vaccines, the demonstration of effective immunity by circulating antibody proves that parenteral vaccination is possible.

Antigen studies

Bacterial pathogens growing *in vivo* in mammalian hosts experience a very hostile environment which in most cases will induce changes in the physiology and structure of the bacterium if it is to survive and multiply. One way in which the presence of novel antigens expressed by bacteria *in vivo* can be detected is by immunoblotting bacterial cells recovered directly without subculture from tissues with convalescent sera.

Sera from the recovered lambs in the convalescent immunity experiments described above were used in immunblots with *P. haemolytica* cells recovered from the pleural fluid of lambs with pneumonic pasteurellosis. (Donachie and Gilmour, 1988). Differences were identified between these *in vivo* grown cells and cells from the same isolate grown *in vitro* in nutrient broth for several passages and subsequent studies indicated that *in vivo* expressed antigens of 70 and 100 kDa were iron-regulated and reacted strongly with immune serum antibodies. The incorporation of iron chelators into laboratory growth media allowed the expression of these proteins in the laboratory. Work by Deneer and Potter (1989), Morck *et al.*, (1991) and Davies *et al.*, (1992) confirmed these findings in A2 and other serotypes and a further iron-regulated protein (IRP) at 35 kDa, not associated with the outer membrane, was reported by Lainson *et al.*, (1991). Murray *et al.*, (1992) showed that the 35, 70 and 100kDa IRPs found in *P. haemolytica* biotype A serotypable isolates were cross reactive in immunoblotting studies using polyclonal and monoclonal antibodies raised against A2 IRPs. This type of cross-reactivity is indicative of potential cross-protective activity *in vivo* and this was tested in an SPF lamb vaccine experiment described below. The IRPs of the T biotype isolates did not cross-react with the antibodies to the same extent as those of the A biotypes and further comparison of the A and T biotype IRPs has revealed that differences in size and antigenicity exist. The main T biotype IRPs have estimated molecular weights of 78 and 100-110 kDa (Deneer and Potter, 1989; Murray *et al.*, 1992).

More detailed examination of the *P. haemolytica* IRPs has revealed that some of them function as transferrin binding proteins (tbps) (Ogunnariwo and Schryvers, 1990). The 110 kDa protein, probably the same protein identified as 100kDa by previous workers, has been shown to have homology with tbps found in other HAP organisms. Studies on the 70kDa protein have revealed that this is not a single protein but a mixture of at least three different proteins 70 a, b and c of which a and b are iron-regulated. The results of analyses on these proteins (Table 1), shows that the 70a protein, the most prominent band in SDS PAGE stained gels, appears to be only weakly immunogenic. The 70b protein is present in smaller amounts but is highly immunogenic and associated with the 110 kDa tbp when the transferrin binding receptor is purified by affinity chromatography (R. Davies, MRI, personal communication). It is probably part of the transferrin binding complex which has been reported to comprise a 110 and a 70 kDa protein. (Ogunnariwo and Schryvers, 1990). The 70c protein which is not iron-regulated, is located in the cytoplasm or the periplasmic space and has an N-terminal amino acid sequence which is similar to the heat shock protein DNAk. Interestingly it was initially identified by its reactivity with a monoclonal antibody raised against A2 IRP cells. This antibody was specific only for A2 cells and antibodies to DNAk and the Mycobacterial homologue do not react with 70c. This suggests species-specific epitopes are found in this highly conserved protein.

Table 1. Characterization of the 70kDa proteins in *P.haemolytica* A2 cells expressing IRPs

	70a	70b	70c
Location in cell	Outer membrane	Outer membrane	Cytoplasm/periplasm
Sarkosyl soluble	-	+	+
Iron regulated	+	+	-
Transferrin associated	-	+	-
Serotype specificity	-	-	+
Reaction with convalescent sera		+	+
Anti-SSE-IRP		+	+

VACCINE STUDIES

IRP vaccines

The findings from laboratory studies suggested that IRPs were involved in the immune response and might contribute to protective immunity in sheep. To test this hypothesis a series of experiments was conducted using SPF lambs immunised with vaccines which either contained or lacked IRP antigens. Table 2 shows the results of an experiment where a cell free sodium salicylate extract (SSE), prepared from the same strain of A2 grown in iron replete (SSE) or iron depleted (SSE-IRP) conditions, was used to vaccinate SPF lambs which were then challenged with homologous *P. haemolytica* A2 cells (Gilmour *et al.*, 1991).

Table 2. Group mean disease scores and percentage protection of SSE and SSE-IRP vaccines in SPF lambs challenged with *P. haemolytica* A2

Group	No. in group	Vaccine antigen	Deaths	Group mean disease score[a]	Protection[b] (%)
1	13	SSE	6	26.7	47.4
2	8	SSE-IRP	0	0.5[c]	99.0
3	7	Unvaccinated	7	50.8	0

[a] Disease score = the sum of clinical findings, consolidated lung lesion area, *P. haemolytica* isolation index and pleurisy scores (Jones *et al.* 1989)

[b] % Protection = $\left(1 - \frac{\text{Group mean score of vaccinate group}}{\text{Group mean score of unvaccinated group}}\right) \times 100$

[c] Significantly different from unvaccinated Group ($p<0.01$) using the Mann-Whitney ranking test

The vaccine containing IRP conferred almost complete protection on SPF lambs. Immunoblotting of the sera from lambs vaccinated with SSE-IRP showed a strong response to the IRPs while other serological tests indicated that there were no leukotoxin neutralising antibodies or significant levels of bactericidal antibodies. Similar results were obtained with a heat-killed bacterin vaccine which offers a more convenient and less costly way to produce a vaccine.

Vaccines containing IRP have also been tested against other serotypes with similar success but it has to be remembered that high levels of protection had already been achieved with SSE vaccines against A1, A6 and A9 (Gilmour *et al.*, 1983). Multivalent IRP vaccines containing these 3 serotypes and A2 have been tested and shown to confer significant protection against each of the homologous strains. To test whether IRP vaccines were cross protective SPF lambs immunised with a multivalent vaccine containing IRP antigens of A1, A2, A6, A7, A9, T3, T4, T10 and T15 were challenged with *P. haemolytica* A12. Significant protection was obtained (98%) compared to unvaccinated controls and immunoblotting with vaccinated lamb sera against A12 envelopes showed strong responses to A12 outer membrane proteins including IRPs (Donachie, unpublished observations).

The efficacy of IRP vaccines against T types has also been demonstrated in SPF lambs. Lambs were vaccinated on two occasions with a multivalent vaccine containing IRPs and challenged with two subcutaneous doses of either *P. haemolytica* T10 or T15 (Hodgson *et al.*, 1993). Table 3 shows the overall efficacy of the vaccines when compared to unvaccinated animals. In all experiments IRP vaccines conferred significant protection against T biotype challenge. All vaccinated animals showed a strong response to IRPs in immunoblotting.

Table 3. Efficacy of T type vaccines

Experiment	Group	Vaccine	Challenge serotype	Deaths/ number infected	Protection[a]
1	1	T10 IRP bacterin+ Lkt	T10	1/10	Yes ($p<0.05$)
	2	Unvaccinated	T10	6/10	
2	1	Multivalent IRP bacterin	T10	4/10	Yes ($p<0.03$)
	2	Unvaccinated	T10	9/10	
3	1	Multivalent IRP bacterin	T15	7/15	Yes ($p<0.05$)
	2	Unvaccinated	T15	13/15	

[a] When compared to unvaccinated control group using the Chi-squared test

Leukotoxin

Leukotoxin has been identified and widely reported as an important virulence factor and vaccine candidate for *P. haemolytica* infections in cattle (Shewen *et al.*, 1988; Jim *et al.*, 1988). Since leukotoxin is produced by all serotypes of *P. haemolytica* recovered from sheep (Sutherland and Donachie, 1985) and neutralising antibodies are present in the sera of experimentally infected sheep (Sutherland, 1989) it can be assumed that it also has great potential as a protective antigen against ovine *P. haemolytica* infections. A crude leukotoxin vaccine, produced by growing *P. haemolytica* A2 in dialysis sacs and concentrating the dialysed cell-free supernatant fluid by lyophilisation, conferred protection on SPF in a homologous challenge experiment (Table 4) (Sutherland *et al.*, 1989).

Table 4. Group means of the experimental disease indices and percentage protection of different *P. haemolytica* vaccines

Group	Vaccine group (n)	Deaths	Disease index	Percent protection	Serum LN (±SEM)	Bactericidal assay (%K) (±SEM)
1	Leukotoxin (7)	1	7	86	15.1 (± 4.3)	100 (± 0.3)
2	SSE (13)	6	26	47	0.1 (±0.1)	77 (± 4.0)
3	Leukotoxin /SSE (14)	0	1	98	38.7 (±10.6)	96 (±2.5)
4	Unvaccinated (7)	7	50	0	0.0	38 (±5.8)

LN - leukotoxin neutralising, %K - percent killing

In addition to the overall disease scores and deaths these results show the degree of anti-leukotoxin and bactericidal antibody response elicited by each of the vaccines. The vaccines containing leukotoxin produced significantly higher leukotxin neutralising and bactericidal antibody titres than those that did not Bactericidal antibodies are reported to be directed against the lipopolysaccharide (LPS) component of *P. haemolytica* A2 (Sutherland, 1988). As this was a crude preparation it is assumed that an antibody response was elicited against cellular components in addition to leukotoxin and LPS such as capsule polysaccharide, outer membrane proteins and other excreted products and their role in immunity needs further consideration. However there was no evidence of antibodies against IRPs.

It has not been possible to purify native leukotoxin in a biologically active form (Himmel *et al.*, 1982; Sutherland and Redmond, 1986; Chang *et al.*, 1987) making it difficult to define its exact role in pathogenesis and immunity. In addition native leukotoxin is produced in relatively small quantities making it necessary to concentrate culture supernatants. Following the identification and isolation of the gene encoding leukotoxin A the close homology of this gene with that encoding the α haemolsin of *Escherichia coli* was noted and subsequent studies led to the designation of the RTX family of toxins (Lo *et al.*, 1985; Lo, 1990). These toxins show a high degree of homology among *P. haemolytica* serotypes and this may explain the cross-neutralization previously reported for the native leukotoxins (Shewen and Wilkie, 1983). This shared anigenicity is useful in the context of sheep vaccines where the inclusion of one leukotoxin may generate cross neutralizing activity against a range of serotypes.

Recombinant leukotoxin (rLKT) from *P. haemolytica* A1 was produced and tested in calves against homologous challenge but shown not to be protective on its own and the authors suggested that other cellular components, most likely at the surface, were required for protection (Conlon *et al.*, 1991). The use of rLKT in sheep vaccines has also proven to be less effective than crude native leukotoxin. rLKT, produced as an insoluble intracellular protein, was combined with a multivalent bacterin-IRP vaccine and used to immunise SPF lambs against A2 challenge. The results (Table 5) show that total disease scores rise with increasing rLKT concentration while the LKT serum neutralizing titer falls.

This suggests that there is a suppression of the immunological response to vaccination by the rLKT component of the vaccine. The rLKT used in the cattle experiments was a soluble, biologically active product which when added to other soluble components of *P. haemolytica* produced enhanced immunity (Conlon *et al.*, 1991). The rLKT used in lambs was an insoluble, biologically inactive product. This may indicate the importance of conformational epitopes. When antisera to native LKT is absorbed with MRI rLKT the unabsorbed antibodies still neutralise native leukotoxin indicating that MRI rLKT is deficient in some conformational epitopes essential for activity and that this insoluble material is not suitable as a vaccine component. It is possible that solubilization and re-naturation of rLKTA may reconstitute such conformational epitopes. More work is needed in this area to determine the best possible formulation for the inclusion of LKT in *Pasteurella* vaccines.

Table 5. Group mean total disease scores and serum LKT neutralizing titers in SPF lambs immunised with bacterin-IRP combined with different doses of rLKT.

Group	Vaccine	Total disease score	P*	Serum LN
1	Multivalent Bacterin	11.8 ± 16.8	0.07	1
2	Multivalent Bacterin + rLKT (0.5μg)	14.6 ± 16.1	0.11	3.8 ± 0.6
3	Multivalent Bacterin + rLKT(5 μg)	16.6 ± 17.4	0.18	2.4 ± 1.4
4	Multivalent Bacterin + rLKT(50 μg)	23.5 ± 18.9	0.55	1.2 ± 0.4
5	Unvaccinated	24.6± 17.9	-	1

p* Mann-Whitney ranking test against the unvaccinated group.

Polysaccharide capsule of *P. haemolytica* A2

The capsule of the A2 cell is composed of colaminic acid (2,8 α linked N-acetyl neuraminic acid) (Adlam *et al.*, 1987) and is identical to that found in *Neisseria meningitidis* Group B isolates, *E. coli* K1 and *Moraxella nonliquafaciens* isolates. The first two organisms are recognised as major pathogens and both cause severe meningitis in human infants. This capsular structure is known to be an extremely poor immunogen and it has been suggested that immune tolerance, due to shared antigenicity with sialic acid residues commonly found in mammalian tissues, loose configuration and host neuraminidases may all contribute to this (Moreno *et al.*, 1985).

P. haemolytica A2 is by far the least immunogenic of the commonly found serotypes and it is likely that the capsule is a major contributor to this characteristic. The importance of anti-capsule antibodies in immunity to A2 infection was indicated in work with monoclonal antibodies raised against *N. meningitidis* group B capsule. These antibodies had been shown to be bactericidal to *N. meningitidis* and conferred passive protection on mice against homologous challenge (Moreno *et al.*, 1983). In our hands the same antibodies reacted strongly in the indirect haemagglutination (IHA) test and were protective against septicaemic A2 infection in mice. That this protection was achieved with an antibody directed at a single epitope on the capsule structure suggests that antibodies to A2 capsule are extremely important in immunity.

Direct immunisation of mice was carried out using a meningococcal vaccine containing group B capsular polysaccharide naturally complexed with outer membrane proteins (CP/OMP) (Moreno *et al.*, 1985). This vaccine conferred protection and elicited antibodies against the capsule of A2 but not the outer membrane proteins confirming the specificity of the protective response. However this vaccine did not elicit anti-capsule antibodies in 8 week old SPF lambs (author, unpublished observations). A CP/OMP vaccine was prepared from *P. haemolytica* A2 following the method described by Moreno *et al.*, (1985) for *N meningitidis*. This was then used to immunise mice and sheep and the results of the mouse immunisation are shown in Table 6. The vaccine gave significant protection compared to whole cells, culture supernatant and crude CP/OMP complex. The level of protection was comparable to that seen with meningococcal vaccine. Adult ewe sheep and their 1 month old lambs were vaccinated with purified and crude CP/OMP complexes and their serum antibody titers measured by IHA and the passive immunisation potential assessed in the mouse model. The results of the IHA test (Figure 2) show that adult sheep responded well to vaccination and antibody titers against capsule were significantly higher in sheep were immunised with the purified complex. Lambs did not respond to immunisation with either vaccine. Mice were passively protected against A2 challenge only with the sera from adult sheep immunised with

Table 6 Counts of *P.haemolytica* A2 in the livers of unvaccinated control mice and mice vaccinated with *P.haemolytica* A2 antigens after challenge[a]

Group	Vaccine	Vaccine dose	Mean count in liver 6 h post challenge (Log_{10} cfu)	Statistical significance[b]
1	Heat killed cells	2.9 x 10^8 cfu/ml	4.5	NS
2	Heated culture extract	2.9 x 10^8 cfu/ml	4.3	NS
3	Crude CP/OMP	10μg[c]	4.1	NS
4	Purified CP/OMP	10μg[c]	1.25	$p<0.002$
5	Meningococcal Group B CP/OMP	1μg total wt	1.8	$p<0.005$
6	Unvaccinated control		5.5	

a Challenge dose = P.haemolytica A2 1×10^7 cfu
b When compared to the unvaccinated control group in the Mann-Whitney Ranking test. NS = not significant
c weight of sialic acid component

the purified PC/OMP complex. These results suggest that antibodies to polysaccharide capsule are crucially important in immunity but that there may be some age restriction since young lambs seem unable to mount an antibody response to vaccination with polysaccharide. Evidence from studies in children have shown that the purified polysaccharides of group A and group C meningococci are ineffective in stimulating immunity in children under 2 years and 6 months respectively (Lifely *et al.*, 1987).

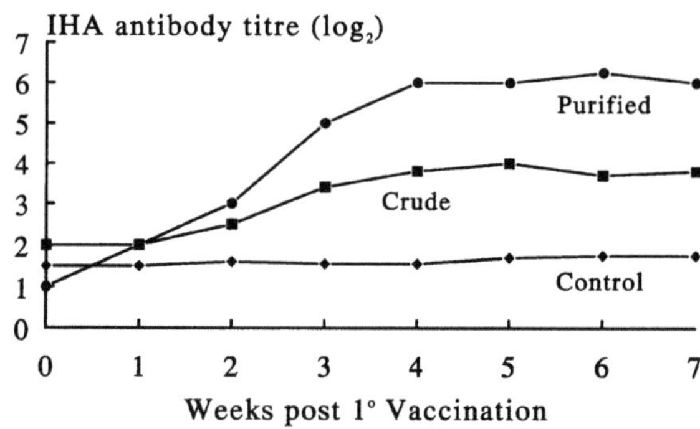

Figure 2. IHA titers of adult ewes vaccinated with purified CP/OMP, crude CP/OMP or sham vaccinated.

CONCLUSIONS

The development of effective vaccines for the prevention of pasteurellosis in sheep has differed from that for cattle disease in that both biotypes and a larger number of serotypes of *P. haemolytica* are involved. The most prevalent is A2 and this has presented the greatest obstacle in the development of effective vaccines. A major advance in the development of vaccines was the analysis of *in vivo* grown bacterial cells and the identification of IRPs. The incorporation of IRPs into vaccines has enhanced protection and given cross protection between serotypes. However there are physical and antigenic differences between A and T biotype IRPs which suggest that cross-protection will not extend to biotype level. Native leukotoxin has been shown to be an effective protective immunogen but more research has to be carried out on the rLKT before it will be a realistic alternative. The A2 capsular polysaccharide is a very effective immunogen in mice and sheep when used as a CP/OMP complex but this is not a commercially attractive option at present. A recombinant product or a peptide mimic may offer a different approach to commercial production. The findings of this development work indicate that the optimum vaccines for sheep pasteurellosis should contain IRPs, leukotoxin and the capsular polysaccharide of A2.

REFERENCES

Adlam, C., 1989, The structure, function and properties of cellular and extracellular components of *Pasteurella haemolytica*, *in* "Pasteurella and Pasteuerellosis", C. Adlam and J.M. Rutter, ed., Academic Press, London, San Diego, pp 223-262.

Adlam, C., Knights, J.M., Mugridge, A., Williams, J.M. and Lindon, J.C., 1987, Production of colominic acid by *Pasteurella haemolytica* serotype A2 organisms. *FEMS-Microbiol. Lett.* 42:23-25

Alley, M.R., (1991), Pneumonia in sheep. Vet. Ann. 31:51-58.

Al-Sultan, I.I. and Aitken, I.D.,1984, Promotion of *Pasteurella haemolytica* infection in mice by iron. *Res. Vet. Sci.* 36:385-386.

Ball,H.J., Connolly, M. And Cassidy,J.,1993, Pasteurella haemolytica serotypes isolated in Northern Ireland during 1989-1991. *Br.Vet. J.* 149:561-570.

Biberstein, E.L., Gills, M. And Knight, H.D., 1960, Serological types of *Pasteurella haemolytica. Cornell Vet.* 50:283-300.

Chang, Y.-F., Young, R., Post, d. and Struck, D.K., 1987, Identification and characterization of the *Pasteurella haemolytica* leukotoxin. *Infect. Immun.* 55:2348-2354.

Conlon, J.A., Shewen, P.E. and Lo, R.Y.C., 1991, Efficacy of recombinant leukotoxin in protection against pneumonic challenge with live *Pasteurella haemolytica* A1. *Infect. Immun.* 59:587-591.

Cullinane,L., Alley, M.R. Marshall,R.B and Manktelow, B.W., 1987, *Bordetella parapertussis* from lambs. *N. Z. Vet. J.* 35:175.

Davies, D.H., Davies, G.B. and Price,M.C, 1980, A longitudinal serological survey of respiratory virus infection in lambs. *N. Z. Vet.J.* 28:125-127.

Davies, D.H., Dungworth, D.L.and Mariassy, A.T.,1981, Experimental adenovirus infection of lambs. *Vet.Micro.* 6:113-128.

Davies, R.L., Parton, R., Coote, J.G. Gibbs, H.A. and Freer, J.H., 1992, Outer-membrane protein and lipopolysaccharide variation in *Pasteurella haemolytica* serotype A1 under different growth conditions. *J. Gen. Micro.* 138:909-22.

Deneer, H.G. and Potter, A.A., Iron-repressible outer-membrane proteins of *Pasteurella haemolytica. J. Gen. Micro.*, 1989, 135:435-443.

Donachie, W., Burrells, C., Sutherland, A.S., Gilmour, J.S. and Gilmour, N.J.L., 1986, Immunity of specific pathogen-free lambs to challenge with an aerosol of *Pasteurella haemolytica* biotype A serotype 2. Pulmonary antibody and cell responses to primary and secondary infections. *Vet. Immun. Immunopath.* 11:265-279.

Donachie W. and Gilmour N.J.L., 1988, Sheep antibody response to cell wall antigens expressed in vivo by *Pasteurella haemolytica* serotype A2. *FEMS Micro. Lett.* 56:271-276.

Donachie, W., Sutherland, A.D. and Jones, G.E., 1986, Assessment of immunity to *Pasteurella haemolytica* in sheep by *in vitro* methods, *Develop.biol. Standard.* 64:63-69.

Dungal, N., 1931, Contagious pneumonia in sheep. *J. Comp. Pathol.* 44:126-143.

Evans, H.B. and Wells, P.W., 1979, A mouse model of *Pasteurella haemolytica* infection and its use in the assessment of the efficacy of *Pasteurella haemolytica* vaccines. *Res. Vet. Sci.* 27:213-217.

Frank,J.H., 1982, Serotypes of *Pasteurella haemolytica* in sheep in the midwestern United States. *Am. J. Vet.Res.* 43:2035-2037.

Fraser, J., Gilmour, N.J.L., Laird, S. and Donachie, W., 1982 Prevalence of *Pasteurella haemolytica* serotypes isolated from ovine pasteurellosis in Britain. *Vet. Record.* 110:560-561.

Gilmour, N.J.L., Donachie, W., Sutherland, A.D., Gilmour, J.S., Jones, G.E. and Quirie, M., 1991. A vaccine containing iron-regulated proteins of *Pasteurella haemolytica* A2 enhances .protection against experimental pasteurellosis in lambs. *Vaccine*. 9:137-140.

Gilmour, N.J.L. and Gilmour, J.S., 1989, Pasteurellosis of sheep, *in* "Pasteurella and Pasteuerellosis", C. Adlam and J.M. Rutter, ed., Academic Press, London, San Diego, pp 223-262.

Gilmour, N.J.L., Martin, W.B., Sharp, J.M., Thompson, D.A., Wells, P.W. and Donachie, W., 1983, Experimental immunisation of lambs against pneumonic pasteurellosis. *Res. Vet. Sci.* 35:80-86.

Gilmour,N.J.L., Quirie, M., Jones, G.E. and Gilmour, J.S., 1988, Metaphylactic use of long-acting oxytetracycline in pasteurellosis in lambs.*Vet. Record.* 123:443-444.

Gilmour, N.J.L., Sharp, J.M. and Gilmour, J.S., 1982, Effect of oxytetracycline therapy in experimentally induced pneumonic pasteurellosis. *Vet. Rec.* 111:97-99.

Hart, R., Mackay, J.M.K., McVittie,C.R. and Mellor, D.J.,1971, A technique for the derivation of lambs by hysterectomy. *Br.Vet. J.*, 127:419-424.

Himmel, M.E., Yates, M.D., Lauerman, L.H. and Sqire, P.G., 1982, Purification and partial characterization of a macrrophage cytotoxin from *Pasteurella haemolytica. Am.J. Vet. Res.* 43:764-767.

Hodgson, J.C., Moon, G.M. Quirie, M. and Donachie, W., 1993, Biochemical signs of endotoxaemia in lambs challenged with T10 strain of *Pasteurella haemolytica* and the effect of vaccination on the host response, *Proc.Sheep. Vet. Soc.* 17:201-204.

Hore, D.E., Stevenson, R.G., Gilmour, N.J.L., Vantsis, J.T. and Thompson, D.A., 1968, Isolation of parainfluenza virus from the lungs and nasal passages of sheep showing respiratory disease. *J.Comp. Path.* 78:259-269.

Jian, Z., Alley, M.R. and Manktelow, B.W., 1991, Experimental pneumonia in mice produced by combined administration of Bordetella parapertussis and Pasteurella haemolytica from sheep. *J.Comp. Pathol.* 104:233-243.

Jim, K., Guichon, T. and Shaw, G., 1988, Protecting feedlot calves from pneumonic pasteurellosis. *Vet. Med.* 83:1084-1087.

Jones,G.E., Donachie, W., Sutherland,A.D., Knox D.P. and Gilmour J.S., 1989, Protection of lambs against experimental pneumonic pasteurellosis by transfer of immune serum. *Vet. Micro.* 20:59-71.

Jones,G.E., Gilmour,J.S., Rae, A.G., McLauchlan, M. and Nettleton, P.F., 1986, A review of experiments on the reproduction of chronic pneumonia in sheep by the use of pneumonic lung homogenate suspensions. *Vet.Bull.* 56:251-263.

Kaehler, K.L., Markham, R.J.F., Muscoplat, C.C. and Johnson, D.W., 1980, Evidence of cytocidal effects of *Pasteurella haemolytica* on bovine peripheral blood mononuclear leukocytes. *Am. J. Vet. Res.* 41:1690-1693.

Lainson, F.A., Harkins,D.C., Wilson, C.F., Sutherland, A.D.,Murray J.E. , Donachie, W. and Baird, G.D., 1991, Identification and localisation of an iron-regulated 35kd protein of *Pasteurella haemolytica* serotype A2. *J. Gen. Micro.* 137:219-226.

LeaMaster, B.R., Evermann, J.F. and Lehmkuhl, H.D., 1987, Identification of ovine adenovirus types five and six in an epizootic of respiratory tract disease in recently weaned lambs. *J. Amer.Vet. Med. Ass.* 190:1545-1547.

Lifely, M.R., Moreno, C and Limdon, J.C., 1987, An integrated molecular and immunological approach towards a meningococcal group B vaccine. *Vaccine*. 5:11-26.

Lo, R.Y.C., 1990, Molecular characterization of cytotoxins produced by *Haemophilus, Actinobacillus, Pasteurella. Can., J. Vet. Res.* 54.S33-S35.

Lo, R.Y.C., Shewen, P.E., Strathdee, C.A. and Greer, C.N., 1985, Cloning and expression of the leukotoxin gene of *Pasteurella haemolytica* A1 in *Eschirichia coli* K-12. *Infect. Immun.* 50:667-671.

Morck, D.W., Ellis, B.D., Domingue, P.A., Olson, M.E. and Costerton, J.W., 1991, *In vivo* expression of iron regulated outer-membrane proteins in *Pasteurella haemolytica*-A1. *Microb-Pathog.*, 11:373-378.

Moreno, C., Hewitt, J., Hastings,K. and Brown, D., 1983, Immunological properties of monoclonal antibodies specific for meningococcal polysaccharides: the protective capacity of IgM antibodies specific for polysaccharide group B. *J.Gen. Micro.* 129: 2451-2456.

Moreno, C., Lifely, M.R. and Esdaile, J., 1985, Immunity and protection of mice against *Neisseria meningitidis* Group B by vaccination, using polsaccharide complexed with outer membrane proteins: a comparison with purified B polysaccharide. *Infect. Immun.* 47: 527-533.

Murray, J.E., Davies, R.C., Lainson, F.A., Wilson, C.F. and Donachie, W., 1992, Antigenic analysis of iron-regulated proteins in *Pasteurella haemolytica* A and T biotypes by immunoblotting reveals biotype specific epitopes. *J. Gen. Micro.* 138:283-288.

Ogunnariwo, J.A.and Schryvers, A.B., 1990, Iron acquisition in *Pasteurella haemolytica*: expression and identification of a bovine-specific transferrin receptor.*Infect. Immun.* 58:2091-2097.

Porter, J. F., Connor, K. and Donachie, W., 1994, Isolation and characterization of *Bordetella parapertussis*-like bacteria from ovine lungs. *Microbiology.* 140:255-261.

Prince, D.V., Clarke, J.K. and Alley, M.R., 1985, Serotypes of *Pasteurella haemolytica* from the respiratory tract of sheep in New Zealand *N. Z. Vet. J.* 33:76-77.

Rimsay, R.L., Coyle-Dennis, J.E, Lauerman, L.H. and Squire, P.G., 1981, Purification and biological characterization of endotoxin fractions from *Pasteurella haemolytica. Am. J. Vet. Res.* 42:2134-2138.

Sharp.J.M., 1991a, Chronic respiratory virus infections, *in* "Diseases of Sheep", W.B. Martin and I.D. Aitken ed., Blackwell Scientific Publications, London.pp 143-150.

Sharp.J.M., 1991b, Acute respiratory virus infections, *in* "Diseases of Sheep", W.B. Martin and I.D. Aitken ed., Blackwell Scientific Publications, London.pp 139-143.

Sharp, J.M., Gilmour, N.J.L.,Thompson, D.A. and Rushton, B., 1978, Experimental infection of specific pathogen-free lambs with parainfluenza virus type 3 and *Pasteurella haemolytica. J. Comp. Path.* 88:237-243.

Shewen, P.E., Sharpe, A. and Wilkie, B. N., 1988, Efficacy testing a P*asteurella haemolytica* extract vaccine. *Vet. Med* 10:1078-1083.

Shewen, P..E., Wilkie, B.N., 1983, *Pasteurella haemolytica* cytotoxin: production by recognized serotypes and neutralization by type-specific rabbit antisera. *Am.J.Vet.Res.* 44:715-719.

Smith, G.R., 1960a, Virulence and toxicity of *Pasteurella haemolytica* in the experimental production of ovine septicaemia. *J.Comp. Pathol.* 70:429-436.

Smith, G.R., 1960b, The pathogenicity of *Pasteurella haemolytica* for young lambs. *J.Comp. Pathol.* 70:326-340.

Smith, G.R., 1964, Production of pneumonia in adult sheep with cultures of Pasteurella haemolytica type-A. *J.Comp. Pathol.* 74:241-249.

Smith, G.R., 1961, The characteristics of two types of *Pasteuerella haemolytica* associated with different pathological conditions in sheep. *J.Path. and Bact.* 81:431-440.

Sutherland, A.D., 1988, A rapid micro-method for the study of antibody-mediated killing of bacteria, with specific application to infection of sheep with *Pasteurella haemolytica. Vet. Microbiol.* 16:263-271.

Sutherland, A.D. and Donachie, W., 1986, Cytotoxic effect of serotypes of *Pasteurella haemolytica* on sheep bronchoalveolar macrophages. *Vet. Micro.* 11:331-336.

Sutherland A.D., Donachie, W., Jones G.E. and Quirie, M., 1988, A crude cytotoxin vaccine protects sheep against experimental *Pasteurella haemolytica* serotype A2 infection.*Vet. Micro.* 19:175-181.

Sutherland, A.D., Gray, E., and Wells, P.W., 1983, Cytotoxic effect of *Pasteurella haemolytica* on ovine bronchoalveolar macrophages in vitro. *Vet. Microbiol.* 8:3-15.

Sneath, P.H.A. and Stevens, M., 1990, *Actinobacillus rossii* sp. nov., *Actinobacillus seminis* sp. nov., nom. rev., *Pasteurella bettii* sp. nov., *Pasteurella lymphangitidis* sp. nov., *Pasteurella mairi* sp. nov., and *Pasteurella trehalosi* sp. *Nov. Int. J. Syst. Bact.* 40: 148-153.

Stamp, J.T., Watt, A.A. and Thomlinson,J.R., 1955, *Pasteurella haemolytica* septicaemia of lambs. *J. Comp. Pathol.* 65:183-196.

VIRAL-BACTERIAL SYNERGISTIC INTERACTIONS/PATHOGENESIS IN CATTLE

L.A. Babiuk, M. Morsy, M. Campos and R. Harland

Veterinary Infectious Disease Organization
124 Veterinary Road
Saskatoon, SK
S7N 0W0

INTRODUCTION

Animals are constantly exposed to microorganisms present in the environment, especially in aerosols which easily enter the respiratory tract, yet pneumonia is a relatively rare event. In fact many organisms associated with lower respiratory tract disease can be part of the commensal bacterial flora of the upper respiratory tract. The normal defence mechanisms of healthy animals maintains the homeostasis between the host and the microorganisms. This implies that very effective defense mechanisms are present to eliminate the vast majority of microorganisms before they establish themselves in the lower respiratory tract and cause clinical disease. However, if there is a disruption of the normal lung defenses, animals fail to clear the infection. Although a number of "stressors" might be involved in disrupting homeostasis, viruses clearly can be critical to enhancing bacterial colonization. It is estimated that individuals suffering from a severe respiratory viral infection have a 40% chance of developing bacterial pneumonia (Jakab, 1982). In cattle, shipping fever or bovine pneumonic pasteurellosis is considered to be one of the major causes of economic loss to the cattle producer. These losses are estimated to be as high as one billion dollars annually. This disease is thought to be caused by the interaction of stressors (management) and infection by viruses and bacteria acting synergistically. Based on these observations a number of viral-bacterial models have been used to reproduce "shipping fever" and to study the interaction between viruses and bacteria in respiratory disease (Babiuk and Acres, 1984). Our laboratory has focused our studies on a bovine herpesvirus-l (BHV-l) model wherein seronegative animals are infected with an aerosol of BHV-l followed four days later by an aerosol of *Pasteurella haemolytica* (Yates *et al.*, 1983; Bielefeldt Ohmann and Babiuk, 1985). Results from this model will be presented to indicate how viral infection of the respiratory tract by BHV-1 alters the respiratory tract environment to allow *Pasteurella haemolytica* to establish itself and cause pneumonia. These events are summarized in Figure 1. Briefly, in addition to altering the epithelial environment, by infecting and causing cell death, BHV-l also influences host specific and nonspecific defense mechanisms. For example, BHV-l has been shown to infect macrophages/monocytes and thereby alter their function, even if cell death does not occur.

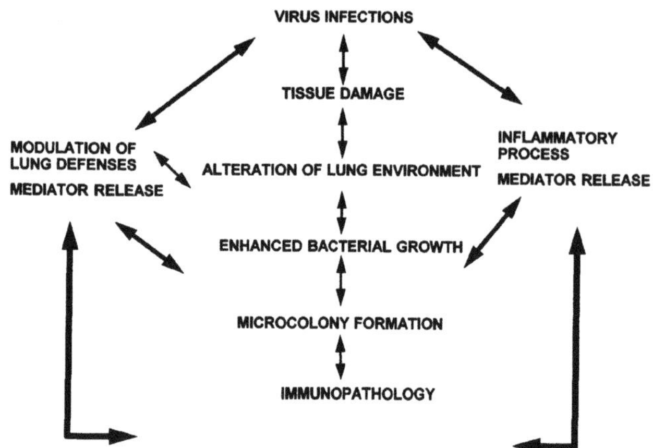

Figure 1. Viral bacterial synergy. Following viral infection, the virus causes tissue damage, induces inflammatory cells and mediator release as well as altering lung defenses. These events alter the environment of the lung allowing bacteria to colonize and produce toxins. These toxins in conjunction with host cell mediators can lead to immunopathology.

Although BHV-1 does not result in a productive infection in bovine lymphocytes it does cause cell death (probably by apoptosis) and consequently there is alteration in responsiveness *in vitro* (Griebel *et al.*, 1990). *In vivo*, lymphocyte reactivity to mitogens is dramatically reduced shortly after viral infection (Bielefeldt Ohmann and Babiuk, 1985), but whether this reduced reactivity is an actual suppression of lymphocyte function or alteration of populations present in the peripheral blood due to lymphocyte trafficking remains to be determined (Griebel *et al.*, 1988). Similarly, polymorphonuclear neutrophil (PMN) functions are also altered following viral infection. Based on these findings our hypothesis is that viruses alter the lung environment and specific and nonspecific defense mechanisms of cattle thereby preventing rapid clearance of the bacteria. The environmental conditions that favor enhanced bacterial growth allow the bacteria to colonize and produce their own products which further reduce immune functions (i.e. leukotoxin). Central to these alterations of immune reactivity is the generation of cytokines. These cytokines are produced by the animal following exposure to viruses as well as bacteria. Evidence will be presented to indicate that cytokines are involved in mediating the acute phase reaction, leukocyte reactivity and by inference can lead to immunopathology.

VIRAL INFECTION AND PULMONARY DEFENSE MECHANISMS

The sequence of events following virus infection of the respiratory tract is similar regardless of whether the virus is an RNA or DNA virus. In bovine respiratory disease it has been clearly shown that a number of viruses including parainfluenza-3, bovine respiratory syncytial virus, bovine herpesvirus and possibly bovine virus diarrhea and bovine adenovirus can all play a major role in enhancing susceptibility of animals to bacterial infection. In each of these virus infections, events begin with the virus initiating contact with the respiratory epithelial cells. This initial contact occurs by aerosolized virus or by direct contact with virus present in droplets or secretions. Immediately upon contact with the respiratory epithelium, the virus interacts with specific host cell receptors and is internalized. Since most respiratory virus infections cause lysis of the cell with the release of a large number of progeny virus there is direct alteration of epithelial cell surface due to cell death. In addition to altering clearance directly by killing ciliated cells, some viruses can alter cellular functions involved in production of secretions. Alteration of the consistency of secretions (more serous) will decrease the efficiency of ciliary clearance of bacteria (Pavia, 1987). In addition to altering secretions, viruses may affect some of the host cells "luxury functions" such that the production of surfactant may be reduced, thereby altering the host nonspecific defense mechanisms. Furthermore, virus infection can alter surface charges on the infected cells which then allow the adherence of bacteria, as the initial stage of colonization. An excellent example of such an event is with influenza viruses where viral infection enhances the ability of bacteria to adhere to virus infected epithelial cells (Nugent and Pesanti, 1983). In the case of bovine respiratory viruses, bovine herpesvirus 1, parainfluenza virus-3 and bovine viral diarrhea virus have been shown to destroy directly ciliary activity whereas bovine respiratory syncytial virus does not (Rossi and Kiesel, 1977).

Defense of the lung against infection is mediated by a plethora of inflammatory and immune effector cells. These act in concert with various soluble factors, especially cytokines which can either up-regulate or down-regulate the activity of specific leukocytes. When some of these mechanisms are altered, due to virus infection or various stressors, this enhances the replication of the bacterium and the end result is pneumonia.

The advent of the bronchial alveolar lavage procedure combined with the availability of various cloned cytokines or cytokine probes has rapidly expanded our understanding, during the past 10 years, of the various secretory and cellular components of the lung and their specific role in interactions amongst themselves as well as with the pathogen. In normal individuals a variety of secretory products are present at various levels of the respiratory tract. In many cases these secretions contain complement components, transferrin, surfactant, fibronectin and immunoglobulins (Woods, 1987; Proctor, 1987; Ramphal *et al.*, 1980; Babiuk *et al.*, 1988). These various humoral factors act in concert with specific immunoglobulin cells to help destroy or reduce the replication of viruses and bacteria. Fibronectin, transferrin and surfactant are three substances which, in addition to being able to potentiate nonimmune opsonic activity in host defense mechanisms are also important in altering the rate of adherence and bacterial replication. For example, fibronectin or fragments thereof have been shown to be functionally important in enhancing macrophage and monocyte phagocytosis of particles both as a conventional opsonin, and also by stimulating macrophages to ingest opsonized particles. Viral infections alter the levels of fibronectin production. Since colonization by Gram-negative bacteria have been correlated with low fibronectin levels (Nugent and Pesanti, 1983; Woods, 1987), this may be a very important factor in colonization. Cell death caused by viruses also increases the release of proteases which can degrade fibronectin on innocent bystander cells, favoring adherence. Thus, reduction of fibronectin can result in poorer clearance as well as enhanced colonization of bacteria. Similarly, surfactant appears to be able to reduce bacterial growth by two mechanisms: by direct killing and by the enhancement of alveolar macrophage activity. Alteration of surfactant levels due to viral infection therefore has a very significant effect on bacterial clearance. Bacteria in turn aid viral replication by producing proteases which may cleave viral glycoproteins and enhance viral infectivity (Kooney and Falkow, 1984). For example, a prerequisite for infection by many myxoviruses or paramyxoviruses is cleavage of the surface glycoproteins. Thus the virus induces surface changes which allow colonization by bacteria which in turn enhance viral replication resulting in reciprocal synergy (Sanford *et al.*, 1978). In summary, the respiratory tract is endowed with many different components all playing a crucial role in clearing various foreign organisms. These factors work in concert to eliminate the organisms, and their relative importance depends on the kinetics of the infection as well as the specific organism involved. Infection can alter some of these factors thereby leading to reduced clearance and pneumonia.

In addition to altering the environment and cellular physiology in the lung, by direct infection of the epithelial cells lining the respiratory tract, viruses can also induce a series of cytokines which attract leukocytes into the site of infection. These cytokines can either enhance or reduce leukocyte activity, alter leukocyte trafficking or provide a secondary cell for the virus to infect. As a result, viruses can either directly or indirectly alter leukocyte activity (Table 1). For example, bovine alveolar macrophages can be infected by bovine respiratory viruses resulting in impairment of immune receptor functions, antibody dependent cell mediated cytotoxicity and cytokine production (Forman and Babiuk, 1982). Virus infection also dramatically influences polymorphonuclear neutrophil (PMN) functions reducing migration and bactericidal activities of PMN (Bielefeldt Ohmann and Babiuk, 1985).

Table 1. Effect of viral infection on leukocyte activity.

Macrophage dysfunction
Chemotaxis
Fc receptors
Phagocytosis
Phagosome-lysosome fusion
Intracellular killing
Decreased lysosome levels
Decreased monokine production
Neutrophil dysfunction
Chemotaxis
Altered maturity
H_2O_2 release
Killing
Chemiluminescence
Lymphocyte dysfunction
Blastogenesis
IL-2 production
Natural killer activity
Altered ratios of various subpopulations

Although many of the mechanisms by which viruses suppress the immune response are not fully understood, virus infection clearly does alter immune reactivity of leukocytes *in vitro* and *in vivo*. Although earlier studies demonstrated that peripheral blood lymphocytes obtained from infected animals had reduced reactivity after infection, it was later determined that some of these alterations in lymphocyte activity could partially be explained by the alteration of leukocyte trafficking, probably mediated by interferon-α (Griebel *et al.*, 1989). More definitive proof was provided by observations that exposure of leukocytes *in vitro* to BHV-l could abolish lymphocyte reactivity (Figure 2). Similarly, preliminary studies demonstrated that modified live BHV-l vaccines interfered with immune response to other vaccines if administered simultaneously (Figure 3). If this occurs with attenuated vaccines then it clearly has implications for increasing susceptibility of animals to other infections following exposure to field strains of virulent virus or even after immunization with modified live virus vaccines.

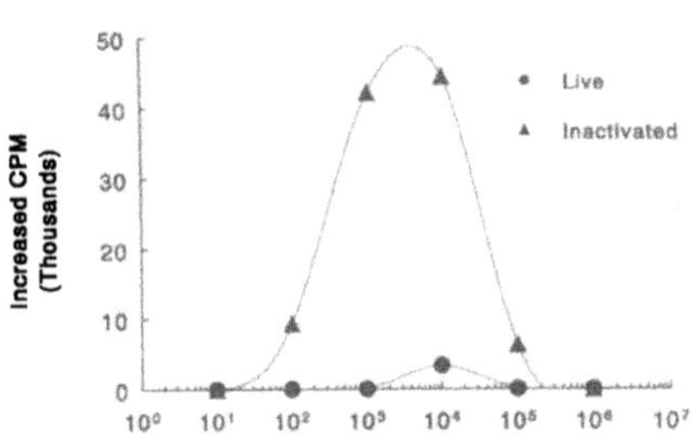

Figure 2. Effect of BHV-1 on lymphoproliferation. Peripheral blood leukocytes from immune animals were cultured in the presence of live or UV inactivated BHV-1 and assayed for proliferation in a ^{3}H-thymidine incorporation assay.

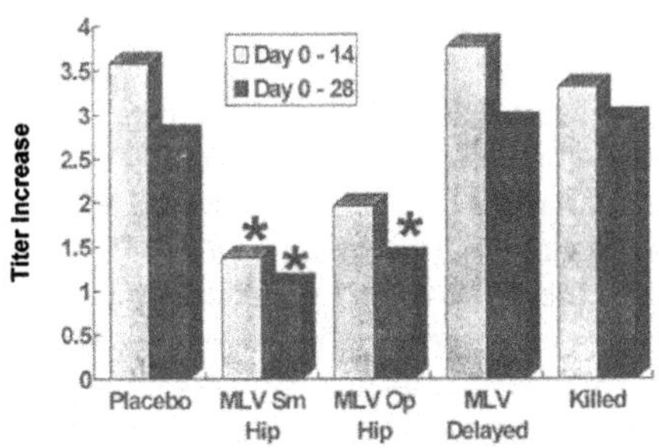

Figure 3. Effect of modified live virus (MLV) vaccination on immune responses to *P. haemolytica* leukotoxin. Conventional calves were immunized with a *Pasteurella haemolytica* leukotoxin containing vaccine alone (Placebo) or in combination with a MLV (containing BHV-1) in the same hip (Sm Hip) or opposite hip (Op Hip). a killed BHV-1 vaccine or MLV was delayed for 2 weeks post leukotoxin vaccination. Antibody titres were tested 14 and 28 days post vaccination significantly different $p < 0.05$ (Tukeys mean test ANOVA).

BACTERIAL ADHERENCE

The binding or attachment of bacteria to various substrates within the environment is an important prerequisite in the process of colonization. Attachment of bacteria to the surface of cells, mediated through adhesins present on the bacterial cell wall, is an important initial step in colonization, and is the basis of infection and production of disease by pathogenic bacteria. Adherence of bacteria to the mucosal surface is a complex phenomenon involving virulence factors, including adhesions (of the pathogenic bacteria) as well as the mucosal surface itself. The expression of these adhesin proteins can be modulated by the environmental condition of the lung and these influence the growth of the bacteria. *Pasteurella haemolytica* has been shown to produce two types of fimbriae which may be involved in adherence as has been shown for many Gram-negative bacteria (Morck *et al.*, 1987; Morck *et al.*, 1988). Whether specific *in vivo* environmental conditions alter the level of expression of these pili is presently unknown but one could speculate that viral infection induces environmental changes conducive to enhanced pili formation and adherence.

Virus infection has been shown to influence adherence and colonization of bacteria. The mechanisms by which viruses alter bacterial adherence are many and include alteration of host cell surface membrane receptors and alteration of the micro-environment where bacterial attachment occurs. The bacteria returns the favor by releasing proteases. The proteases cleave viral glycoproteins and enhance viral infectivity (Nugent and Pesanti, 1983; Sanford *et al.*, 1978; Kooney and Falkow, 1984; Davison and Sanford, 1981). Furthermore, these proteases cleave fibronectin, thereby increasing the ability of the bacteria to adhere. Proteases produced by viral infected cells or bacteria can also decrease the effectiveness of some of the normal lung defenses. For example proteases can cleave and inactivate secretory immunoglobulin that would normally reduce viral or bacterial growth (Kooney and Falkow, 1984). Another important factor for bacterial growth is the availability of iron. If a virus infects the respiratory tract and induces extensive cell damage and haemorrhages, this will increase the level of iron and other divalent cations, which will favor both bacterial adhesion and growth. In many cases iron is an essential element for bacterial growth and iron regulated outer membrane proteins are involved in modulating virulence. Furthermore, iron bound to transferrin, and other divalent cations such as zinc, have been shown to be critical in immune regulation. Thus, the level of iron and zinc can affect not only bacterial adhesion and growth, but also the host's ability to respond to the infection (Sugarman, 1980; Sugarman *et al.*, 1982).

Following the burst of bacterial growth, the bacteria often form micro-colonies which can then enter the lower respiratory tract by aspiration or by "creeping" colonization of the respiratory tract due to reduced mucocilliary clearance following virus infection. Large micro-colonies are difficult to phagocytoze and in fact cause phagocytic cells to release their enzymes, causing tissue damage. In many cases bacteria such as *P. haemolytica* produce a leukotoxin which kills leukocytes, further improving the environment for bacterial replication and adhesion (Benson *et al.*, 1978; Kaeler *et al.*, 1980; Shewen and Wilkie, 1982). Leukotoxin has clearly been shown to be central in *P. haemolytica* induced disease. It is produced only during the logarithmic growth phase, similar to what is observed in cattle where there is an "explosion" of bacterial growth once colonization occurs. This toxin (101 - 105Kd protein) specifically kills leukocytes that are attempting to clear the bacteria. Thus once the bacteria are established it is almost impossible to eliminate them because the effector cells are incapacitated. Indeed, bacteria that are phagocytozed by leukocytes can also kill the leukocytes by leukotoxin produced internally. Although the exact mechanism of cellular killing by leukotoxin has not been elucidated, animals with high levels of antibody to leukotoxin are resistant to disease (Confer *et al.*, 1988). Based on this observation a vaccine containing leukotoxin is being marketed.

CYTOKINES IN BACTERIAL CLEARANCE AND IMMUNOPATHOLOGY

Cytokines can be divided into early cytokines (IL-l, TNFα, IL-6) and late cytokines (IL-2, IL-4, IFNγ). In most instances early cytokines are involved in modulating nonspecific immune responses, chemotaxis and migration of cells to the site of infection and subsequent activation of cells for antigen processing and presentation. In contrast the late cytokines are primarily involved in clonal expansion and fine tuning of the immune response including: differentiation, cytotoxic T cell responses, and isotype switching. Furthermore, cytokines can be involved in maintaining a balance between a THl and TH2 immune response (Mosmann *et al.*, 1986). When these cytokines are produced at the appropriate times and in appropriate concentrations the end result is clearance of the infection and concomitant immunity. It has clearly been shown that certain disease situations arise as a result of an inappropriate response. For example, Leishmania resistant mice develop a THl response whereas susceptible animals develop a TH2 response, due to different cytokine profiles induced early in the infection process (Heinzel *et al.*, 1989). Similarly in respiratory infections there may be an imbalance of cytokines because one pathogen overstimulates the production of a specific cytokine or inhibits cytokine production by an other pathogen. This initial response can then influence many subsequent events including leukocyte trafficking and cytokine production. The level of interferon production is probably a reflection of the degree of virus replication and correlates with 2-5A synthetase. Following superinfection with *P. haemolytica* there is a dramatic increase in 2-5A synthetase activity, suggesting that the bacteria enhances viral replication, or at least increases interferon production, a stimulator of 2-5A synthetase (Figure 4). Results also indicate an inverse relationship between 2-5A synthetase levels and survival (Figure 5) suggesting that an imbalance of cytokine production may be involved in viral bacterial synergy.

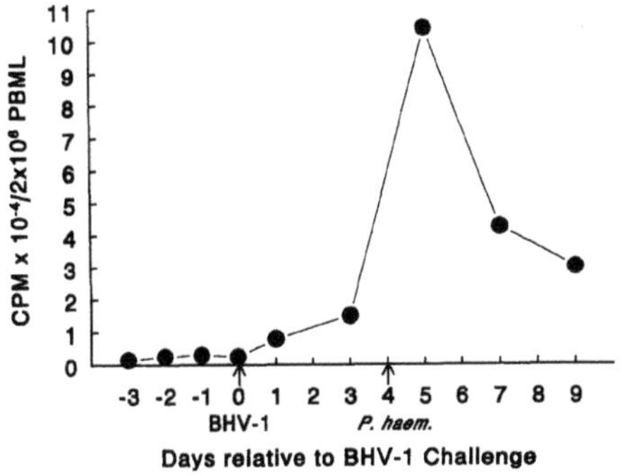

Figure 4. 2-5A synthetase activity in PBML following BHV-1 and *P. haemolytica* challenge.

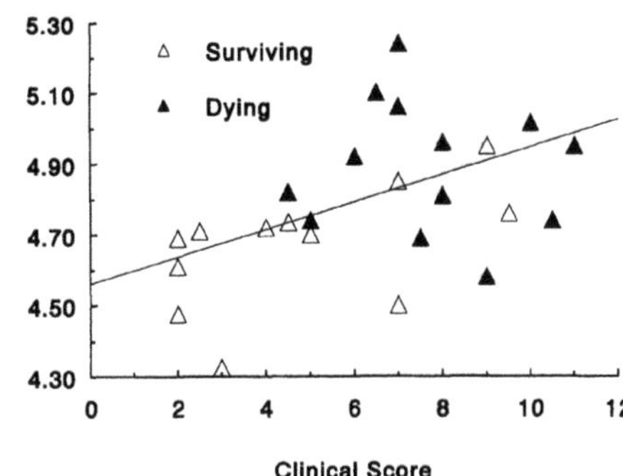

Figure 5. Correlation of 2-5A synthetase activity in PBML 5 days after BHV-1 infection (1 day post *P. haemolytica*) with survival.

Studies have clearly shown that neutrophils are involved in lung injury and that neutrophil depletion reduces pulmonary injury and immunopathology associated with pneumonic pasteurellosis (Slocombe *et al.*, 1985). Based on these studies we tested the hypothesis that following infection with virus or *P. haemolytica*, cytokines could trigger the rapid influx of neutrophils into the lung to either kill the bacteria or induce lung damage. Since endothelial cell damage occurs in the region of neutrophil attachment and neutrophil adherence to endothelial cells is a requirement for migration of neutrophils into the site of bacterial infection, we treated bovine endothelial cells with different cytokines to determine their role in enhancing adherence of neutrophils to endothelial cells. Tumor necrosis factorα clearly enhanced neutrophil attachment to endothelial cells (Warren, 1994). Viral or bacterial infections can induce cytokine production, which will allow neutrophils to attach to endothelial cells and migrate to the site of infection. We hypothesized that the pulmonary macrophage may be one of the initial cells that induces cytokine production and facilitates neutrophil infiltration into the lung. To test this hypothesis we isolated bovine alveolar macrophages and exposed them to *P. haemolytica* and tested their ability to produce chemotactic and immune regulatory cytokines. Within 30-60 min post exposure of alveolar macrophages to *P. haemolytica* large quantities of mRNA for IL-8, TNFα and IL-1 were produced. In fact *P. haemolytica* was more efficient at producing IL-8 mRNA. cytokines than was *E. coli* (Figure 6, 7). P. *haemolytica* also induced the secretion of high amounts of IL-1, TNFα and IL-8 proteins from bovine alveolar macrophage (Figure 8, 9, 10).

Since TNFα was produced at high levels, and it is known to have multiple biological effects *in vivo* we assessed its affect on the acute phase protein response in cattle. TNFα levels of <1 μg/kg were extremely efficient in inducing a rapid (within 8 hr post exposure) acute phase response. These acute phase response levels were comparable to those observed in bovine respiratory disease. In addition to testing the acute phase response in animals suffering from bovine respiratory disease we also correlated the levels of acute phase proteins (haptoglobin) with the animal's ability to recover from infection. Concomitant with an increase in haptoglobin there was decrease in plasma zinc and survival (Figure 11). Whether this is a cause or an effect is not presently known, but is well established that zinc is important in maintaining development of the immune response and in colonization and growth of the bacteria. The role of zinc in the immune response has been recognized in numerous populations suffering from various diseases and has been proposed to play a central regulatory role in immunity (Keen and Gershwin, 1990). Our results demonstrate that cytokine combinations (TNFα and IFN-γ) can act synergistically to reduce zinc levels. By combining IFN-γ with TNFα the equivalent reduction in zinc levels could be achieved with a 1000-fold less TNFα. These results clearly indicate the complexity of interactions between various cytokines and their role in immunopathology in the lung.

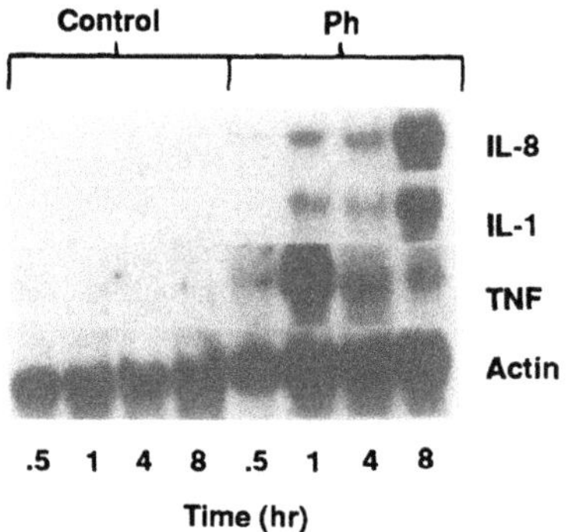

Figure 6. Induction of cytokine mRNA synthesis in BAM. BAM cells ($5x10^7$) were cultured in 75 cm^2 flasks in 30 ml of minimum essential medium and stimulated with heat-killed *P. haemolytica* at 1:1 cell ratio for the indicated time points. Levels of IL-8, IL-1β, TNFα and actin mRNAs were determined by Northern blot analysis of total RNA. Identical results were obtained in experiments with BAM cells from four different animals.

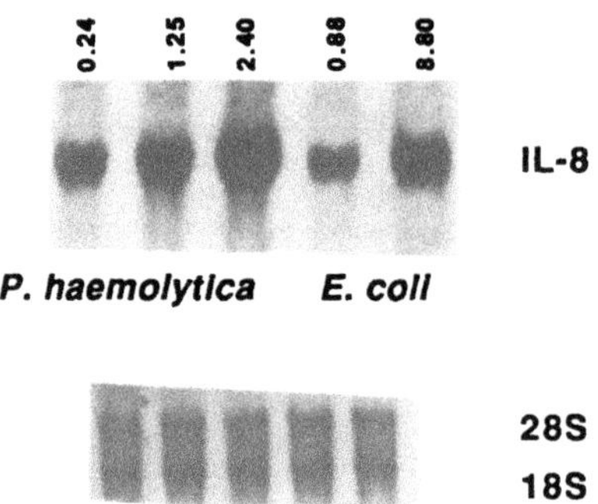

Figure 7. Dose response of *P. haemolytica* and *E. coli* endotoxins. BAM were stimulated for 8 hr with the indicated concentrations of heat-killed *P. haemolytica* or *E. coli* endotoxin. The levels of IL-8 mRNA increased with increased endotoxin concentrations. Low levels of *P. haemolytica* LPS induced higher levels of IL-8 mRNA than those induced by much higher concentrations of *E. coli* LPS.

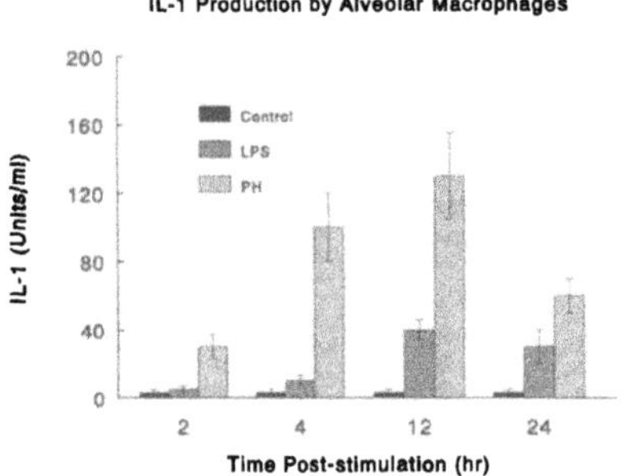

Figure 8. IL-1 secretion by BAM. BAM cells ($5x10^7$) were cultured in 75 cm^2 flasks in 30 ml of minimum essential medium and stimulated either with heat-killed *P. haemolytica* at 1:1 cell ratio or *E. coli* LPS (20 μg/ml). Tissue culture supernatants were harvested at the indicated time points and assayed for IL-1β bioactivity. Data are the means ± SEM of triplicate samples from 6 different animals.

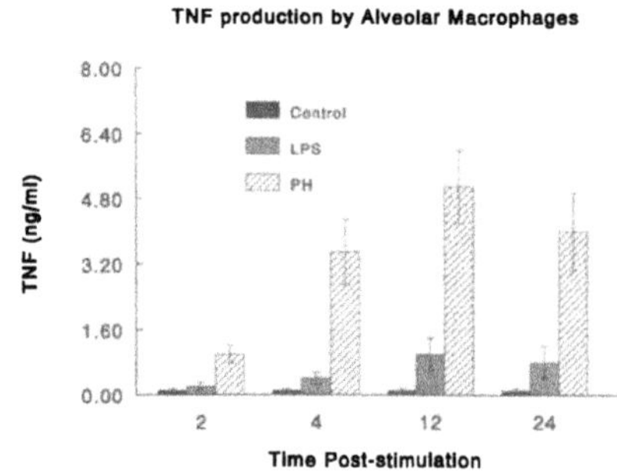

Figure 9. TNFα secretion by BAM. BAM cells ($5x10^7$) were cultured in 75 cm^2 flasks in 30 ml of Minimum essential medium and stimulated either with heat-killed *P. haemolytica* at 1:1 cell ratio or *E. coli* LPS (20 μg/ml). Tissue culture supernatants were harvested at the indicated time points and assayed for TNFα bioactivity. Data are means ± SEM of triplicate samples from 6 different animals.

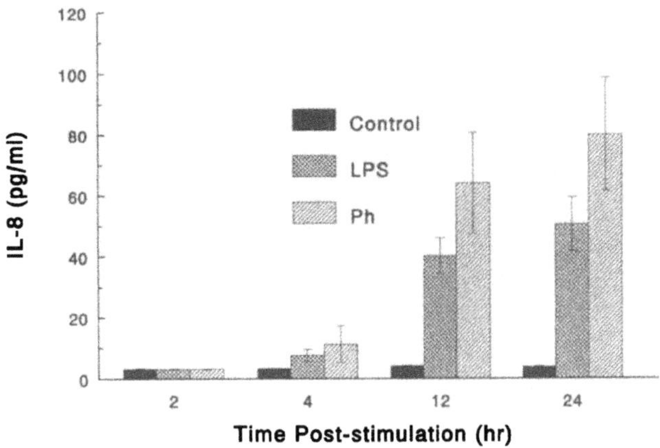

Figure 10. IL-8 secretion by BAM. BAM cells (5×10^7) were cultured in 75 cm^2 in 30 ml of minimum essential medium and stimulated either with heat-killed *P. haemolytica* at 1:1 cell ratio or *E. coli* LPS (20 μg/ml). Tissue culture supernatants were harvested at the indicated time points and used for detection of IL-8 by ELISA. Data are the means ± SEM of triplicate samples from 6 different animals.

In summary, viruses often enter the host and alter both the environmental conditions in the lung which favor bacteria replication as well as modulate the host defense mechanism involved in clearing the bacteria. In many instances, these interactions are extremely complex, but are probably similar for many viral-bacterial interactions in the respiratory tract. Some of the main interactions involve the host response to the infection by providing a variety of mediators which can alter both the ability of the pathogen to replicate as well as the host response. Replication of the virus also induces specific by-products which in turn release other mediators which can further modify lung defenses and encourage bacterial adherence and growth. The bacterium, in turn can induce chemotactic factors (cytokines) which encourage infiltration of specific effector cells into the lung as well as leukotoxin which kills effector cells. These effector cells release mediators which cause tissue damage and immunopathology, which further encourage bacterial growth and may result in the death of the animal. In order to be able to control this complicated scenario it is important to either prevent the initial infection by vaccination against the viral and bacterial pathogens or to ensure that the appropriate constellation of cytokines and mediators are present that will aid in clearing the infection.

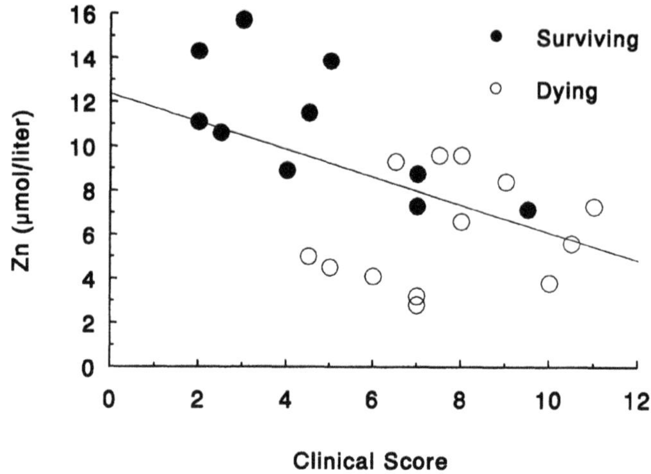

Figure 11. Plasma zinc levels 5 days post BHV-1 infection.

REFERENCES

Babiuk, L.A. and Acres, S.D., 1984, Experimental models for bovine respiratory diseases, *in:* "Bovine Respiratory Disease: A Symposium", R.W. Loan, ed., Texas A & M Univ. Press, College Station.

Babiuk, L.A., Lawman, M.J.P. and Bielefeldt Ohmann, H., 1988, Viral-bacterial synergistic interactions in respiratory disease, *Adv. in Virus Res.* 35:219.

Benson, M.L., Thomson, R.G. and Valli, V.E.O., 1978, The bovine alveolar macrophage II: *in vitro* studies with *Pasteurella haemolytica*, *Can. J. Comp. Med.* 42:368.

Bielefeldt Ohmann, H. and Babiuk, L.A., 1985, Viral bacterial pneumonia in calves: effect of bovine herpesvirus-1 on immunological functions, *J. Infect. Dis.* 151:937.

Confer, A.W., Panciera, R.J. and Mosier, D.A., 1988, Bovine pneumonic pasteurellosis: immunity to *Pasteurella haemolytica*, *J. Am. Vet. Med. Assoc.* 193:1308.

Davison V.E. and Sanford, B.A. , 1981, Adherence of *Staphylococcus aureus* to influenza A virus-infected MDBK cells, *Infect. Immun.* 32:1118.

Forman, A.J. and Babiuk, L.A., 1982, The effect of infectious bovine rhinotracheitis virus infection on bovine alveolar macrophage function, *Infect. Immun.* 35:1041.

Griebel, P.J., Bielefeldt Ohmann, H., Campos, M., Qualtiere, L., Davis, W.C., Lawman, M.J.P. and Babiuk, L.A., 1989, Bovine peripheral blood leukocyte population dynamics following treatment with recombinant bovine interferon α, *J. Interferon Res.* 9:241.

Griebel, P.J. Bielefeldt Ohmann, H., Lawman, M. and Babiuk, L.A., 1990, The interaction between bovine herpesvirus-1 and activated bovine T cells, *J. Gen. Virol.* 77:369.

Griebel, P.J., Qualtiere, L., Davis, W.C., Lawman, M.J.P. and Babiuk, L.A., 1988, Bovine peripheral blood leukocyte subpopulation dynamics following a primary bovine herpesvirus type-1 infection, *Virol. Immunol.* 4:267.

Heinzel, F.P., Sadick, M.D., Holaday, B.J., Coffman, R.L. and Locksley, R.M., 1989, Reciprocal expression of IFNγ or interleukin-4 during the resolution or progression of murine leishmaniasis, *J. Exp. Med.* 169:59.

Jakab, G.J., 1982, Immune impairment of alveolar macrophage phagocytosis during influenza virus pneumonia, *Am. Rev. Resp. Dis.* 126:776.

Kaehler, K.L., Markham, R.J.F., Muscoplat, C.C. and Johnson, D.W., 1980, Evidence of species specificity in cytocidal effects of *Pasteurella haemolytica*, *Infect. Immun.* 30:615.

Keen, C.L. and Gershwin, M.E., 1990, Zinc deficiency and immune function, *Ann. Rev. Nutr.* 10:415.

Kooney, J.M. and Falkow, S., 1984, Nucleotide sequence homology between the immunoglobulin A1 protease of *Neisseria gonorrhoeae, Neisseria meningitidis* and *Haemophilus influenza, Infect. Immun.* 43:101.

Morck, D.W., Raybould, T.J.G., Acres, S.D., Babiuk, L.A., Nelligan, J. and Costerton, J.W., 1987, Electron microscopic description of glycocalyx and fimbria on the surface of *Pasteurella haemolytica* A1, *Can. J. Vet. Res.* 51:83.

Morck, D.W., Watts, T.C., Acres, S.D. and Costerton, J.W., 1988, Electron microscopic examination of cells of *Pasteurella haemolytica* A1 in experimental infected cattle, *Can. J. Vet. Res.* 52:343.

Mosmann, T.R., Cherwinski, H., Bond, M.W., Giedlin, A. and Coffman, R.L., 1986, Two types of murine helper T cell clone. I. definition according to profiles of lymphokine activities and secreted proteins, *J. Immunol.* 136:2348.

Nugent K.M. and Pesanti, E.L., 1983, Tracheal function during influenza infections, *Infect. Immun.* 42:1102.

Pavia, D., 1987, Acute respiratory infection and mucociliary clearance, *Eur. J. Respir. Dis.* 71:219.

Proctor, R.A., 1987, Fibronectin: a brief review of its structure, function and physiology, *Rev. Infect. Dis.* 9(Suppl4):317.

Ramphal, R., Small, D.M., Shands, J.W., Fischlschweiger, W. and Small, P.A., 1980, Adherence of *Pseudomonas aeruginosa* to tracheal cells injured by influenza infection or by endotracheal intubation, *Infect. Immun.* 27:614.

Rossi, C.R. and Kiesel, G.K., 1977, Susceptibility of bovine macrophages and tracheal-ring cultures to bovine viruses, *Am. J. Vet. Res.* 38:1705.

Sanford, B.A., Shelokov, A. and Ramsay, M.A., 1978, Bacterial adherence to virus-infected cells: a cell culture model of bacterial superinfection, *J. Infect. Dis.* 137:176.

Shewen, P.E. and Wilkie, B.N., 1982, Cytotoxin of *Pasteurella haemolytica* acting on bovine leukocytes, *Infect. Immun.* 35:91.

Slocombe, R.F., Malark, J. and Ingersoll, R., 1985, Importance of neutrophils in the pathogenesis of pneumonic pasteurellosis in calves, *Am. J. Vet. Res.* 46:2253.

Sugarman, B., 1980, Effect of heavy metals on bacterial adherence, *J. Med. Microbiol.* 13:351.

Sugarman, B., Epps, L.R. and Stenback, W.A., 1982, Zinc and bacterial adherence, *Infect. Immun.* 37:1191.

Warren, L.W., 1994, The effect of bovine herpesvirus-1 on pulmonary endothelial cell function, M.Sc. Thesis, Univ. of Sask., p. 127.

Woods, D.E., 1987, Role of fibronectin in the pathogenesis of gram negative pneumonia, *Rev. Infect. Dis.* 9(Suppl4):386.

Yates, W.D.G., Babiuk, L.A. and Jericho, K.W.F., 1983, Viral-bacterial pneumonia in calves: duration of the interaction between bovine herpesvirus 1 and *Pasteurella haemolytica, Can. J. Comp. Med.* 47:257.

PATHOGENESIS AND VIRULENCE OF *PASTEURELLA HAEMOLYTICA* IN CATTLE: AN ANALYSIS OF CURRENT KNOWLEDGE AND FUTURE APPROACHES

A.W. Confer, K.D. Clinkenbeard, and G.L. Murphy

Department of Veterinary Pathology
College of Veterinary Medicine
Oklahoma State University
Stillwater, OK 74078 USA

INTRODUCTION

Pasteurella haemolytica is the cause of a severe fibrinous pleuropneumonia of cattle termed Shipping Fever (Frank, 1989). Serovar A1 is isolated most often from bovine pneumonic pasteurellosis. The pathogenesis of pneumonic pasteurellosis relies on several bacterial-host interactions: nasopharyngeal colonization, inhalation, pulmonary alveolar colonization, host response to colonization, and bacterial evasion of host defense (Whiteley *et al.*, 1992) .

Several *P. haemolytica* virulence factors may influence the outcome of bacterial-host interactions. The pathogenic roles of endotoxin, leukotoxin (LKT), and capsular polysaccharide (CPS) are documented (Confer *et al.*, 1990). Other potential virulence factors whose roles are less well documented or not defined include: fimbriae (pili, Morck *et al.*, 1987), outer membrane proteins (Davies *et al.*, 1992), iron-regulated proteins (Murray *et al.*, 1992; Yu *et al.*, 1992), serotype-specific agglutinating antigen (Gonzalez-Rayos *et al.*, 1986), neuraminidase (Straus *et al.*, 1993), and neutral glycoprotease (sialoglycoprotease, Abdullah *et al.*, 1992) .

Our understanding of the potential pathogenic roles of these virulence factors has been developed through *in vitro* experiments characterizing the actions of those factors on isolated cell populations such as leukocytes and endothelial cells. However, it has been difficult to integrate our understanding of the action of LKT, CPS, and endotoxin *in vitro* with specific pathogenic events *in vivo*. This is due to the complexity of the bacterial-host interactions in pneumonic pasteurellosis, the lack of a suitable laboratory animal model that mimics bovine pneumonic pasteurellosis, and the lack of protection of cattle after vaccination with purified factors (CPS or LKT). Therefore, future studies of the pathogenesis of pneumonic pasteurellosis must involve development of mutant strains of *P. haemolytica* that are deficient in individual virulence factors for use in experimental infection studies to identify the specific role of each virulence factor in pathogenesis. *In vivo* testing of those bacteria both as vaccines and challenge strains will further allow us to determine the role(s) of individual virulence factors in infection and immunity within the complexity of the living animal.

This review will discuss current knowledge and understanding of bacterial-host interactions in the nasopharynx and pulmonary alveoli that lead to pneumonic pasteurellosis

Haemophilus, Actinobacillus, and Pasteurella
Edited by W. Donachie *et al.*, Plenum Press, New York, 1995

and the pathogenic roles that known and potential virulence factors of *P. haemolytica* play or potentially play in those interactions. Finally, we will provide an overview of current knowledge of and potential uses for mutant strains of *P. haemolytica.*

NASOPHARYNGEAL COLONIZATION

Several recent discoveries on the *P. haemolytica*-host interaction have increased our understanding of the early phases of pathogenesis of pneumonic pasteurellosis. It has been well established that *P. haemolytica* A1 is carried in the nasopharynx and tonsils in low, often undetectable numbers (Frank, 1989). Viral infections and stress result in the proliferation of *P. haemolytica* A1 with subsequent inhalation of bacteria into the lung. However, we are only beginning to understand some of the cellular and molecular events in the nasopharynx which allow *P. haemolytica* to proliferate (Table 1).

Table 1. Possible factors and mechanisms enhancing nasopharyngeal colonization by *P. haemolytica.*

Factor	Mechanism
Bacterial Factors	
Fimbriae	adherence to epithelium
Capsule	adherence to mucus or epithelial surface
Outer membrane proteins	adherence to mucus or epithelial surface
Lipopolysaccharide	adherence to mucus or epithelial surface
Neuraminidase	alteration of mucus or epithelial surface
Neutral glycoprotease	alteration of mucus or epithelial surface
Host Factors	
Elastase	secreted after viral infection reducing epithelial fibronectin
Environmental Factors	
Temperature	alter expression of surface adhesins
Oxygen concentration	alter expression of surface adhesins
Serum proteins	alter expression of surface adhesins

A necessary requirement for nasopharyngeal colonization is adherence of bacteria to the epithelial surface (Mason *et al.*, 1993). Adherence usually occurs via bacterial fimbriae-associated adhesins which have high affinity for mucosal glycolipids or glycoproteins present on epithelial cells. It is uncertain whether *P. haemolytica* adheres to the mucosal epithelium or proliferates within the mucous layer. Morck *et al.*, (1987) and Potter *et al.*, (1988) demonstrated fimbriae on *P. haemolytica*, though specific adhesins were not identified. Gonzalez and Maheswaren (1993) and Murphy (1992, unpublished data) failed to demonstrate fimbriae in numerous *P. haemolytica* A1 isolates from the lung or upper respiratory tract. Those results could indicate that expression of fimbriae is not a universal phenomenon among *P.haemolytica* isolates. However, phase variation of fimbriae expression has been documented in bacteria (Mason *et al.*, 1993) and could account for negative results obtained in various laboratories. If *P. haemolytica* fimbriae are readily expressed *in vivo*, they are probably important in attachment to the nasopharyngeal epithelial cell surface. However, if they are not expressed, how might adherence occur?

Not all bacteria that adhere to mucosal surfaces have demonstrable fimbriae or adhesins (Jann and Hoschutzky, 1991). Bacteria can nonspecifically adhere to epithelial surfaces via capsular or other bacterial surface structures or remain within the mucous layer. Various pathogenic bacteria have been shown to adhere to the carbohydrate chains of mucins. Uhlich *et al.*, (1993) found that *P. haemolytica* A1 adhered to bovine nasal mucus *in vitro*, and adhesion could be altered by enzymatic degradation of both protein and carbohydrate components of mucus. Botcher *et al.*, (1993) found that OMP preparations, capsular extracts, and whole cell suspensions of *P. haemolytica* bound to preparations of bovine tracheal mucus. Therefore, *P. haemolytica* seems capable of adhering to the mucous lining of the respiratory tract. Those findings, however, may only account for the state in which *P. haemolytica* exists in low numbers prior to the bacterial proliferation that precedes pneumonia. In addition, adherence to mucous should enhance clearance of bacteria, *P. haemolytica*-derived neuraminidase has been shown to be produced *in vivo* (Straus *et al.*, 1993; Straus and Purdy, 1994). Neuraminidase may exert a pathogenic role by removing sialic acid from mucus or cell surface glycoproteins, reducing the protective effects of mucus while enhancing *P. haemolytica* adherence to mucosal epithelium.

Fibronectin is a large glycoprotein on the surface of epithelial cells. It has been hypothesized that loss of nasal fibronectin through increased elastase activity causes enhanced bacterial adhesion through exposure of epithelial receptors (Dal Nogare *et al.*, 1987). Briggs and Frank (1992) found that intranasal inoculation of cattle with infectious bovine rhinotracheitis virus caused a marked increase in elastase activity in nasal mucus. Studies in human surgical patients suggested that stress may be responsible for reduced epithelial-cell associated fibronectin, allowing bacteria such as *Pseudomonas aeruginosa* to readily colonize those patients (Mason *et al.*, 1993). Concurrent viral infection and/or stress in shipped cattle could contribute to loss of fibronectin and increased *P. haemolytica* adhesion; however, the definitive studies have yet to be reported.

With the exception of putative fimbriae, the surface structures of *P. haemolytica* potentially responsible for adherence are not known. Capsular polysaccharides, lipopolysaccharide (LPS), outer membrane proteins, and serotype-specific agglutinating antigen may function as adhesins in the absence of fimbriae. Variations in culture conditions *in vitro* modify expression of *P. haemolytica* LPS and various proteins, including OMPs, iron-regulated proteins, LKT, and a heat-shock protein (HSP60) (Confer and Durham, 1992; Davies *et al.*, 1992; Mosier *et al.*, 1993). It seems likely that stress-related changes in the nasal cavity, such as temperature, oxygen concentration, host enzyme activity, or serum exudation, may modulate expression of bacterial surface proteins, enzymes, or LPS thereby enhancing nasopharyngeal colonization.

PULMONARY ALVEOLAR COLONIZATION AND HOST RESPONSES

Once *P. haemolytica* gain access to the pulmonary alveoli via inhalation from the nasopharynx, the bacteria must overcome host defense mechanisms including pulmonary surfactant, alveolar macrophages, the cough reflex and mucociliary apparatus for pneumonia to result. Numerous bacterial factors stimulate an intense acute inflammatory response in pulmonary alveoli. If that intense inflammatory response is coupled with an appropriate humoral immune response, the bacteria are destroyed, pulmonary damage is minimal, and the infection is averted. However, when the immune response is insufficient or overcome, bacterial virulence factors and the host inflammatory response (particularly leukocyte products) induce localized tissue damage, allowing extension of infection to adjacent tissue, and inciting systemic changes associated with the acute inflammatory process. Several recent findings have both clarified and complicated our understanding of the bacterial-host interaction at the pulmonary alveolar level.

LKT and endotoxin are the best characterized *P. haemolytica* virulence factors with respect to their effects within alveoli. Since its initial discovery, LKT has been the subject of intense investigation relative to its molecular nature and pathogenic and immunogenic potential (Table 2). Clinkenbeard *et al.*, (1989a,b,c) found that *P. haemolytica* LKT was a pore-forming cytolysin indicating that it injures target cells by binding to the plasma membrane and creating transmembrane pores. Redistribution of ions across the membrane results in colloid osmotic swelling and cell lysis with leakage of intracellular macromolecules out of the cell.

Table 2. Biological effects of *P. haemolytica* leukotoxin (LKT).

High concentrations of LKT - cytolytic for ruminant leukocytes and platelets

Low concentrations of LKT - noncytolytic but modify leukocyte function

- Neutrophils
  - activation as evidenced by
    - stimulation of the oxidative burst and release of oxygen free radicals
    - cytoskeletal changes
    - release of secondary granules
    - release of arachidonic acid metabolites
- Mononuclear phagocytes
  - stimulates IL-1 and TNFα release
- Mast cells
  - stimulates histamine release
- Lymphocytes
  - reduction in mitogen-mediated blastogenesis

P. haemolytica LKT's target specificity is for ruminant leukocytes and platelets (Shewen and Wilkie, 1982; Clinkenbeard and Upton, 1991). The mechanism conferring LKT's species specific leukotoxic activity has not been determined. One possible mechanism of target cell specificity is that susceptible target cells express LKT receptors that mediate LKT binding, whereas cells not susceptible to LKT lack or have too few LKT receptors to mediate LKT sensitivity. Using an indirect binding assay, Clinkenbeard and co-workers (1993) found that LKT bound to membranes of two LKT susceptible cell types (bovine lymphoma cells and bovine platelets). LKT also bound to a cell type that is not LKT susceptible (bovine red blood cells). However, the susceptible cell types bound more LKT than did the non-susceptible cell type. Similarly, Bauer and Welch (1994) recently reported that LKT bound to a LKT susceptible cell line (bovine lymphoma cells) as well as to a LKT non-susceptible cell line (human lymphoma cells). Determination of the mechanism conferring LKT target cell specificity will require further experimentation using radiolabeled, purified LKT in direct binding assays.

Cytolysis is a Ca^{2+} dependent process, and neutrophils are more susceptible to injury than are monocytes (Stevens and Czuprynski, 1994). In calves experimentally challenged with *P. haemolytica*, Whiteley *et al.*, (1990) found immunohistochemically that LKT was associated with degenerating leukocytes within the pulmonary alveoli. Those findings, as well as the demonstration that cattle develop systemic antibodies to LKT following natural exposure, substantiate that LKT is produced *in vivo* and lend support to its pathogenic role.

LKT derived from *P. haemolytica* grown in *vitro* is highly aggregated (≈ 8,000 kDa) and has low leukotoxic activity (Waurzyniak *et al.*,1994). LKT can be disaggregated by 0.5 % bovine serum albumin (BSA) or 3M guanidine resulting in enhanced leukotoxic activity up to 20-fold. In recent studies (Clinkenbeard, 1994 unpublished data), other chaotropic agents such as 6M urea and 1M sodium thiocyanate or detergents (1% Tween) resulted in only partial disaggregation and less enhancement of leukotoxic activity. The *in vitro* phenomenon of aggregation and disaggregation of LKT could occur during pulmonary infection. One of the first events to occur within alveoli after *P. haemolytica* colonization is increased capillary permeability with serum flooding of alveoli (Weiss *et al.*,

1991). Exudation of serum into alveolar spaces may be to the host's advantage should that serum be rich in LKT-neutralizing antibodies. However, in the absence of such antibodies, BSA could disaggregate LKT resulting in enhanced cytolytic activity. This event would place the resident macrophages and immigrating neutrophils at greater risk of cytolysis, reducing phagocytic potential within the alveoli, and inducing pulmonary injury through release of leukocytic products into the alveoli.

Whereas high concentrations of LKT cause cytolysis of leukocytes, low concentrations of LKT activate bovine neutrophils as demonstrated by stimulation of an oxidative burst (luminol-dependent chemiluminescence), cytoskeletal alterations and release of secondary granules (Czuprynski *et al.*, 1991). Ortiz-Carranza and Czuprynski (1992) found that LKT-induced activation of neutrophils was Ca^{2+} dependent, as is neutrophil activation by other methods. Maheswaren *et al.*, (1992) found that low concentrations of LKT induced nearly immediate release of products of the respiratory burst, i.e. the oxygen free radical superoxide anion (O_2^-) and hydrogen peroxide. The toxic oxygen products generated by the respiratory burst preceded degranulation and cytolysis as measured by lactate dehydrogenase release. Sublethal concentrations of LKT also caused release of the lysosomal enzymes arylsulfatase B and myeloperoxidase. It is likely that these oxygen products and proteases released from lysosomal granules cause tissue necrosis. In addition, toxic oxygen free radicals can directly damage endothelium and inactivate antiproteases such as alpha-1 antitrypsin, thus enhancing or prolonging protease activity and tissue damage.

Which is more detrimental to the host, low concentrations of LKT with neutrophil activation or high concentrations with cytolysis? Perhaps disaggregation of LKT by BSA could result in enhanced cytolysis but less pulmonary injury, whereas aggregated LKT causes neutrophil activation, release of toxic oxygen products and proteases and more pulmonary injury. Clarification of these concepts awaits further investigation.

Besides the leukocytic products released by neutrophils described above, LKT affects the inflammatory response through stimulation of release of other mediators, IL-1, TNFα, arachidonic acid metabolites, and histamine. Recently, Stevens and Czuprynski (1993) reported that sublethal concentrations of *P. haemolytica* LKT stimulated release of IL-1 and TNFα from bovine mononuclear phagocytes and that neutralization of LKT-induced cytolysis with a monoclonal antibody (MAb) enhanced cytokine release. In addition, Hendricks *et al.*, (1992) reported that LKT stimulated bovine neutrophils to release the arachidonic acid metabolites via the lipoxygenase pathway, leukotriene B4 (LTB_4) and 5-hydroxyeicosatetraenoic acid. Clinkenbeard, Clarke and co-workers (1994) confirmed that LKT stimulated synthesis of LTB_4 by bovine neutrophils, but could not detect synthesis of another important arachidonic acid metabolite, thromboxane B2. In those studies LTB_4 synthesis from endogenous arachidonic acid required LKT-induced plasma membrane damage and lagged behind LKT-induced cytolysis. However, LTB_4 synthesis from exogenous arachidonic acid (available in abundance in inflammatory sites) preceded neutrophil lysis and occurred in the absence of plasma membrane damage. Finally, Adusu *et al.*, (1994) recently reported that *P. haemolytica* LKT induced histamine release from isolated bovine pulmonary mast cells within 5 minutes after exposure. LKT-induced histamine release could be blocked by incubation with LKT neutralizing antibodies, a β-adrenoceptor antagonist (propranolol) and a phospholipase A_2 inhibitor (p-bromophenacyl bromide). Similarly, mast cells isolated from the pneumonic areas of lungs from calves experimentally challenged with *P. haemolytica* had less histamine than did those cells isolated from normal areas of the lung. The biologic effects of these LKT-induced inflammatory mediators can be to increase vascular permeability, stimulate chemotaxis of neutrophils, promote leukocyte adhesion to endothelium, stimulate synthesis of platelet activating factor, and increase procoagulant properties of endothelium. Because histamine is generally an early inflammatory mediator, it could account for the rapid influx of serum proteins and neutrophils seen in *P. haemolytica*-induced pneumonia.

The final effect of *P. haemolytica* LKT which we wish to comment on is that sub-lethal concentrations of LKT inhibit mitogen-induced blastogenesis of bovine lymphocytes *in vitro* (Majury and Shewen, 1991; Czuprynski and Ortiz-Carranza, 1992). In those studies, blastogenic responsiveness could be restored by addition of partially purified IL-1, recombinant IL-2 or LKT-neutralizing antibody to the cultures. The biological relevance of this phenomenon is not known, but a reduction in peripheral blood lymphocytes with sequestration and infiltration of lymphocytes into the lung microvasculature and pulmonary alveoli has been described after experimental *P. haemolytica* infection (Whitely *et al.*, 1992). Therefore, the potential exists for intra-alveolar LKT to come in contact with lymphocytes and reduce the immune-mediated defenses against *P. haemolytica*. Also, if components of LKT responsible for attenuating lymphocyte responsiveness gain access to pulmonary lymph, abrogation of an immune response could result. Perhaps an LKT-induced reduction in lymphoblastogenesis would be particularly detrimental by reducing the rate at which a potentially immune animal could mount an anamnestic antibody response after natural challenge.

Although neutrophil products are credited with much of the pulmonary necrosis that accompanies pneumonic pasteurellosis, one cannot overlook that in pasteurellosis severe vascular thrombosis results in ischemia and is probably also responsible for substantial tissue damage. Numerous recent *in vitro* studies implicate *P. haemolytica* LPS as a major cause of vascular necrosis and thrombosis (Whiteley *et al.*, 1992).

The polysaccharide component of *P. haemolytica* LPS has been well characterized chemically and immunologically in recent years (Confer, 1993; Davies *et al.*, 1992). Numerous effects of *P. haemolytica* lipid A, the endotoxin component of LPS, have been demonstrated *in vitro* suggesting a pathogenic role for LPS in pneumonic pasteurellosis. Whiteley *et al.*, (1990) demonstrated in experimentally challenged calves that *P. haemolytica* LPS was present within the cytoplasm of neutrophils, alveolar macrophages, endothelial cells, and pulmonary intravascular macrophages and on epithelial surfaces. The widespread distribution of LPS throughout the pulmonary alveolar lumens and capillaries suggests that LPS could potentially affect various cellular components of the host (Table 3).

Several investigators have documented that *P. haemolytica* endotoxin is directly toxic to endothelium *in vitro* (Paulsen *et al.*, 1989; Breider *et al.*, 1990). Because of the extensive microvascular thrombosis that occurs in pneumonic pasteurellosis, endotoxin-induced endothelial cell damage is probably of key importance in activating the coagulation cascade and stimulating platelet aggregation. Dose-dependent endothelial cell degeneration occurs as characterized by shape changes, cell retraction, and increased membrane permeability culminating in cell death. Breider *et al.*, (1991) found that *P. haemolytica* endotoxin-induced endothelial cell damage could be prevented with immune serum or incubation with neutrophils, whereas alveolar macrophages enhanced *P. haemolytica*-induced endothelial cell damage (Sharma *et al.*, 1992a). Recently, Bienhoff *et al.*, (1992) found that bovine alveolar macrophages stimulated with *P. haemolytica* endotoxin released TNFα. Sharma *et al.*, (1992b) demonstrated that TNFα and IL-1 synergistically enhanced *P. haemolytica* endotoxin-induced damage to endothelial cells. Pace *et al.*, (1993) found serum TNFα concentrations increased between 2 and 8 hours after experimental challenge of cattle with *P. haemolytica*. Espinasse *et al.*, (1993), however, failed to demonstrate circulating TNFα in calves experimentally infected with *P. haemolytica*. One could speculate that early in infection *P. haemolytica* endotoxin and/or LKT could stimulate alveolar macrophages to secrete TNFα which in turn augments endotoxin-induced endothelial damage. In addition to endothelial cell degeneration and necrosis, *P. haemolytica* endotoxin can induce expression of the coagulation factor thromboplastin (factor III) on endothelial cells (Breider and Yang, 1994). Thromboplastin is a cell membrane-associated glycoprotein that initiates the extrinsic coagulation cascade and stimulates microvascular thrombotic events. Therefore, *P. haemolytica* endotoxin could enhance thrombosis even in the absence of endothelial cell degeneration.

Table 3. Biological effects of *P. haemolytica* endotoxin (LPS).

Directly toxic to endothelial cells
- Endothelium is protected by neutrophils or immune sera
- Endothelial damage enhanced by macrophages, TNFα, or IL-1
Induces expression of thromboplastin on endothelial cells
Stimulates release of TNFα and IL-1 by pulmonary alveolar macrophages

MUTANT STRAINS OF *P. HAEMOLYTICA*

Future determinations of the role of *P. haemolytica* virulence factors in the pathogenesis of pneumonic pasteurellosis of cattle lie in the development of mutant strains of the bacterium for *in vitro* and *in vivo* assessment of virulence. Spontaneous mutants of *P. haemolytica* which have reduced virulence have not been described. Chemical-induced mutagenesis of *P. haemolytica* is possible, but those mutations are randomly produced and secondary mutations can minimize the value of this approach (Table 4).

A LKT-deficient mutant of *P. haemolytica* was developed using nitrosoguanidine mutagenesis (Froshauer *et al.*, 1993). Use of that mutant provides a mechanism for studying *P. haemolytica*-leukocyte interactions without the complication of LKT inducing cell injury, death or activation. In *in vitro* studies, the LKT-deficient *P. haemolytica* was readily phagocytized by bovine neutrophils and macrophages. Bacterial killing was less for *P. haemolytica* than for other bacteria investigated. Co-infection of phagocytes with other bacteria indicated that normal neutrophil bactericidal activity had not been compromised. Inoculation of cattle with the LKT-deficient *P. haemolytica* induced lung lesions. They were reported, however, to be less extensive than those produced by a wild-type strain, thus substantiating the importance of LKT and other virulence factors in the pathogenesis of pneumonic pasteurellosis.

Once it is determined that this mutant is not deficient in other virulence factors, the LKT-deficient mutant offers additional possibilities for study. Activation and phagocytic potential of and LKT-deficient *P. haemolytica* killing by bovine neutrophils and macrophages could be studied with regard to the interactions of various concentrations of recombinant LKT added to culture media along with the mutant. This would enhance our understanding of the role of LKT in bacterial-phagocyte interactions. *In vivo* studies comparing LKT-deficient and wild-type *P. haemolytica* could be expanded using dose-response challenge studies with a variety of challenge methods to better understand the pathogenic role of LKT. Use of viral-bacterial and bacteria alone challenge methods might help us understand the role of *P. haemolytica* LKT in producing lesions with and without viral associated injury. A sequential time-dose light and electron microscopic study comparing lungs infected with either mutant or wild-type strains should help in determining the extent of necrosis induced by LKT versus the role of other virulence factors and ischemia in producing necrosis.

Table 4. Mutant strains of P. *haemolytica.*

LKT-deficient mutant - nonisogenic
- Readily phagocytized by neutrophils and macrophages
- Less readily killed by phagocytes than wild-type
- Induced less severe lung lesions than did wild-type

Lipoprotein (*Lpp*)-deficient mutant - isogenic
- Enhanced iron-regulated protein production
- Less virulent for mice than wild-type
- More susceptible to complement-mediated killing than wild-type
- Altered membrane protein composition

aroA-deficient mutant - isogenic
- Attenuated - requiring supplementation of growth media with aromatic amino acids

Development of isogenic *P. haemolytica* mutants through molecular manipulation of specifically targeted genes seems the most logical approach to study the influence of individual bacterial components on bacterial growth and virulence and bacterial-host interactions. Progress in the development of genetic systems for analysis of virulence factors in *P. haemolytica* has been slow. Efforts to introduce foreign DNA into *P. haemolytica* have often not been successful. Briggs *et al.*, (1994) recently identified a restriction endonuclease, *Pha*I, in *P. haemolytica* that may contribute to the difficulties that researchers have had in attempting to introduce exogenous DNA. However, methods for introduction of plasmid DNA into *P. haemolytica* by electroporation have been developed (Briggs *et al.*, 1994; Craig *et al.*, 1989; Murphy and Whitworth, 1994). Also broad host range shuttle vectors (pJFF224-NX and pJFF224-XN) were recently derived from the plasmid RSF1010 (Frey, 1992). In that study, those vectors allowed stable transformation and expression of cloned genes in *P. haemolytica* and *Actinobacillus pleuropneumoniae.*

Murphy and co-workers have constructed isogenic mutants of *P. haemolytica* Al that no longer synthesize three membrane lipoproteins *(Lppl, Lpp2, and Lpp3)* with molecular mass of approximately 28 kDa (Murphy and Whitworth, 1994). The genes for those proteins were originally cloned by Craven *et al.*, (1991). In that study, a potential biological relevance was attached to those proteins because a significant correlation was found between antibody responses to one of the lipoproteins and resistance to experimental challenge. Sequence data (Murphy and Whitworth, 1993), later confirmed by Cooney and Lo (1993), indicated that these lipoproteins are the products of three tandem genes transcribed from a single promoter. The determined amino acid sequences of the proteins encoded by these genes are similar to a 28 kDa inner membrane lipoprotein from *Escherichia coli* and a 28 kDa lipoprotein of *Haemophilus influenzae* type b. In the latter bacterium, the lipoprotein is associated with virulence in a rat challenge model. Dabo *et al.*, (1994) cloned the *Lppl, Lpp2, Lpp3* genes separately into an expression vector and recombinant forms of the three proteins were purified after expression in *E. coli.* Using densitometric analyses of Western blots of the three proteins, a significant correlation was seen between antibodies to *Lpp3* and resistance of cattle to experimental *P. haemolytica* challenge. Therefore, the importance of these lipoproteins in pasteurellosis seems probable.

Murphy and co-workers have recently completed several studies looking at growth of the *Lpp*-deficient mutant, production of virulence factors, and potential virulence. *In vitro* growth of the *Lpp*-deficient mutant and LKT production appear similar to the parent strain. To assess production of iron-regulated OMPs (IROMPs), the mutant and parent strains were incubated in BHI broth with or without the iron chelator 2,2-dipyridyl. Outer membranes were prepared using N-lauroylsarcosine (Sarkosyl, Squire *et al.*, 1984). Analysis of OMP- and IROMP-enriched fractions by SDS-PAGE indicated that by as early as 2 hours of growth, the 70 and 100 kDa IROMPs were seen in both the wild-type and mutant strains incubated with dipyridyl and with the mutant grown in normal BHI broth. The intensity of the bands was less for the mutant in normal BHI broth than for those grown with the chelator. Similar changes were seen at 6 and 18 hours of growth. The 77 kDa IROMP was expressed only in mutant and wild-type bacteria grown for at least 6 hours with the iron chelator. Therefore, the absence of *Lppl, Lpp2,* and *Lpp3* could modify the envelope such that iron uptake is reduced causing the bacterium to express IROMPs under iron-sufficient conditions.

To compare the LD_{50}s for the *Lpp*-deficient mutant and wild-type *P. haemolytica,* young Swiss-Webster mice were inoculated intraperitoneally with either organism suspended in hog gastric mucin (Gentry *et al.*, 1987). The mutant had a 40% greater LD_{50} (2.0 x 10^6 CFU/ml) than did the wild-type (1.2 x 10^6). Although, intraperitoneal inoculation of mice is probably not a good model for bovine pasteurellosis, these results suggest that the *Lpp*-deficient mutant probably has reduced ability to grow *in vivo.*

The ability of bovine serum to induce complement-mediated killing of the *Lpp*-deficient mutant and parent strains was also compared. Sera from calves vaccinated with live *P. haemolytica,* with OMPs, and with saline were compared for complement-mediated killing (Chae *et al.*,, 1990). Sera from *P. haemolytica*-vaccinated calves caused approximately 29% killing of the wild-type and 66-69% killing of the mutant. Similarly, sera from the saline vaccinates plus complement failed to kill the wild-type but resulted in a 51% killing of the mutant. Even when saline was substituted for sera, there was 27% killing of the mutant, but the wild-type grew. Perhaps, loss of *Lppl, Lpp2,* and *Lpp3* expression enhances cell envelope fragility allowing the cells to be more readily lysed or allows for expression, greater concentrations or clustering of C3 receptors on the cell envelope.

Recently, Tatum *et al.*, (1994) described construction of an *aroA*-deficient mutant of *P. haemolytica* A1. The *aroA*-deficient mutant has a defective aromatic amino acid biosynthetic pathway requiring supplementation of growth medium with aromatic amino acids. The *aroA*-deficient mutants constructed for several pathogenic Gram negative bacteria are attenuated. Therefore, the authors propose its use as a live vaccine.

In the future, development of a universal set of *in vitro* and *in vivo* assays for virulence assessment of *P. haemolytica* mutants would aid in comparisons of the role of various virulence factors in the pathogenesis of bovine pneumonic pasteurellosis. This is probably particularly true for comparative *in vivo* studies in cattle.

ACKNOWLEDGEMENTS

This work supported in part by grants 90-34116-5353 and 92-37204-7773 from the USDA-CSRS and by projects OKL01438 and OKL02179 of the Oklahoma Agricultural Experiment Station. Published as J94-016 from the College of Veterinary Medicine and Oklahoma Agricultural Experiment Station.

REFERENCES

Abdullah, K.M., Udoh, E.A., Shewen, P.E., and Mellors, A., 1992. A neutral glycoprotease of *Pasteurella haemolytica* A 1 specifically cleaves O-sialoglycoproteins. *Infect. Immun.* 60:56-62.

Adusu, T.E., Conlon, P.D., Shewen, P.E., and Black, W.D., 1994. *Pasteurella haemolytica* leukotoxin induces histamine release from bovine pulmonary mast cells. *Can. J. Vet. Res.* 58:1-5.

Bauer, M. and Welch, R.A., 1994. Association of RTX toxin with target cells. *Proc. 94th Ann. Mtg. Amer. Soc. Microbiol.*, Las Vegas. B-113

Bienhoff, S.E., Allen, G.K., and Berg, J.N., 1992. Release of tumor necrosis factor-alpha from bovine alveolar macrophages stimulated with bovine respiratory viruses and bacterial endotoxins. *Vet. Immunol. Immunopathol.* 30:341-357.

Botcher, L., Stoter, I., and Hellmann, E., 1993. In vitro adharenz von *Pasteurella haemolytica* an trachealmucin und einer trachealepithelzellprapation vom rind. Berl. Munch. *Tierarztl. Wochenschr.* 106:333-336.

Breider, M.A. and Yang, Z., 1994. Tissue factor expression in bovine endothelial cells induced by *Pasteurella haemolytica* lipopolysaccharide and interleukin-1. *Vet. Pathol.* 31:55-60.

Breider, M.A., Kumar, S. and Corstvet, R.E., 1990. Bovine pulmonary endothelial damage mediated by *Pasteurella haemolytica* pathogenic factors. *Infect. Immun.* 58: 1671-1677.

Breider, M.A., Kumar, S. and Corstvet, R.E., 1991. Protective role of bovine neutrophils in *Pasteurella haemolytica-mediated* endothelial cell damage. *Infect. Immun.* 59:4570-4575 .

Briggs, R.E. and Frank, G.H., 1992. Increased elastase activity in nasal mucus associated with nasal colonization by *Pasteurella haemolytica* in infectious bovine rhinotracheitis virus-infected calves. *Am. J. Vet. Res.* 53:631-635.

Briggs, R.E., Tatum, F.M., Casey, T.A., and Frank, G.H., 1994. Characterization of a restriction endonuclease, *PhaI*, from *Pasteurella haemolytica* serotype A1 and protection of heterologous DNA by a cloned *PhaI* methyltransferase gene. *Appl. Environ. Microbiol.* 60:2006-2010.

Chae, C.H., Gentry, M.J., Confer, A.W. and Anderson, G.A., 1990. Resistance to host immune defense mechanisms afforded by capsular material of *Pasteurella haemolytica*, serotype 1. *Vet. Microbiol.* 25:241-251.

Clinkenbeard, K.D. and Upton, M.L., 1991. Lysis of bovine platelets by *Pasteurella haemolytica* leukotoxin. *Am. J. Vet. Res.* 52:453-457.

Clinkenbeard, K.D., Mosier, D.A., and Confer, A.W., 1989a. Effects of *Pasteurella haemolytica* leukotoxin on isolated bovine neutrophils. *Toxicon.* 27:797-804.

Clinkenbeard, K.D., Clarke, C.R., Hague, C.M., Clinkenbeard, P.A., Srikumaran, S., and Morton, R.J., 1994. *Pasteurella haemolytica* leukotoxin-induced synthesis of eicosanoids by bovine neutrophils *in vitro*. *J. Leukocyte Biol.*, 56:644-649.

Clinkenbeard, K.D., Mosier, D.A., and Confer, A.W., 1989b. Transmembrane pore size and role of cell swelling in cytotoxicity caused by *Pasteurella haemolytica* leukotoxin. *Infect. Immun.* 57:420-425.

Clinkenbeard, K.D., Mosier, D.A., Timko, A.L., and Confer, A.W., 1989c. Effects of *Pasteurella haemolytica* leukotoxin on cultured bovine lymphoma cells. *Am. J. Vet. Res.* 50:271-275.

Clinkenbeard, K.D., Clinkenbeard, P., and Waurzyniak, B., 1993. Binding of *Pasteurella haemolytica* leukotoxin to isolated bovine platelet and RBC ghosts and cultured bovine lymphoma cell plasma membranes. *Proc. 74th Ann. Mtg. Conf. Res. Work. Ani. Dis.*, Chicago, 80.

Confer, A.W., 1993. Immunogens of Pasteurella. *Vet. Microbiol.* 37:353-368.

Confer, A.W. and Durham, J.A., 1992. Sequential development of antigens and toxins of *Pasteurella haemolytica* serotype 1 grown in cell culture medium. *Am. J. Vet. Res.* 53:646-652.

Confer, A.W., Panciera, R.J., Clinkenbeard, K.D. and Mosier, D.A., 1990. Molecular aspects of virulence of *Pasteurella haemolytica*. *Can. J. Vet. Res.* 54 Suplmt:S48-S52.

Conlon, J.A. and Shewen, P.E., 1991a. Efficacy of recombinant leukotoxin in protection against pneumonic challenge with live *Pasteurella haemolytica* A1. *Infect. Immun.* 59:587-591 .

Cooney, B.J. and Lo, R.Y.C., 1993. Three contiguous lipoprotein genes in *Pasteurella haemolytica* A1 which are homologous to a lipoprotein gene in *Haemophilus influenzae* type b. *Infect. Immun.* 61:4682-4688.

Craig, F.F., Coote, J.G., Parton, J.G., Freer, J.H., and Gilmour, N.J.L., 1989. A plasmid which can be transferred between *Escherichia coli* and *Pasteurella haemolytica* by electroporation and conjugation. *J. Gen. Microbiol.* 135:2885-2890.

Craven, R.C., Confer, A.W., Gentry, M.J., 1991. Cloning and expression of a 30 kDa surface antigen of *Pasteurella haemolytica*. *Vet. Microbiol.* 27:63-78.

Czuprynski, C.J. and Ortiz-Carranza, O., 1992. *Pasteurella haemolytica* leukotoxin inhibits mitogen-induced bovine peripheral blood mononuclear cell proliferation *in vitro*. *Microbial Pathog.* 12:459-463 .

Czuprynski, C.J., Noel, E.J., Ortiz-Carranza, and Shrikumaran, S., 1991. Activation of bovine neutrophils by partially purified *Pasteurella haemolytica* leukotoxin. *Infect. Immun.* 59:3126-3133.

Dabo, S.M., Confer, A.W., Styre, D., and Murphy, G.L., 1994. Expression, purification, and immunologic analysis of three *Pasteurella haemolytica* Al 28-30 kDa lipoproteins. *Microbial Pathog.*, in press.

Dal Nogare, A.R., Toews, G.B., and Pierce, A.K., 1987. Increased salivary elastase precedes gram-negative bacillary colonization in postoperative patients. *Am. Rev. Respir. Dis.* 135:671-678.

Davies, R.L., Parton, R., Coote, J.G. Gibbs, H.A., and Freer, J.H., 1992. Outermembrane protein and lipopolysaccharide variation in *Pasteurella haemolytica* serotype A1 under different growth conditions. *J. Gen. Microbiol.* 138:909-922.

Espinasse, J., Peel, J.E., Voirol, M.J., Schelcher, F., and Valarcher, J.F., 1993. Absenceof circulating TNF~ in experimental bovine pneumonic pasteurellosis. *Vet. Rec.* 132:303-304.

Frank, G.H., 1989. Pasteurellosis of cattle. In: C. Adlam and J.M. Rutter (Editors), *Pasteurella* and Pasteurellosis. *Academic Press,* London, pp. 197-222.

Frey, J., 1992. Construction of a broad host range shuttle vector for gene cloning and expression in *Actinobacillus pleuropneumoniae* and other *Pasteurellaceae. Res. Microbiol.* 143:263-269.

Froshauer, S.A., Ricketts, A.P., Harrigan, M.B., Mann, D.W., and Natoli. Interaction of *Pasteurella haemolytica* with bovine defense cells. *Proc. 74th Annu. Mtg. Conf. Res. Work. Anim. Dis.,* Chicago. 236.

Gentry, M.J., Confer, A.W., and Craven, R.C., 1987. Effect of repeated in vitro transfer of *Pasteurella haemolytica* A1 on encapsulation, leukotoxin production, and virulence. *J. Clin. Microbiol.* 25:142-145.

Gonzalez, C.T. and Maheswaran, S.K., 1993. The role of induced virulence factors produced by *Pasteurella haemolytica* in the pathogenesis of bovine pneumonic pasteurellosis: Review and hypothesis. *Brit. Vet. J.* 149:183-193.

Gonzalez-Rayos, C., Lo, R., Shewen, P.E., and Beveridge, T.J., 1986. Cloning of a serotype-specific antigen from *Pasteurella haemolytica Al. Infect. Immun.* 53:505-510.

Hendricks, P.A.J., Binkhorst, G.J., Drijver, A.A., and Nijkamp, F.P., 1992. *Pasteurella haemolytica* leukotoxin enhances production of leukotriene B4 and 55-hydroxyeicosatetraenoic acid by bovine polymorphonuclear leukocytes. *Infect. Immun.* 60:3238-3243.

Jann, K. and Hoschutzky, H., 1991. Nature and organization of adhesins. Curr. Top. *Microbiol. Immun.* 151:55-62.

Maheswaran, S.K., Weiss, D.J., Kannan, M.S., Townsend, E.L., Reddy, K.R., Whitely, L.O., and Srikumaran, S., 1992. Effects of *Pasteurella haemolytica* Al leukotoxin on bovine neutrophils: degranulation and generation of oxygen-derived free radicals. *Vet. Immunol. Immunopathol.* 33:51-68.

Majury, A.L. and Shewen, P.E., 1991a. The effect of *Pasteurella haemolytica* A1 leukotoxic culture supernate on the in vitro proliferative response of bovine lymphocytes. *Vet. Immunol. Immunopathol.* 29:41-56.

Majury, A.L. and Shewen, P.E., 1991b. Preliminary investigation of the mechanism of inhibition of bovine lymphocyte proliferation by *Pasteurella haemolytica* A1 leukotoxin. *Vet. Immunol. Immunopathol.* 29:57-68.

Mason, C.M., Nelson, S., and Summer, W.R., 1993. Bacterial colonization: Pathogenesis and clinical significance. *Immun. Allergy Clinics N. Amer.* 13:93-108.

Morch, D.W., Raybould, T.J.G., Acres, S.D., Babiuk, L.A., Nelligan, J. and Costerton, J.W., 1987. Electron microscopic examination of cells of *Pasteurella haemolytica*Al. *Can. J. Vet. Res.* 51:83-88.

Mosier, D.A., Simons, K.R., Leedle, J.A.Z., Hart, and Chegappa, M.M., 1993. The effect of atmospheric conditions and temperature on antigen expression by *Pasteurella haemolytica. Proc. 74th Annu. Meet. Conf. Res. Work. Anim. Dis.,* Chicago, 239.

Murphy, G.L. and Whitworth, L.C., 1993. Analysis of tandem, multiple genes coding 30kDa membrane proteins in *Pasteurella haemolytica* A1. *Gene.* 129:107-111.

Murphy, G.L. and Whitworth, L.C., 1994. Construction of isogenic mutants of *Pasteurella haemolytica* by allelic replacement. *Gene,* 129:107-111.

Murray, J.E., Davies, R.C., Lainson, F.A., Wilson, C.F., Donachie, W., 1992. Antigenic analysis of iron-regulated proteins in *Pasteurella haemolytica* A and 1 biotypes by immunoblotting reveals biotype-specific epitopes. *J. Gen. Microbiol.* 138:283-288.

Ortiz-Carranza, O. and Czuprynski, C.J., 1992. Activation of bovine neutrophils by *Pasteurella haemolytica* leukotoxin is calcium dependent. *J. Leukocyte Biol.* 52:558-564.

Pace, L.W., Kreeger, J.M., Bailey, K.L., Turnquist, S.E. and Fales, W.H., 1993. Serum levels of tumor necrosis factor-α in calves experimentally infected with *Pasteurella haemolytica* A1. *Vet. Immunol. Immunopathol.* 35:353-364.

Paulsen, D.B., Mosier, D.A., Clinkenbeard, K.D. and Confer, A.W., 1989. The direct effects of *Pasteurella haemolytica* lipopolysaccharide on bovine pulmonary endothelial cells *in vitro. Am. J. Vet. Res.* 50:1633-1637.

Potter, A.A., Ready, K., and Gilchrist, J., 1988. Purification of fimbriae from *Pasteurella haemolytica* A1. *Microbial Pathogen.* 4:311-316.

Sharma, S.A., Olchowy, T.W.J., and Breider, M.A., 1992a. Alveolar macrophage and neutrophil interactions in *Pasteurella haemolytica-induced* endothelial cell injury. *J. Infect. Dis.* 165:651-657.

Sharma, S.A., Olchowy, T.W.J., Yang, Z., and Breider, M.A., 1992b. Tumor necrosis factor-α and interleukin 1 enhance lipopolysaccharide-mediated bovine endothelial cell injury. *J. Leukocyte Biol.* 51:579-585.

Shewen, P.E. and Wilkie, B.N., 1982. Cytotoxin of *Pasteurella haemolytica* acting on bovine leukocytes. *Infect. Immun.* 35:911-914.

Squire, P.G., Smiley, D.W., and Croskell, R.B., 1984. Identification and extraction of *Pasteurella haemolytica* membrane proteins. *Infect. Immun.* 45:667-673.

Stevens, P. and Czuprynski, C., 1993. *Pasteurella haemolytica* leukotoxin modulation of bovine mononuclear phagocytes in vitro. *Proc. 74th Annu. Mtg. Conf. Res. Work. Animal Dis.*, Chicago, 235.

Stevens, P. and Czuprynski, C., 1994. Flow cytometric investigation of *Pasteurella haemolytica* leukotoxin mediated killing of bovine leukocytes. *Proc. Virulence Mech. Bact. Pathogens*, Ames, 22.

Straus, D.C. and Purdy, C.W., 1994. In vivo production of neuraminidase by *Pasteurella haemolytica* A1 (Ph A1) in goats following transthoracic bacterial challenge. *Proc. 94th Ann. Meet. Amer. Soc. Microbiol.*, Las Vegas, B-128.

Straus, D.C., Unbehagen, P.J., Purdy, C.W., 1993. Neuraminidase production by a *Pasteurella haemolytica* A1 strain associated with bovine pneumonia. *Infect. Immun.* 61:2553-259.

Tatum, F.M., Briggs, R.E., and Halling, S.M., 1994. Molecular cloning and nucleotide sequencing and construction of an *aroA* Mutant of *Pasteurella haemolytica* serotype A1. *Appl. Environ. Microbiol.* 60:2011-2016.

Uhlich, G.A., Mosier, D.A., and Simons, K.R., 1993. An enzyme-linked immunosorbent assay for detection of adhesion of *Pasteurella haemolytica* to bovine nasal mucus. *Proc. 74th Ann. Mtg. Conf. Res. Work. Anim. Dis.*, Chicago, 24.

Waurzyniak, B.J., Clinkenbeard, K.D., Confer, A.W., and Srikumaran, S., 1994. Enhancement of *Pasteurella haemolytica* leukotoxic activity by bovine serum albumin. *Am. J. Vet. Res.*, 55:1267-1274.

Weiss, D.J., Bauer, M.C., Whiteley, L.O., Maheswaran, S.K., and Ames, T.R., 1991. Changes in blood and bronchoalveolar lavage fluid components in calves with experimentally induced pneumonic pasteurellosis. *Am. J. Vet. Res.* 52:337-344.

Whiteley, L.O., Maheswaran, S.K., Weiss, D.J., and Ames, T.R., 1990. Immunohistochemical localization of *Pasteurella haemolytica* A1-derived endotoxin, leukotoxin, and capsular polysaccharide in experimental bovine pasteurella pneumonia. *Vet.Pathol.* 27:150-161.

Whiteley, L.O., Maheswaran, S.K., Weiss, D.J., Ames, T.R. and Kannan, M.S., 1992. *Pasteurella haemolytica A1* and bovine respiratory disease: pathogenesis. *J.Vet. Int. Med.* 6:11-22.

Yu, R.H., Gray-Owen, S.D., Ogunnariwo, J. and Schryvers, A.B., 1992. Interaction of ruminant transferrins with transferrin receptors in bovine isolates of *Pasteurella haemolytica* and *Haemophilus somnus*. *Infect. Immun.* 60:2992-2994.

HAEMOPHILUS SOMNUS: ANTIGEN ANALYSIS AND IMMUNE RESPONSES

L.B. Corbeil,[1] R.P. Gogolewski,[2] L.R. Stephens,[3] and T.J. Inzana[4]

Department of Pathology, University of California, San Diego 92103-8416,[1]
P.O. Box 135, Ingleburn, NSW 2565, Australia,[2]
Victorian Institute of Animal Science, Attwood, Victoria 3049 Australia,[3]
The Center for Molecular Medicine and Infectious Diseases, Virginia Polytechnic Institute and State University, Blacksburg, VA 24061 [4]

INTRODUCTION

Haemophilus somnus causes several disease syndromes in cattle and sheep but is also carried asymptomatically, especially on the genital mucosa. The factors involved in determining whether disease or asymptomatic carriage result are only beginning to be defined. Both host and bacteria contribute to the outcome of the dynamic interaction. To gain some insight into *H. somnus*/host interactions, it is necessary to consider what is known of the disease or asymptomatic conditions seen as well as current information on virulence factors and host immune response. This review constitutes the biased synthesis of the reviewers' thoughts on available data relative to *H. somnus* infection.

NATURAL HISTORY OF INFECTION

In cattle, *H. somnus* was first noted to cause thrombotic meningoencephalitis (TME) (Stephens *et al.*, 1981). Now it is known to also be involved in the etiology of upper respiratory infection, pneumonia, septicemia, abortion, arthritis, myocarditis and infertility (Corbeil, 1990; Harris and Janzen, 1989, Humphrey and Stephens 1983, Kwiecien and Little, 1991; Miller *et al.*, 1983). It is also carried asymptomatically on the prepuce of most bulls (Humphrey *et al.*, 1985; Slee and Stephens, 1985; Ward *et al.*, 1983) and the vagina of some cows (Corbeil, 1986; Kwiecien and Stephens, 1991; Ward *et al.*, 1983). An upper respiratory carrier state has also been reported (Humphrey and Stephens, 1983) but appears to be much more rare than the genital carrier state. However, it may be that the organism persists longer in the lower respiratory tract than in the upper tract. In a study of experimental pneumonia with 5 calves, we showed that clinical pneumonia was present for less than a week but the organisms could be isolated from bronchial lavage fluids for 5 to 10 weeks or more, even when nasal swabs were culturally negative (Gogolewski *et al.*, 1989).

Sheep also may carry *H. somnus* in the genital tract generally without clinical signs, or may have *H. somnus* induced disease (Lees *et al.*, 1994). Ovine *H. somnus*, has been called *Histophilus ovis* or *Haemophilus agni* in earlier literature but several recent reports indicate that *Histophilus ovis*, *H. agni*, and *H. somnus* should be considered as one taxon (McGillivery *et al.*, 1986; Stephens *et al.*, 1983; Walker *et al.*, 1985). The carrier state is primarily preputial (Lees *et al.*, 1990; Walker and LeaMaster, 1986) but the organism may also be found in the vagina of ewes (Higgins *et al.*, 1981). Epidydimitis in young rams is one of the more frequently reported disease syndromes (Bulgin and Anderson, 1983; Walker *et al.*, 1986 and 1988). In addition, vulvitis (Ball *et al.*, 1991), lower reproductive rates (Lees *et al.*, 1990), mastitis (Beauregard and Higgins, 1983), septicemia, arthritis, meningitis and pneumonia (Rahaley, 1978) have been reported. Rahaley (1978) produced septicemia by intravenous or intranasal inoculation of isolates from the ovine vagina or from lamb septicemia. Recently we reproduced meningitis (Lees *et al.*, 1994) with isolates from septicemia and preputial carriers by intracisternal inoculation, but not all preputial isolates were capable of producing disease.

VIRULENCE ATTRIBUTES

Several virulence factors have been reported which may help to explain why disease occurs in some cases and not others. Obviously host factors may also be involved but this will be discussed later. Attachment to host cells is one of the bacterial functions involved in the first step of pathogenesis. *Haemophilus somnus* has been shown to adhere to several host cell types including bovine turbinate cells (Ward *et al.*, 1983) and bovine endothelial cells (Thompson and Little, 1981). More recently, we (Stephens LR and Corbeil LB, unpublished data) showed H. *somnus* attaches in large numbers to bovine vaginal epithelial cells (VEC). It may be significant that the organisms adhered primarily to the large flat more differentiated cells (Figure 1). Most of these cells were covered with *H. somnus*, reminiscent of clue cells which are characteristic of human bacterial vaginosis. Since human bacterial vaginosis is associated with abnormal outcome of pregnancy (McDonald *et al.*, 1991, Holst *et al.*, 1994), the finding that *H. somnus* covers epithelial cells in clue cell like fashion may be significant. In our experience, nearly all of more than 100 *H. somnus* strains tested adhered in large numbers to primary bovine VEC. Isolates of other bacterial species did not adhere in large numbers to VEC. Most but not all of the adherent isolates of *H. somnus* also autoagglutinated. Thus auto-agglutination and adherence appear to be linked but probably not closely. Although the molecular basis of adherence was not defined, hydrophobic interactions were implicated. Identification of the molecules involved in adherence could be a fruitful approach to an understanding of virulence mechanisms and as a target for immunoprophylaxis.

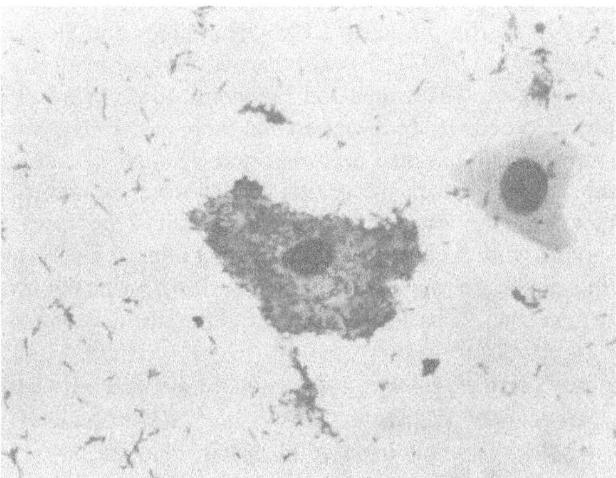

Figure 1. Attachment of *Haemophilus somnus* to bovine vaginal epithelial cells. Note that the larger cell is almost covered with *H. somnus* whereas the smaller cell has no more *H. somnus* than in the background.

Cytotoxicity or interference with host cell function is another factor associated with virulence that is often crucial to pathogenesis. In our studies of experimental pneumonia, calves were killed at 24 hours after inoculation of 10^7 *H. somnus, so* that early steps in the pathogenesis could be identified (39). The most prominent finding on histopathologic examination of severely affected lung was marked degeneration of macrophages and abundant extracellular bacteria (Figure 2). The *in vitro* studies of Lederer *et al.*, (1987) indicated that *H. somnus* was not killed by bovine mononuclear cells. These findings are consistent with our findings *in vivo* (Gogolewski *et al.*, 1987b) and suggest that *H. somnus is* not killed by macrophages because the phagocytes are degenerate, not because *H. somnus* survives in healthy macrophages. Vasculitis characterized by degeneration of endothelial cells and transmural neutrophil infiltration is also a prominent finding in *H. somnus* induced disease (Gogolewski *et al.*, 1987b; Humphrey and Stephens, 1983). These endothelial changes seen *in vivo* are similar to those described in the *in vitro* studies of Thompson and Little (1981) on the interactions of *H. somnus* and bovine endothelial cells from arterial explants. Not only did *H. somnus* attach to endothelial cells but this attachment induced endothelial cell contraction and exposure of the basement membrane. Endothelial cell toxicity, therefore, may be a first step in the pathogenesis of the neutrophilic vasculitis typical of acute *H. somnus* disease. Neutrophils, too, are inefficient at killing *H. somnus* according to the *in vitro* studies of Czuprynski and Hamilton 1985. This may be due, at least in part, to secretion of factors by *H. somnus* which suppress neutrophil function. Interestingly, Chiang *et al.* (1986) identified these factors as purine and pyrimidine bases as well as ribonucleotides and a ribonucleoside. Others have demonstrated the reduced activity of neutrophils obtained from calves infected with *H. somnus* compared with neutrophils from healthy calves (Pfeifer *et al.*, 1992).

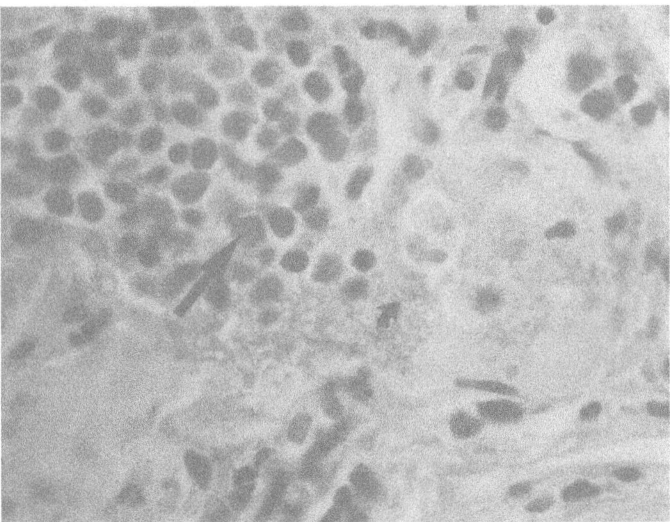

Figure 2. Abundant *Haemophilus somnus* in the alveoli (small arrow), associated with degenerate macrophages (larger arrow) twenty four hours after intrabronchial inoculation. Note that the stromal cells are not degenerate, indicating that fixation was adequate (H&E staining, lOOX magnification).

Resistance to killing by complement is a virulence attribute important in survival after invasion since complement components are abundant in blood and extracellular fluids. In secretions, less is known about the abundance and activity of complement components but during inflammation and exudation, at least, complement mediated killing in secretions could be important. Some time ago, we investigated the susceptibility of *H. somnus* isolates to killing by bovine serum complement. Killing was defined as a one log decrease, or more, in colony counts after incubation in fresh normal bovine serum for 1 hour. Essentially all isolates that were associated with disease were serum resistant whereas approximately 25% of preputial isolates from normal bulls were serum sensitive (Corbeil *et al.*, 1985). The killing was predominantly by the classical complement pathway and lysozyme did not play a major role (Corbeil *et al.*, 1985).

From the above discussion it is obvious that many virulence factors are involved in several steps of pathogenesis. For example, adherence is likely to be important in colonization, complement resistance in survival in the circulation or inflammatory sites and cytotoxicity in evading killing by phagocytes and in initiation of vasculitis as well as invasion through the endothelium. The host damage which occurs as a result may be further exacerbated by inflammatory mediators released by the host in response to *H. somnus*.

MOLECULAR CHARACTERIZATION OF VIRULENCE FACTORS

The molecules involved in virulence are only beginning to be understood. The outer membrane has received most attention since *H. somnus* does not have flagella and appears to be nonencapsulated (Stephens and Little, 1981). The endotoxin of the outer membrane has been characterized as a lipooligosaccharide (LOS) without long O side chains (Inzana *et al.*, 1988). Toxicity similar to other endotoxins was demonstrated (Inzana *et al.*, 1988). The molecular composition of *H. somnus* LOS may partially explain the differences among strains in resistance to killing by bovine complement and in ability to cause septicemia. Most serum sensitive preputial strains (Corbeil *et al.*, 1985) examined in silver stained SDS PAGE gels had one or two major LOS bands less than Mr 3800. In contrast, serum resistant virulent strains had higher molecular mass LOS bands (Mr 4000 to 5000) that are lacking in the LOS of preputial isolates (Figure 3).

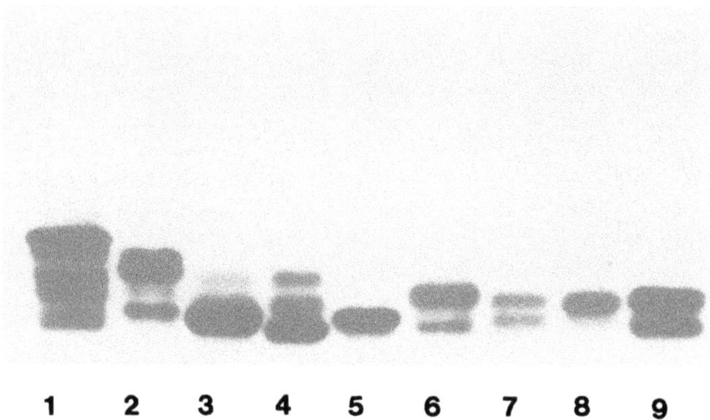

Figure 3. Silver stained SDS-PAGE gel of H. *somnus* LOS from serum resistant strains 649, 1297, 2336 and 8025 as well as serum sensitive preputial strains IP, 24P, 124P, 127P, 129Pt (left to right). Note that the preputial strains have fewer and/or lower molecular mass bands than the disease isolates.

Further studies showed that a serum resistant abortion isolate (strain 649) has several of the higher molecular mass LOS bands greater than Mr 4000 and is capable of causing haematogenous abortion in mice. In contrast, serum sensitive preputial strain 127P has one predominant LOS band less than Mr 3800 and one minor band of about Mr 4500 (Figure 3) and does not cause murine abortion (Inzana and Todd, 1992). We had previously shown that strain 649 caused experimental bovine abortion (Widders *et al.*, 1986) but had not tested the ability of preputial strain 127P to cause abortion in cattle. However Little and coworkers have tested the ability of several strains to cause experimental TME or pneumonia, and found isolates from TME or pneumonia caused disease but preputial strains were less virulent (Groom *et al.*, 1988; Humphrey and Stephens, 1983). The ability of isolates from the female genital tract to cause disease was more varied. For example, of 14 isolates from the female reproductive tract inoculated intracisternally into 14 calves (one isolate per calf), 6 caused neurologic disease and 8 did not (Kwiecien and Little, 1992). Whether these strains had differences in LOS composition is not known.

The differences in LOS composition between isolates obtained from disease and the preputial strains is more complex than first appreciated. Although the number and Mr of LOS bands appears to be important in serum sensitivity, antigenic variation also appears to play a role in host-parasite interactions. The LOS of disease isolates undergoes alteration in electrophoretic profile and antigenic reactivity as they are reisolated from mice and their aborted fetuses after intravenous challenge of pregnant mice (Inzana and Todd, 1992), from cattle after intrabronchial challenge, or after *in vitro* passage (Inzana *et al.*, 1992). No variation in LOS profile occurs in strain 127P following challenge of pregnant mice or after *in vitro* passage (Inzana and Todd, 1992, Inzana *et al.*, 1992). This phase variation in the LOS of virulent isolates may represent a mechanism of evasion of host defense against *H. somnus*. Evidence to support this hypothesis was gained from the *in vivo* experiments. In the calf challenge study, there was a correlation between reactivity of serum antibodies with LOS phase variants throughout the duration of the infection. That is, more of the LOS phase variants reacted with antibodies from the calves by immunoblotting with serum collected near the end of the infection than that collected near the beginning of infection. Calves with antibodies to all the LOS variants cleared their infections, whereas at least one calf that had antibodies reactive with only half of the isolates' LOSs still had a substantial bacterial burden. This suggested that when antibody responses had specificity for all variants, the host cleared the infection.

A second factor involved in serum resistance came to our attention serendipitously. When Sue Dowling detected high background in ELISA assays for bovine antibody to *H. somnus* whole cells, she hypothesized that the organisms had adsorbed antibody while growing on blood agar plates. Shortly thereafter a postdoctoral fellow, Phil Widders, joined the laboratory and suggested that the binding of antibody may be by the Fc fragment. Using papain digestion to generate Fc fragments as well as in competitive inhibition studies with affinity purified bovine anti-DNP, he then showed that bovine immunoglobulin did bind to *H. somnus* via the Fc fragment (Widders *et al.*, 1988). Michele Yarnall *et al.*, (1988 a and b) subsequently identified and characterized the immunoglobulin binding proteins (IgBPs) of *H. somnus*. Two types of IgBPs were detected (Yarnall *et al.*, 1988b). A 41kDa IgBP weakly bound bovine IgG1, IgG2, IgA and IgM. This IgBP co-migrates with or is identical with, the major outer membrane protein (MOMP) of *H. somnus*. The other type of IgBP may play a greater role in virulence since the Ig binding is much stronger. This IgBP is detected on Western blots as a series of bands of approximately 350, 270 and 120kDa, as well as other bands in this molecular weight range (Yarnall *et al.*, 1988b). This group of IgBPs binds predominantly IgG2 (Yarnall *et al.*, 1988b) and is secreted into the culture medium (Yarnall *et al.*, 1988a). Interestingly, the 41kDa IgBP (p41) is not recognized by convalescent serum although bovine antibody to it can be prepared by inoculation of the isolated, denatured molecule in adjuvant (Yarnall *et al.*, 1988a). The high molecular weight IgBP (designated p270) elicits an antibody response during infection (Yarnall *et al.*, 1988a) and can be used diagnostically if antigen concentrations are used in immunodots which are lower than amounts needed to detect IgBP activity (Yarnall and Corbeil, 1989). Under these conditions p270 detects antibody in vaccinated cattle or those with disease but not in normal animals that are culture negative or asymptomatic carriers (Yarnall and Corbeil, 1989).

The role of the IgBPs in virulence has been especially interesting. After testing the IgBP binding capacity of many isolates from diseased or carrier animals, we found that all isolates bound IgG-Fcs except for a few serum susceptible preputial isolates (Widders *et al.*, 1989a). These isolates lacked the high molecular weight IgBPs and had a 33kDa MOMP rather than the characteristic 41kDa MOMP (Widders *et al.*, 1989a). Thus the presence of IgBPs was associated with serum resistance. Later, IgBPs were also shown to be associated with resistance to phagocytosis by bovine PMNs (Alexander-Hanley and Widders, 1989). Further characterization of the high molecular weight IgBPs was undertaken by molecular genetic analysis. We had already cloned the gene for the 120kDa IgBP (pl20), when we determined that IgBPs were associated with serum resistance (Corbeil *et al.*, 1985; Widders *et al.*, 1989a). This IgBP was shown to be genetically linked to a 76 K (p76) surface protein of H. *somnus* (Cole *et al.*, 1992; Corbeil *et al.*, 1988) which is recognized by convalescent serum (Corbeil *et al.*, 1987). Both p76 and pl20 reacted with antiserum raised to p270 but DNA probes derived from inserts expressing pl20 or p76 bound to different restriction fragments from *H. somnus* (Cole *et al.*, 1992). Thus we speculated that pl20 and p76 constituted different components of p270 (Cole *et al.*, 1992). Further evidence to support this hypothesis was gained when we demonstrated that serum sensitive isolates lacking p270, also lacked pl20 and p76 (Cole *et al.*, 1992). In fact, the four serum sensitive isolates tested, not only lacked pl20 and p76 but also lacked the 13.4 kb DNA sequence which encodes these two proteins (Cole *et al.*, 1992). Sequence analysis of the insert encoding p76 revealed two 1.2 kb tandem direct repeats (DRs). The 5'DR included the start site for p76 and the 3'DR had a flanking inverted repeat, suggestive of an insertion-like sequence (Cole *et al.*, 1993). There was 65% identity between the two DRs at the protein level, suggesting that if insertion and duplication occurred, the events were long enough ago to allow for substantial divergence between the two DRs. The presence of the DRs in serum resistant but not serum susceptible strains, however, suggests that they may be necessary for expression of proteins involved in resistance to complement mediated killing.

Another surface antigen which appears to be important in protection is the 40kDa (p40) immunodominant protein. We showed that convalescent phase serum, which was passively protective against experimental pneumonia, recognized primarily 78kDa (p78) and p40 outer membrane proteins (OMPs) (Gogolewski *et al.*, 1987a). Bovine antiserum was then prepared against each of these OMPs and the immunoglobulin fractions tested in passive protection experiments (Gogolewski *et al.*, 1988). Antibody to p40 protected but antibody to p78 did not (Gogolewski *et al.*, 1988). The failure of anti-p78 may have been due to factors other than specificity as discussed below. However, protection by anti-p40 was promising. Thus we further defined the antigen migrating at ~40kDa in gradient (7 1/2 - 17 1/2%) SDS-PAGE gels. When 8 or 10% polyacrylamide gels were run and blotted, two bands were detected in the area where only one was detected on gradient gels. The band reacting with the protective anti-p40 retained the p40 designation and the band not reacting with the protective antibody was designated p39 (Corbeil *et al.*, 1991). The latter protein was recognized by a monoclonal antibody (3G9) which did not react with p40. Both antigens were conserved in all isolates of *H. somnus* tested (Corbeil *et al.*, 1991). In addition, antigens cross-reacting with p40 were detected in a few other bacterial species, especially those in the family Pasteurellaceae. Since the cross-reacting antigens were surface exposed in *Pasteurella haemolytica,* we hypothesized that anti-p40 also may be protective against pasteurellosis. Monoclonal antibody to p39, on the other hand, only reacted with *H. somnus* from cattle and sheep, suggesting it may be a useful diagnostic antigen (Corbeil *et al.*, 1991). Subsequently two genes for antigens of ~40 K have been cloned and characterized by Theisen et al. (1992 and 1993). Both antigens were lipid modified. Therefore they have been designated LppA (Theisen *et al.*, 1992) and LppB (Theisen et al. 1993). Protein LppB was shown to bind Congo red (Theissen *et al.*, 1993), a factor which is strongly associated virulence in other bacteria. It would be interesting to determine whether either LppB or LppA is similar to the p40 OMP which we showed to be potentially protective.

Transferrin receptors of *H. somnus* have been characterized recently (Ogunnariwo *et al.*, 1990; Yu *et al.*, 1992). Bovine transferrin (but not porcine, human or chicken transferrin) was shown to serve as an iron source for bovine *H. somnus* (Ogunnariwo *et al.*, 1990). Two iron-regulated OMPs (105 and 73kDa) were isolated by transferrin affinity purification methods (Ogunnariwo *et al.*, 1990). This suggests that a receptor mediated iron uptake mechanism is involved. Another report by the same group of investigators (Yu *et al.*, 1992) shows that bovine *H. somnus* binds bovine transferrin but not ovine transferrin. It is tempting to speculate that this is the reason for the reported species specificity of bovine *H. somnus* for cattle but not sheep (Biberstein 1981; Young and Hoerlein, 1970). This is especially suggestive since we recently showed that isolates of ovine *H. somnus* pathogenic for sheep had the same antigens we had previously identified in bovine *H. somnus* (Lees *et al.*, 1994): p270, p78, p76, p40 and p39. This observation of conservation of key antigens in both bovine and ovine *H. somnus,* although not explaining species specificity, may indicate that vaccines containing these antigens (especially p40) would protect against both bovine and ovine disease. Some evidence for this hypothesis is gained from the observations of Bulgin (1990) that use of a commercially available bovine *H. somnus* vaccine appears to reduce the incidence of *H. somnus* induced disease in sheep.

Other antigens of *H. somnus* have been characterized by several groups. Tagawa et al. (1993d) purified the MOMP of *H. somnus* and determined it to have an apparent molecular mass of 40kDa. The amino-terminal sequence of this MOMP showed considerable homology to porins of other gram-negative bacteria. Antigenic analysis revealed three surface exposed epitopes and two non exposed epitopes (Tagawa *et al.*, 1993a). In our studies the MOMP was designated as a 41kDa protein. Since there are many proteins that migrate very close to one another in SDS PAGE gels, it is likely that Tagawa's 40kDa OMP is the 41kDa OMP in our studies. In addition, Tagawa *et al.*, (1993b) have characterized a heat modifiable OMP of *H. somnus* which has an apparent molecular mass of 28kDa when solubilized at 60°C and a mass of 37kDa when solubilized at 100°C. This OMP reacted intensely with convalescent phase serum and had both surface exposed and non-surface exposed epitopes. Antibodies to this OMP reacted with all *H. somnus* strains tested. It cross-reacted with and had N terminal sequence homology with *E. coli* OmpA. Whether this OMP is related to any of the OMPs we studied is unclear. Lastly, Tagawa *et al.*, (1993b) described a 17.5kDa OMP which was also conserved in all *H. somnus* strains tested. In 1993, Won and Griffith (1993) identified a 31kDa protein recognized by bovine antibody to *H. somnus*. The gene for this protein was cloned and expressed in *E. coli.* Recombinants lysed bovine erythrocytes and deletion of the open reading frame resulted in loss of hemolytic activity. Interestingly a deduced amino acid sequence shared homology with *E. coli* OMPA. Since the expressed protein in E. *coli* conferred hemolytic activity, this may be a virulence factor of *H. somnus* also.

Commercially available killed whole cell vaccines have been available for some time. Vaccines clearly protect against TME (Stephens *et al.*, 1982) but protection against pneumonia is more controversial (Groom and Little, 1988; Humphrey and Stephens, 1983; Martin *et al.*, 1984) and little if any data has been reported on protection against reproductive failure. Studies of both antigenic specificity and immunologic functions involved in protection are necessary not only because of the questions about protection against pneumonia and reproductive failure but also because adverse reactions against whole cell vaccines sometimes occur. Stephens *et al.*, (1984) showed that anionic membrane antigens protect against TME and the same group later demonstrated that this was not due to LOS (Silva and Little, 1990). Whether the anionic membrane preparation was rich in p40 is not known but it would be interesting to determine its p40 content since our data suggests this is a potentially protective antigen (Gogolewski *et al.*, 1988) as described above.

Immune functions necessary in protection have not been well studied. Because *H. somnus* appears to be an extracellular pathogen (Corbeil, 1990), we focused on mechanisms of humoral immunity. Cell mediated immune mechanisms may also be involved but antibody should be especially important. This was confirmed by our studies which showed that preincubation of *H. somnus* for 15 minutes in subagglutinating concentrations of antibody before inoculation of 10^7 organisms intrabronchially, protected against pneumonia (Gogolewski *et al.*, 1987a and 1988). Several lines of evidence suggested that IgG2 antibody may be most important. Early studies of Nanssen showed that cattle deficient in IgG2 were more susceptible than normal cattle to pyogenic infections including pneumonia (Nanssen, 1972). We later demonstrated that low IgG2 antibody levels were associated with onset of calfhood pneumonia (Corbeil *et al.*, 1984). In studies of *H. somnus* infection, we found that cattle with low IgG2 antibody titers to *H. somnus* were most susceptible to experimental *H. somnus* abortion (Widders *et al.*, 1989b). In other studies of the immune response to *H. somnus*, IgG2 titers increased the most after infection (Widders *et al.*, 1986 and 1989). Furthermore anti-p40 antibodies which protected against experimental pneumonia contained both IgG1 and IgG2 antibodies to p40 whereas antiserum to another surface exposed OMP, p78, which was not protective, contained only IgG1 antibodies (Gogolewski *et al.*, 1988). These observations led us to hypothesize that IgG2 antibodies to p40 would protect against *H. somnus* induced pneumonia better than IgG1 antibodies to p40. To test this hypothesis, we purified IgG1 and IgG2 antibodies separately from the polyclonal antibody preparation to p40 which we had previously shown to be passively protective (Gogolewski *et al.*, 1988). As before (Gogolewski *et al.*, 1987a and 1988), passive protection experiments were done by preincubating *H. somnus* strain 2336 in anti-p40 but this time IgG1 and IgG2 anti-p40 were tested individually. The same isotypes isolated from pre-immunization serum were inoculated into the contralateral lobes as controls. The protective capacity of anti-p40 IgG 1 and anti-p40 IgG2 were evaluated in separate experiments using four calves for each isotype. Twenty-four hours after inoculation, the calves were necropsied. The lungs were fixed by intra-vascular perfusion of fixative, sectioned at 3 mm intervals in the dorsoventral plane and volumes of pneumonia were determined by computer-assisted image analysis (Gogolewski *et al.*, 1987b). The mean difference in volume of pneumonic lung between pre- and postimmunization antibody was greater after inoculation of *H. somnus* preincubated with antip40 IgG2 (30 cm^3) than with anti-p40 IgG1 (16 cm^3) although this difference was not statistically significant. Expressed another way, the mean lesion size after preincubation of *H. somnus* with IgG1 anti-p40 was 66% of the IgG1 control whereas the mean lesion size after preincubation of *H. somnus* with IgG2 anti-p40 was 47% of the IgG2 control. This trend suggests that IgG2 anti-p40 is more protective than IgG1 anti-p40, which is consistent with other data mentioned above indicating that IgG2 is important in protection against *H. somnus* infection.

REFERENCES

Alexander-Hanley, C., Widders, P., 1989, Serum killing and phagocytosis of *Haemophilus somnus* isolates that differ in Fc receptor activity, ASM meeting, Abstract B-174.

Ball, H.J., Kennedy, S., Ellis, W.A., 1991, Experimental reproduction of ovine vulvitis with bacteria of the haemophilus/histophilus group, *Res. Vet. Sci.* 50:81-85.

Beauregard, M., and Higgins, R., 1983, Ovine mastitis due to *Histophilus ovis, Can. Vet.* J. 24:284-286.

Biberstein, E., 1981, *Haemophilus somnus* and *Haemophilus agni. in:* "Haemophilus, Pasteurella and Actinobacillus", M. Kilian, W. Frederiksen, E. Biberstein, eds., p. 125, Acad Press, Inc, London.

Bulgin, M.S., 1990, Epididymitis in rams and lambs, *Adv. in Sheep and Goat Med.* 6:683-690.

Bulgin, M.S., and Anderson, B.C., 1983, Association of sexual experience with isolation of various bacteria in cases of ovine epididymitis, J. *Am. Vet. Med. Assoc.* 182:372-374.

Chiang, Y.W., Kaeberle, M.L., and Roth, J.A., 1986, Identification of suppressive components in *Haemophilus somnus* fractions which inhibit bovine polymorphonuclear leucocyte function., *Infect. Immun.* 52:792-797.

Cole, S.P., Guiney, D.G., and Corbeil, L.B., 1992, Two linked genes for outer membrane proteins are absent in four non-disease strains of *Haemophilus somnus, Molec. Microbiol.* 6:1895-1902.

Cole, S.P., Guiney, D.G., and Corbeil, L.B., 1993, Molecular analysis of a gene encoding a serum-resistance-associated 76 kDa surface antigen of *Haemophilus somnus.* J. *Gen. Microbiol.* 139:2135-2143.

Corbeil, L.B., 1990, Molecular aspects of some virulence factors of *Haemophilus somnus, Can.* J. *Vet. Res.* 54:S57-S62.

Corbeil, L.B., Arthur, J.E., Widders, P.R., Smith, J.W., and Barbet, A.F., 1987, Antigenic specificity of convalescent serum from cattle with *Haemophilus somnus* induced experimental abortion, *Infect. Immun.* 55:1381-1386.

Corbeil, L.B., Blau, K., Prieur, D.J., and Ward, A.C.S., 1985, Serum susceptibility of *Haemophilus somnus* from bovine clinical cases and carriers, J. *Clin. Micro.* 22:192-198.

Corbeil, L.B., Chikami, G., Yarnall, M., Smith, J., and Guiney, D.G., 1988, Cloning and expression of genes encoding for *Haemophilus somnus* antigens, *Infect. Immun.* 56:2736-2742.

Corbeil, L.B., Kania, S.A., and Gogolewski, R.P., 1991, Characterization of immunodominant surface antigens of *Haemophilus somnus, Infect. Immun.* 59:4295-4301.

Corbeil, L.B., Watt, B., Corbeil, R.R., Betzen, T.G., Brownson, R.K., and Morrill, J.L., 1984, Immunoglobulin concentrations in serum and nasal secretions of calves at the onset of pneumonia, *Am.* J. *Vet. Res.* 45:773-778.

Czuprynski, C.J., and Hamilton, H.L., 1985, Bovine neutrophils ingest but do not kill *Haemophilus somnus* in vitro, *Infect. Immun.* 50:431-436.

Gogolewski, R.P., Kania, S.A., Inzana, T.J., Widders, P.R., Liggitt, H.D., and Corbeil, L.B., 1987a, Protective ability and specificity of convalescent serum from *Haemophilus somnus* pneumonia, *Infect. Immun.* 55:1403-141 1.

Gogolewski, R.P., Kania, S.A., Liggitt, H.D., and Corbeil, L.B., 1988, Protective ability of monospecific sera against 78 kDa and 40-kDa outer membrane antigens of *"Haemophilus somnus", Infect. Immun.* 56:2301-2316.

Gogolewski, R.P., Leathers, C.W., Liggitt, H.D., and Corbeil, L.B., 1987b, Experimental *Haemophilus somnus* pneumonia in calves and immunoperoxidase localization of bacteria, *Vet. Pathol.* 24:250-256.

Gogolewski, R.P., Schaefer, D.C., Wasson, S.K., Corbeil, R.R., and Corbeil, L.B., 1989, Pulmonary persistence of *Haemophilus somnus* in presence of specific antibody, J. *Clin. Microbiol.* 27:1767-1774, 1989.

Groom, S.C., and Little, P.B., 1988, Effects of vaccination of calves against induced *Haemophilus somnus* pneumonia, *Am. J. Vet. Res.* 49:793-800.

Groom, S., Little, P., and Rosendal, S., 1988, Virulence differences among three strains of H. *somnus* following intratracheal inoculation of calves, *Can. J. Vet. Res.* 52:349-354.

Harris, F.W., and Janzen, E.D., 1989, The *Haemophilus somnus* disease complex (Hemophilosis): A review., *Can. Vet. J.* 30:816-822.

Higgins, R., Godbout-deLasalle, Messier, S., Couture, Y., and Lamothe, P., 1981, Isolation of *Histophilus ovis* from vaginal discharge in ewes in Canada, *Can. Vet. J.* 22:395-396.

Holst, E., Goffeng, A.R., and Andersch, B., 1994, Bacterial vaginosis and vaginal microorganisms in idiopathic premature labor and association with pregnancy outcome. *J. Clin. Microbiol.* 32:176-186.

Humphrey, J.D., *et al.*, 1985, Prevalence and distribution of *Haemophilus somnus* in the male bovine reproductive tract, *Am. J. Vet. Res.* 43:792-795.

Humphrey, J., and Stephens, L., 1983, *Haemophilus somnus:* A review, *Vet. Bull.* 53:987-1004.

Inzana, T.J., Gogolewski, R.P., and Corbeil, L.B., 1992, Phenotypic phase variation in *Haemophilus somnus* lipooligosaccharide during bovine pneumonia and after *in vitro* passage, *Infect. Immun.* 60:2943-2951.

Inzana, T.J., Iritani, B., Gogolewski, R.P., Kania, S.A., Corbeil, L.B., 1988, Purification and characterization of lipooligosaccharides from four strains of *Haemophilus somnus*, *Infect. Immun.* 56:2830-2837.

Inzana, T.J., and Todd, J., 1992, Immune response of cattle to *Haemophilus somnus* lipid A-protein conjugate vaccine and efficacy in a mouse abortion model. *Am. J. Vet. Res.* 53:175-179.

Kwiecien, J.M., and Little, P.B., 1991, *Haemophilus somnus* and reproductive disease in the cow, *Can. Vet. J.* 32:595-601.

Kwiecien, J.M., and Little, P.B., 1992, Isolation of pathogenic strains of *Haemophilus somnus* from the female bovine reproductive tract, *Can. J. Vet. Res.* 56:127-134.

Lederer, J.A., Brown, J.F., and Czuprynski, C.J., 1987, *Haemophilus somnus*, a facultative intracellular pathogen of bovine mononuclear phagocytes, *Infect. Immun.* 55:381-387.

Lees, V.W., Meek, A.H., and Rosendal, S., 1990, Epidemiology of *Haemophilus somnus* in young rams, *Can. J. Vet. Res.* 54:331-336.

Lees, V.W., Yates, W.D.G., and Corbeil, L.B., 1994, Ovine *Haemophilus somnus:* Experimental intracisternal infection and antigenic comparison with bovine *Haemophilus somnus*, *Can. J. Vet. Res.* 58:202-210.

Martin, W., Acres, S., Janzen, E., Wilson, P., and Allen, B., 1984, A field trial of preshipment vaccination of calves, *Can. Vet. J.* 25:145-147.

McDonald, H.M., O'Loughlin, J.A., Jolley, P., Vigneswaran, R., and McDonald, P.J., 1991, Vaginal infection and preterm labour, *J. Obstet. Gynecol.* 98:427-435.

McGillivery, D.J., Webber, J.J., and Dean, H.F., 1986, Characterisation of *Histophilus ovis* and related organisms by restriction endonuclease analysis, *Aust. Vet. J.* 63:389-393.

Miller, R.B., Lein, D.H., McEntee, K.E., Hall, C.E., Shaw, S. 1983, *Haemophilus somnus* infection of the reproductive tract of cattle: A review, JAVMA 182:1390-1392.

Nanssen, P., 1972, Selective immunoglobulin deficiency in cattle and susceptibility to infection, *Acta Pathol. Microbiol. Scand. Sect. B.* 80:49-54.

Ogunnariwo, J.A., Cheng, C., Ford, J., and Schryvers, A.B., 1990, Response of *Haemophilus somnus* to iron limitation: expression and identification of a bovine-specific transferrin receptor, *Microbial Pathogen.* 9:397-406.

Rahaley, R.S., 1978, Pathology of experimental *Histophilus ovis* infection in sheep, *Vet. Pathol.* 15:631-637.

Pfeifer, C.G., Campos, M., Beskorwayne, T., Babiuk, L.A., and Potter, A.A., 1992, Effect of *Haemophilus somnus* on phagocytosis and hydrogen peroxide production by bovine polymorphonuclear leukocytes, *Microbial Pathogen.* 13:191-202.

Silva, S.V.P.S., and Little, P.B., 1990, The protective effect of vaccination against experimental pneumonia in cattle with *Haemophilus somnus* outer membrane antigens and interference by lipopolysaccharide, *Can. J. Vet. Res.* 54:326-330.

Slee, KJ., and Stephens, L.R., 1985, Selective medium for isolation of *Haemophilus somnus* from cattle and sheep, *Vet. Res.* 116:215-217.

Stephens, L.R., and Little, P.B., 1981, Ultrastructure of *Haemophilus somnus*, causative agent of bovine infectious thromboembolic meningoencephalitis, *Am. J. Vet. Res.* 43:1638-1640.

Stephens, L.R., Humphrey, J.D., Little, P.B., and Barnum, D.A., 1983, Morphological, biochemical, antigenic and cytochemical relationship among *Haemophilus somnus*, *Haemophilus agni*, *Haemophilus haemoglobinophilus*, *Histophilus ovis* and *Actinobacillus seminis*, *J. Clin. Microbiol.* 17:728-737.

Stephens, L.R., Little, P.B., Humphrey, J.D., Wilkie, B.N., and Barnum, D.A., 1982, Vaccination of cattle against experimentally induced thromboembolic meningoencephalitis with a *Haemophilus somnus* bacterin, *Am. J. Vet. Res.* 43:1339-1342.

Stephens, L.R., Little, P.B., Wilkie, B.N., Barnum, D.A., 1981, Infectious thromboembolic meningoencephalitis in cattle: a review, *J. Am. Vet. Med. Assoc.* 178:378-384.

Stephens L.R., Little, P.B., Wilkie, B.N., and Barnum, D.A., 1984, Isolation of *Haemophilus somnus* antigens and their use as vaccines for prevention of bovine thromboembolic meningoencephalitis. Am J Vet Res 45:234-239.

Stephens, L.R., Slee, K.J., Poulton, P., Larcombe, M., and Kosior, E., 1986, Investigation of purulent vaginal discharge in cows, with particular reference to *Haemophilus somnus*, *Aust. Vet. J.* 63:182-185.

Tagawa, Y., Haritani, M., Ishikawa, H., and Yuasa, N., 1993a, Antigenic analysis of the major outer membrane protein of *Haemophilus somnus* with monoclonal antibodies, *Infect. Immun.* 61:2257-2259.

Tagawa, Y., Haritani, M., Ishikawa, H., and Yuasa, N., 1993b, Characterization of a heat-modifiable outer membrane protein of *Haemophilus somnus*, *Infect. Immun.* 61:1750-1755.

Tagawa, Y., Haritani, M., and Yuasa, N., 1993c, Characterization of an immunoreactive 17.5-kilodalton outer membrane protein of *Haemophilus somnus* by using a monoclonal antibody, *Infect. Immun.* 61:4153-4157.

Tagawa, Y., Ishikawa, H., and Yuasa, N., 1993d, Purification and partial characterization of the major outer membrane protein of *Haemophilus somnus*, *Infect. Immun.* 61:91-96.

Theisen, M., Rioux, C.R., and Potter, A.A., 1992, Molecular cloning, nucleotide sequence, and characterization of a 40,000-molecular-weight lipoprotein of *Haemophilus somnus*, *Infect. Immun.* 60:826-831.

Theisen, M., Rioux, C.R., and Potter, A.A., 1993, Molecular cloning, nucleotide sequence, and characterization of *LppB*, encoding an antigenic 40-kilodalton lipoprotein of *Haemophilus somnus*, *Infect. Immun.* 61:1793-1798.

Thompson, K.G., and Little, P.B., 1981, Effect of *Haemophilus somnus* on bovine endothelial cells in organ culture, *Am. J. Vet. Res.* 42:748-754.

Walker, R.L., Biberstein, E.L., Pritchett, R.F., and Kirkham, C., 1985, Deoxyribonucleic acid relatedness among *"Haemophilus somnus," "Haemophilus agni," "Histophilus ovis," "Actinobacillus seminis,"* and *Haemophilus influenzae*, *Int. J. Sys. Bacteriol.* 35:46-49.

Walker, R.L., and LeaMaster, B.R., 1986, Prevalence of *Histophilus ovis* and *Actinobacillus seminis* in the genital tract of sheep, *Am. J. Vet. Res.* 47:1928-1930.

Walker, R.L., LeaMaster, B.R., Biberstein, E.L., and Stellflug, J.N., 1988, Serodiagnosis of *Histophilus* ovis-associated epididymitis in rams, *Am. J. Vet. Res.* 49:208-212.

Walker, R.L., LeaMaster, B.R., Stellflug, J.N., and Biberstein, E.L., 1986, Association of age of ram with distribution of epididymal lesions and etiologic agents, *J. Am. Vet. Med. Assoc.* 188:393-396.

Ward, A.C.S., and Corbeil, L.B., 1983, A selective medium for Gram-negative pathogens from bovine respiratory and reproductive tracts. *A. A. Vet. Lab. Diagn.* 26:103-112.

Ward, G.E. Nivard, J.R., and Maheswaran, S.K., 1984, Morphologic features, structure and adherence to bovine turbinate cells of three *Haemophilus somnus* variants, *Am. J. Vet. Res.* 45:336-338.

Widders, P.R., Dorrance, L.A., Yarnall, M., and Corbeil, L.B., 1989a, Immunoglobulin-binding activity among pathogenic and carrier isolates of *H. somnus*, *Infect. Immun.* 57:639-642.

Widders, P.R., Dowling, S.C., Gogolewski, R.P., Smith, J.W., and Corbeil, L.B., 1989b, Isotypic antibody responses in cattle infected with *Haemophilus somnus*, *Res. Vet. Sci.* 46:212-217.

Widders P.R., Smith, J.W., Yarnall, M., McGuire, T.C., Corbeil, L.B., 1988, Non-immune immunoglobulin binding of *Haemophilus somnus*, *J. Med. Micro.* 26:307-311.

Widders, P.R., Paisley, L.G., Gogolewski, R.P., Evermann, J.F., Smith, J.W., and Corbeil, L.B., 1986, Experimental abortion and the systemic immune response in cattle to *Haemophilus somnus*, *Infect. Immun.* 54:555-560.

Won, J., and Griffith, R.W., 1993, Cloning and sequencing of the gene encoding a 31-kilodalton antigen of *Haemophilus somnus*, *Infect. Immun.* 61:2813-2821.

Yarnall, M., Gogolewski, R.P., and Corbeil, L.B., 1988a, Characterization of two *Haemophilus somnus* Fc receptors, *J. Gen. Microbiol.* 134:1993-1999.

Yarnall, M., Widders, P.R., and Corbeil, L.B., 1988b, Isolation and characterization of Fc receptors from *Haemophilus somnus*, *Scand. J. Immunol.* 28:129-137.

Yarnall, M., and Corbeil, L.B., 1989, Antibody response to an *Haemophilus somnus* Fc receptor, *J. Clin. Microbiol.* 27:111-117.

Young, S., and Hoerlein, A.B., 1970, Experimental reproduction of thromboembolic meningoencephalitis in calves, *Can. Vet. J.* 11:46.

Yu, S., Gray-Owen, S.D., Ogunnariwo, J., and Schryvers, A.B., 1992, Interaction of ruminant transferrins with transferrin receptors in bovine isolates of *Pasteurella haemolytica* and *Haemophilus somnus*, *Infect. Immun.* 60:2992-2994.

LIPOSACCHARIDES AND CAPSULES OF THE HAP GROUP BACTERIA

B. Fenwick

Department of Pathology & Microbiology
College of Veterinary Medicine
Kansas State University
Manhattan, KS 66506

INTRODUCTION

As with other Gram-negative bacteria, the liposaccharides (LS) and capsules of the organisms of the *Haemophilus*, *Actinobacillus*, and *Pasteurella* (HAP) group serve as the principal interface between the organism and the environment. As such, they play a fundamental role in determining an organism's pathogenic potential. Understanding the interactions between the bacterium and its environment at a structure/function level as well as those factors which provide a protective rather than destructive host response is important in improving our ability to limit these organisms' potential to cause disease.

The purpose of this paper is to review the structure/function relationships involving the LS and capsules of the HAP group bacteria in relation to virulence and protective immunity. The goal is to review recent developments in these specific areas since the last reviews were published (Inzana 1990; Fenwick 1990) and, in a small way, perhaps help chart what appear to be particularly rewarding challenges for the future.

LIPOSACCHARIDES

Structure

Conserved as well as unique antigenic epitopes are present in the lipopolysaccharides (LPS) and lipooligosaccharides (LOS) of the bacteria in the HAP group. Recently, the structural composition and consequently the antigenic characteristics of these complexes have been found to be less stable than previously thought. In addition to qualitative variability and random antigenic drift, it is now clear that modifications occur in relation to stage of growth, environmental conditions, and even host induced modification. In terms of virulence and successful avoidance of host defenses, the advantage to the bacteria of being able to produce an assorted collection of related yet immunologically unique surface antigens as circumstances require is obvious.

Structural diversity in the LS may be related to host specificity as well as differences in pathogenic potential among closely related bacteria. This diversity may in part be the basis for difficulties in developing accurate serodiagnostic tests, the ability to serotype specific organisms, differences in immunity provided by various vaccines, and contradictory experimental results involving what are incorrectly believed to be the same strain of organism.

Haemophilus, Actinobacillus, and Pasteurella
Edited by W. Donachie *et al.*, Plenum Press, New York, 1995

Identification of the spontaneous occurring mutants as well as experimentally induced modifications or deletions of specific components has begun to provide unequivocal data as to the role of LS in virulence and protective immunity for the HAP group bacteria.

Haemophilus. Detailed studies of the structure of the LOS of *H. ducreyi* by nuclear magnetic residence demonstrate subtle differences between strains and rather large differences between *H. ducreyi* and *H. influenzae* (Melaugh *et al.*, 1992; Schweda *et al.*, 1994). Conserved as well as unique antigen epitopes are present in the LOS of *H. influenzae* (Campagnari *et al.*, 1990). In contrast, the LOS of *H. ducreyi* appear to be organism specific (Alfa *et al.*, 1993). At least some of the genes involved in the synthesis of *H. influenzae* LOS have been identified (Spinola *et al.*, 1990; Cope *et al.*, 1991; Abu-Kaik *et al.*, 1991).

The LOS of *H. somnus*, *H. influenzae*, *H. influenzae* (biogroup *aegyptius*) can undergo both compositional and antigenic variations (Mertsola *et al.*, 1991; Inzana *et al.*, 1992a; Rubin and St. Geme, 1993). These phase variations occur in association with growth *in vivo* and following passage *in vitro*. There is evidence that LOS variants are related to changes in virulence and may represent a means of evading host immunity. In fact, host mediated modifications of LOS appear to provide a means to mimic host antigens (Mandrell and Apicella, 1993). There is even evidence to indicate *H. influenzae* can express more than one LOS type at the same time (Patrick *et al.*, 1989). It is critical to consider that the antigenic character of the LOS expressed *in vivo* is considerably different from that expressed under laboratory conditions. The specific structural changes associated with these shifts in LOS type as well as the mechanisms involved have not as yet been described fully.

Selection for LOS variants by detection of mutations in the terminal sugar of *H. ducreyi* LOS by pyocin-mediated lysis provides a means to evaluate the contribution of various LOS types to virulence and protective immunity (Campagnari *et al.*, 1994). The ability to quantify critically the ribitol-ribose-phosphate portion of polysaccharide and protein conjugate vaccines for *H. influenzae* type b is an important step towards evaluating the antigenicity of the conjugate's components (Tsai *et al.*, 1994).

Actinobacillus. The structure of the LPS of *A. pleuropneumoniae* has been characterized in detail (Fenwick and Osburn, 1986a; Byrd and Kadis, 1989; Altman *et al.*, 1989; Altman *et al.*, 1990; Beynon *et al.*, 1991a; Beynon *et al.*, 1991b; Beynon *et al.*, 1992; Perry, 1990). Smooth, semi-rough, and rough LPS types occur and tend to be serotype specific (Fenwick and Osburn, 1986b; Byrd and Kadis, 1989). Growth stage dependent differences in exposure of LPS core antigens have been demonstrated in *A. pleuropneumoniae* (Fenwick *et al.*, 1986a). Based on structural similarities between serotypes and data from antibody adsorption studies, the capsule provides serotype specificity with cross reactions being due to the similar LPS structure and common antigens among the various serotypes. Additional studies using immune serum and monoclonal antibodies confirmed these results, but also indicated the presence of LPS serotype specific epitopes (Fenwick and Osburn 1986b, Nakai *et al.*, 1991; Nakai *et al.*, 1992). A more precise serotyping scheme for *A. pleuropneumoniae* based on both capsular and LPS antigens has been proposed by Beynon et al. (1992).

Pasteurella. Studies on the structure of the LPS of *P. haemolytica* have been completed (Richards and Leitch, 1989; Severn and Richards, 1993). A rather high degree of variability in the LPS within serotypes of *P. haemolytica* has been demonstrated (Davies *et al.*, 1991; Ali *et al.*, 1992; Utley *et al.*, 1992; McCluskey *et al.*, 1994). Based on the frequency of occurrence of certain LPS types in a given species, the suggestion has been made that certain LPS types might be related to host specificity (Ali *et al.*, 1992). In addition, structural studies have provided a basis for serologic cross reactions between the various serotypes of *P.*

haemolytica (Rimler, 1990). LPS antigens are exposed on the surface of encapsulated strains of *P. haemolytica* A1 (Wilson *et al.*, 1992).

The media used as well as the time in culture influence the endotoxin activity of culture supernatants of *P. haemolytica* (Confer and Durham, 1992). Growth under anaerobic conditions alters the low-molecular weight components of the LPS in comparison to aerobically grown cells (Davies *et al.*, 1992). Growth of *P. multocida* of avian origin under iron-limiting conditions reduces the amount of LPS in culture supernatants (Ficken *et al.*, 1992a). On the other hand, growth conditions did not alter the amount of capsule or LPS produced by bovine strains (Confer *et al.*, 1992). The SDS-PAGE profile of the LPS seems to be stable throughout the growth cycle of *P. haemolytica* (Davies *et al.*, 1991). However, sub-inhibitory concentration of penicillin influence the LPS profile of some strains of *P. multocida* (Lebrun *et al.*, 1992).

Virulence

While the inflammatory potential of bacterial cell wall peptidoglycans and protein exotoxins have been documented, their pathogenic potential is overshadowed by both the magnitude and variety of host reactions which can be induced by LS in general and lipid-A (endotoxin) specifically. The endotoxins of the HAP group bacteria acting in concert (additively or synergistically) with other virulence factors (e.g. protein exotoxins) clearly have the ability to enhance the organism's pathogenic potential.

The relative importance of endotoxin as a virulence factor is paradoxical. Few compounds have greater pathogenic potential than endotoxin in terms of inducing a systemic inflammatory response. Yet, in large part it is the ability of endotoxin to nonspecifically activate host defenses which ultimately allows the host to overcome an acute Gram-negative bacterial infection. On the other hand, the inflammatory response and antibiotic therapy may contribute to the clinical severity of an infection by killing the organism with subsequent rapid release of large amounts of endotoxin. The processes by which the host metabolizes, detoxifies, and regulates the response to endotoxin are just beginning to be understood.

On a quantitative basis, the amount of endotoxin between avirulent and pathogenic Gram-negative bacteria is similar, yet slight structural differences in the LS and differences in host reactions to endotoxin may explain rather profound differences in virulence between strains of closely related bacteria. Immunologic, enzymatic, and biochemical strategies to limit the pathogenic potential of endotoxins hold the promise of being able to define better the contribution of endotoxin to virulence. Perhaps more importantly, these efforts may provide more effective treatments for Gram-negative infections as well as reduce the occurrence of adverse reactions associated with vaccines containing endotoxin.

Haemophilus. Mutations in the LOS of *H. influenzae* have been used to evaluate the role in LOS in virulence. The results provide convincing evidence that LOS plays a significant role in serum resistance and tissue invasiveness (Cope *et al.*, 1990). The LOS of *H. influenzae* appears to work synergistically with other cell wall components in the induction of meningitis (Burroughs *et al.*, 1992). More precisely, endotoxin-free peptidoglycan of *H. influenzae* can induce a significant inflammatory reaction (Burroughs *et al.*, 1993). The peptidoglycan can also interfere with neutrophil function (Cundell *et al.*, 1993). The endotoxin of *H. influenzae* can enhance the release of histamine from basophils that have been stimulated by IgE or by non-immunologic means (Clementsen *et al.*, 1990). Depending on the strain of *H. influenzae* and antibiotics used, more endotoxin can be released following therapy, and thus contribute to the disease severity (Arditi *et al.*, 1989; Arditi *et al.*, 1993; Binger *et al.*, 1992). Oligosaccharide-conjugate vaccines for *H. influenzae* have been remarkably effective, both in reducing the occurrence of clinical disease and carriers (Barbour *et al.*,

1993). The immune response of cattle to a *Haemophilus somnus* lipid A-protein conjugates vaccine as well as its efficacy in preventing abortions in mice have been evaluated (Inzana and Todd, 1992).

The LOS of *H. ducreyi* has been implicated as the cause of chancroid as both live and killed cells placed intradermally cause ulceration in mice in the absence of any demonstrable exotoxin or proteolytic activity (Abeck and Korting, 1992). The LOS of *H. ducreyi* is more toxic in the rabbit intradermal toxicity test than an equivalent amount of LOS from *H. influenzae* or LPS from *E. coli* (Campagnari *et al.*, 1991).

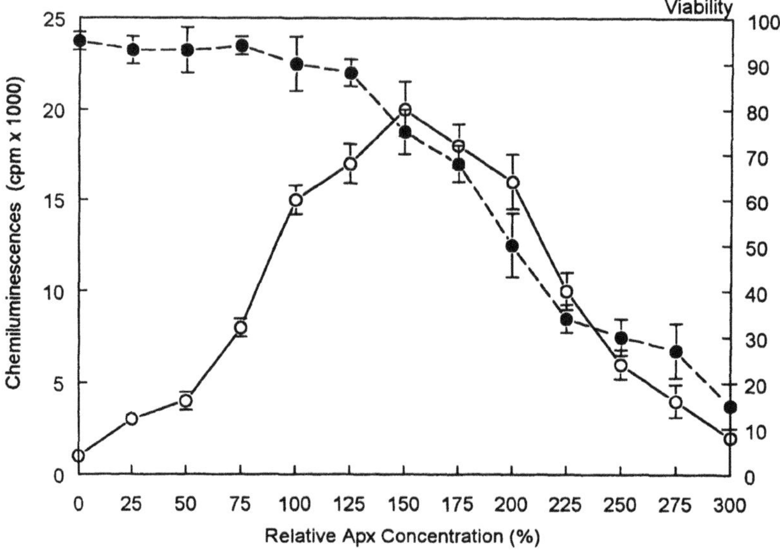

Figure 1. Effect of ApxI and II on isolated porcine neutrophils *in vitro*. Increasing concentrations of endotoxin-free preparations of ApxI from *A. pleuropneumoniae* were mixed with isolated porcine neutrophils for 90 minutes, and the resultant chemiluminescence (open dots) and cell viability (closed dots) determined.

Actinobacillus. The LPS of *A. pleuropneumoniae* has the potential to cause significant tissue damage, yet the lesions induced by *A. pleuropneumoniae* LPS are not typical of the infection (Fenwick and Osburn 1986a, Fenwick and Osburn 1986b). On the other hand, the LPS of *A. pleuropneumoniae* binds proteins in the respiratory tract mucus of pigs (Belanger *et al.*, 1994) and the ability of the organism to bind to porcine tracheal rings is related to the LPS type; smooth-type LPS strains bind better than strains with semi-rough-type LPS (Belanger *et al.*, 1990). Binding is related also to the amount of capsule produced which is believed to block the interaction between the LPS and mucus (Belanger *et al.*, 1992). Following the success with *H. influenzae*, immune responses to LPS-protein conjugate vaccines appear to be a promising new method of providing protective immunity (Fenwick and Osburn, 1986c; Byrd and Kadis, 1992; Byrd *et al.*, 1992).

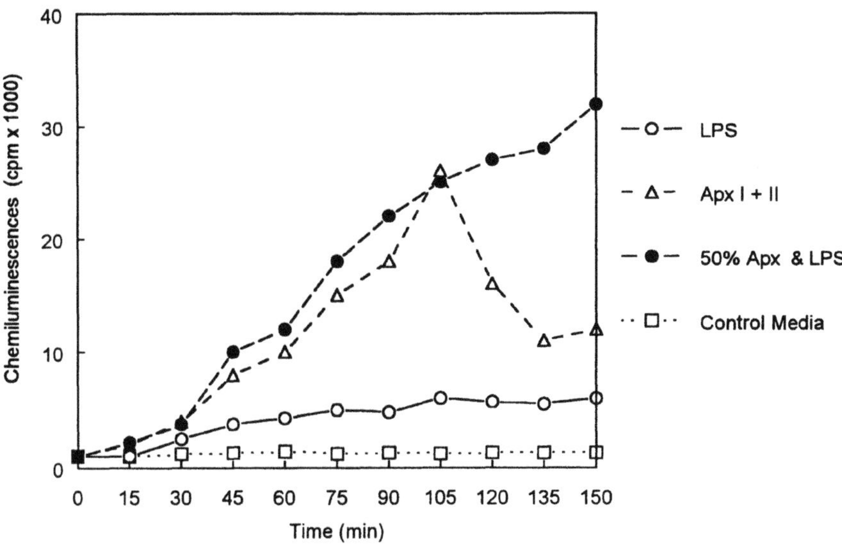

Figure 2. Effect of ApxI and II with and without LPS on activation of porcine neutrophils over time. *A. pleuropneumoniae* culture supernatants containing ApxI, ApxII, and endotoxin were either heat treated to destroy exotoxin activity (Apx I + II) or the LPS removed using a polymyxin B column. These two preparations were also mixed in equal parts (50% Apx & LPS). Culture media was used as a negative control (Control Media).

The interaction between the LPS and exotoxins of *A. pleuropneumoniae* (ApxI, ApxII, and ApxIII) is just beginning to be appreciated. To evaluate this interaction, a number of experiments were conducted. In the first experiment (Figure 1), porcine neutrophils were incubated with increasing concentrations of an endotoxin-free ApxI and ApxII. As has been documented with leukotoxin of *P. hemolytica*, at low concentrations ApxI activated the cells while at higher concentrations the cells were killed. In the second experiment (Figure 2), the interactive effect of ApxI and ApxII with LPS was examined. The data indicate that the ability of Apx to activate porcine neutrophils is both dependent on concentration and time. In addition, the presence of LPS prevents, or at least delays, the cytotoxic effect of ApxI and II. When tested *in vivo,* the same preparations demonstrate an additive effect (Figure 3).

Pasteurella. Using immunohistochemistry, *P. haemolytica* LPS can be found within neutrophils, alveolar macrophages, endothelial cells, and pulmonary intravascular macrophages of calves following experimental inoculation (Whiteley *et al.*, 1990). It seems clear that *P. haemolytica* LPS has a direct cytotoxic effect on bovine endothelial cells (Breider *et al.*, 1990; Paulsen *et al.*, 1989). The toxic effect of LPS on endothelial cells can be enhanced by macrophage-produced cytokines (Sharma *et al.*, 1992). Neutrophils provide a degree of protection against this effect (Breider *et al.*, 1991a; Breider *et al.*, 1991b; Kumar *et al.*, 1991). On the other hand, *P. haemolytica* leukotoxin rather than LPS mediates neutrophil related killing of endothelial cells (Maheswaran *et al.*, 1992; Maheswaran *et al.*, 1993). *P. haemolytica* LPS induces arachidonic acid production by bovine endothelial cells which can augment the adherence of neutrophils to endothelial cells (Paulsen *et al.*, 1990). The LPS of *P. haemolytica* is more potent than the LPS of *E. coli* in inducing procoagulant activity in pulmonary alveolar macrophages (Car *et al.*, 1991; Slocombe *et al.*, 1990).

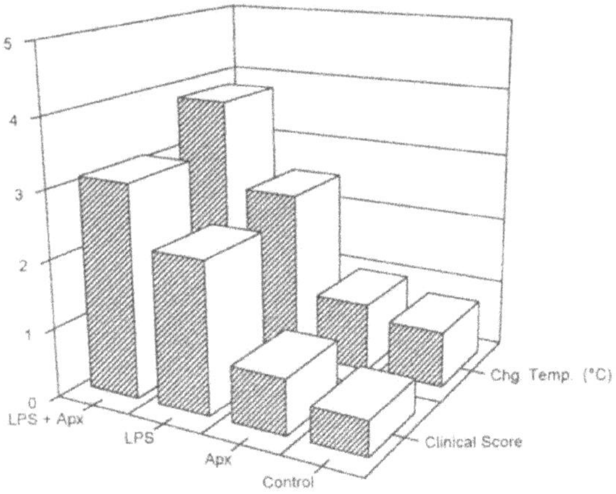

Figure 3. Effect of Apx and LPS in pigs. Pigs were injected with culture supernatant of *A. pleuropneumoniae* (LPS + Apx) or the same amount of supernatant in which the endotoxin had been removed or the Apx activity destroyed by heating. The clinical reaction to these preparations were determined 4 hours post injection by clinical score (scale of 0 to 5) and change in temperature.

The LPS of *P. haemolytica* induces the production of interleukin-1 by blood monocytes, but not alveolar macrophages (Czuprynski *et al.*, 1991). The interactions between LPS and leukotoxin are not well defined. The LPS and leukotoxin produced by *P. haemolytica* have an additive effect on changes in coagulation and peripheral leukocytes following intraduodenal injection in calves (Bowersock *et al.*, 1990). An interesting suggestion is that subcytotoxic concentrations of leukotoxin may enhance the ability of neutrophils to protect against LPS mediated killing of endothelial cells by increased LPS receptor expression or increased production of endotoxin deacylating enzyme(s) (McDermot and Fenwick, 1992).

Endotoxin appears to be a necessary component of vaccines designed to protect turkeys from *P. multocida* (Ficken *et al.*, 1992b). Anti-idiotypic antibodies which mimic *P. multocida* LPS are capable of providing protection from disease following challenge with strains of homologous LPS type (Sutherland *et al.*, 1993). Antibodies to either the capsule or LPS of *P. haemolytica* A1 fail to provide antibody-dependent complement-mediated killing, but can facilitate phagocytosis (Wilson *et al.*, 1992). Complement causes the release of LPS from avian strains of *P. multocida* with virulent strains able to retain their viability while vaccines strains are killed (Lee *et al.*, 1992).

CAPSULE

Structure

The mechanisms of bacterial encapsulation are rather highly conserved as demonstrated by a high degree of homology of the genes involved in the transport of capsule to the cell surface of Gram-negative bacteria (Frosch *et al.*, 1991; Kroll *et al.*, 1990). Similarities in capsular composition have been used to suggest common evolutionary origins (Frosch *et al.*, 1991). The ability to modify genetically the structure and antigenic character of specific polysaccharide antigens holds the promise of being able to determine more precisely their role in bacterial virulence as well as in protective immunity (Sato *et al.*, 1992).

Haemophilus. Growth conditions influence the amount of capsule (possibly LOS also) produced by *H. influenzae* (Kuratana and Anderson, 1991). The genes of *H. influenzae* that code for capsule production are duplicated. This duplication seems to be the basis for the spontaneous occurrence of capsule-deficient mutants *via rec*˙ dependent loss of one copy (Brophy *et al.*, 1991). On the other hand, the duplicate set of genes also allow for overproduction of capsule which may provide a means to increase capsule under appropriate conditions (Corn *et al.*, 1993).

Actinobacillus. Serotype-dependent differences, which range from a continuous capsular layer to patchy coverage with the underlying outer membrane exposed, have been found in the amount of capsule produced by *A. pleuropneumoniae* (Steffens *et al.*, 1990). The amount of capsule appeared to be inversely related to the number of sialylate glycoconjugates. Using freeze-substitution, the capsule of *A. pleuropneumoniae* appeared to have a more random arrangement than other Gram-negative bacteria (Graham *et al.*, 1991). Capsular antigens are of value in the serodiagnosis of *A. pleuropneumoniae* infection (Inzana *et al.*, 1990; Inzana *et al.*, 1992b). Using the *H. influenzae* capsule genes, the genes involved with capsule production by *A. pleuropneumoniae* appear to have been identified (Ward and Inzana, 1993).

Pasteurella. Shared polysaccharide antigen(s) have been recognized in *P. haemolytica* and *P. multocida* (Austin *et al.*, 1992; Austin and Corstvet, 1993). Structure rather than immunologic characteristics have been used to distinguish various serogroups of *P. multocida* (Rimler, 1994). Growth conditions influence capsule production of *P. multocida.* Under iron limited conditions *P. multocida* produces significantly less capsule, binds more polymyxin, and shows increased affinity for respiratory tract mucus than cells grown under iron-sufficient conditions (Jacques *et al.*, 1994). Sub-inhibitory concentrations of penicillin influence the production of capsule by *P. multocida* (Jacques *et al.*, 1991).

Virulence

The role of bacterial capsules in avoiding host defensive systems (interference with opsonization and phagocytosis, complement resistance, etc.) is well documented. It is also clear that specific capsular types confer particular virulence attributes. Capsules by their nature are poor immunogens, but once recognized immunologically by the host can provide an effective target for antibodies, and thus some degree of protection. Promising new methods to enhance the immunogenicity of capsular antigens are being investigated. The relative importance in terms of protection provided by mucosal versus systemic immunity and by various antibody isotypes directed against the capsule is now only beginning to be appreciated.

Haemophilus. Capsulation of *H. influenzae* decreases polymorphonuclear leukocyte activation as measured by organism-induced chemiluminescence (Cook *et al.*, 1989). Capsular type (a,b,c,e, or f) appears not to influence this effect. The data suggest that the relative increased virulence of type b capsulation is not associated with an increased ability to block opsonisation or resist intracellular killing (Zwahlen *et al.*, 1989, Cook *et al.*, 1989, Swift *et al.*, 1991). The surface characteristics of *H. influenzae* type b influence the organism's ability to activate neutrophils, yet the serotype b capsular type does not appear to play a role in avoiding opsonisation (Cook *et al.*, 1989). Isogenic capsule-deficient mutants of *H. influenzae* show increased adherence and invasion compared to their encapsulated parents (St. Geme and Falkow, 1991; Virji *et al.*, 1991; Read *et al.*, 1992). In other words, capsule appears to block rather than enhance bacterial adherence and reduces rather than increases the ability to invade tissues.

Actinobacillus. Porcine mucus contains proteins which bind *A. pleuropneumoniae* LPS, yet capsule appears to block organism adhesion (Belanger *et al.*, 1994). Immune response to *A. pleuropneumoniae* capsule is dependent on the route of immunization (oral versus aerogenous) and significantly less when killed cells as opposed to viable cells were used (Hensel *et al.*, 1994). Multiple *in vitro* passages of *A. pleuropneumoniae* cause a significant reduction of capsule which is associated with a significant reduction of virulence in relation to the original strain that was passed only once (Rosendal and MacInnes, 1990). In this case, changes in the outer membrane proteins, LPS, or exotoxins were not detected. Ethyl methanesulfate mutagenesis of *A. pleuropneumoniae* was used to induce a capsular deficient mutant which was apparently unchanged in terms of LPS, proteins, and production of exotoxins (Inzana *et al.*, 1993). The capsular deficient *A. pleuropneumoniae* mutant did not cause disease at dosages well above the lethal dose of the parent strain.

Pasteurella. Encapsulated avian strains of *P. mutocida* possess an A-type capsule which provides serum resistance. The capsule appears to block the effectiveness of complement rather than inhibit the generation of membrane attack complexes (Hansen and Hirsh, 1989). Variability in the amount of capsule between strains has been noted (Harmon *et al.*, 1991). Using an experimental *P. multocida* septicemia model in turkeys, encapsulation influences the ability to kill the organism but does not change organism clearance from the blood by the liver or spleen (Tsuji 1989). Using a mouse model, the amount of capsule produced by avian strains of *P. multocida* was not always related to virulence (Dubreuil *et al.*, 1992). Based on serologic evidence, immunity to capsular antigen was best correlated to protection from hemorrhagic septicemia in cattle (Afzal *et al.*, 1992).

In porcine strains of *P. multocida*, encapsulation is associated with increased virulence in terms of being able to produce turbinate lesions, yet capsule appears to interfere with binding to respiratory tract mucus *in vitro* (Jacques *et al.*, 1993). Adherence appears to be related to LPS as noncapsulated strains readily bind to mucus which can be blocked by addition of LPS.

While the capsule of *P. haemolytica* does not by itself mediate tissue damage, it can form complexes with pulmonary surfactant thus facilitating attachment to alveolar walls (Brogden *et al.*, 1989). In contrast to previous studies, it is now clear that the ability of *P. haemolytica* capsule to stimulate macrophages and monocytes is due to small amounts of contaminating endotoxin (Czuprynski *et al.*, 1991).

CONCLUSION

Efforts to characterize the biochemical composition and related functions of the LS and capsules of the HAP group bacteria have provided insight into broad differences between related organisms, but have failed to provide detailed data on fine differences in the secondary and tertiary structure. The application of more advanced biochemical characterizations including nuclear magnetic resonance, laboratory synthesis, selective mutation, and antigenic characterization using monoclonal antibodies have produced some exciting and on occasion unexpected findings.

During the past few years, perhaps the most important finding is that the LS and capsule of the HAP group bacteria are far from static. The high degree of variability between strains of the same species suggest they are contributing factors to host specificity and strain dependent difference in virulence. In addition, environmentally induced changes in structure and antigenic character are more common than previously recognized. Studies designed to reveal the fundamental mechanisms associated with these modifications as well as their role in virulence and protective immunity hold considerable promise.

REFERENCES

Abeck, D., and Korting, H.C., 1992, Mechanisms of skin adherence, penetration and tissue necrosis production by *Haemophilus ducreyi*, the causative agent of chancroid, *Acta. Derm. Venereol. Suppl. Stockh.* 174:1.

Abu-Kaik, Y., McLaughlin, R.E., Apicella, M.A., and Sinola, S.M., 1991, Analysis of *Haemophilus influenzae* type b lipooligosaccharide-synthesis gene that assemble or expose a 2-keto-3-deoxyoctulosonic acid epitope, *Mol. Microbiol.* 5:2475.

Afzal, M., Muneer, R., and Akhtar, S., 1992, Serological evaluation of *Pasteurella multocida* antigens associated with protection in buffalo calves, *Rev. Sci. Tech.* 11:917.

Alfa, M.J., Olson, N., Degagne, P., Plummer, F., Namaara, W., Maclean, I., and Ronald, A.R., 1993, Humoral immune response of humans to lipooligosaccharide and outer membrane proteins of *Haemophilus ducreyi*, *J. Infect. Dis.* 167:1206.

Ali, Q., Davies, R.L., Parton, R., Coote, J.G., and Gibbs, H.A., 1992, Lipopolysaccharide heterogeneity in *Pasteurella haemolytica* isolates from cattle and sheep, *J. Gen. Microbiol.* 138:2185.

Altman, E., Perry, M.B., and Brisson, J.R., 1989, Structure of the lipopolysaccharide antigenic O-chain produced by *Actinobacillus pleuropneumoniae* serotype 4, *Carbohydr. Res.* 191:295.

Altman, E., Brisson, J.R., and Perry, M.B., 1990, Structural characterization of the capsular polysaccharide and O-chain of the lipopolysaccharide of *Actinobacillus pleuropneumoniae* serotype B (strain 405), *Can. J. Chem.* 68:329.

Arditi, M., Ables, L., and Yogev, R., 1989, Cerebrospinal fluid endotoxin levels in children with *H. influenzae* meningitis before and after administration of intravenous ceftriaxone, *J. Infect. Dis.* 160:1005.

Arditi, M., Kabat, W., and Yogev, R., 1993, Antibiotic-induced bacterial killing stimulates tumor necrosis factor-alpha release in whole blood, *J. Infect. Dis.* 167:240.

Austin, F.W., Corstvet, R.E., Schnorr, K.L., and Todd, W.J., 1992, A characterization of monoclonal antibodies prepared against *Pasteurella haemolytica* serotype 1 surface antigens, *Vet. Microbiol.* 32:327.

Austin, F.W., and Corstvet, R.E., 1993, Purification of a *Pasteurella haemolytica* serotype 1-specific polysaccharide epitope by use of monoclonal antibody immunoaffinity, *Am. J. Vet. Res.* 54:695.

Barbour, M.L., Booy, R., Crook, D.W., Griffiths, H., Chapel, H.M., Moxon, E.R., and Mayon-White, D., 1993, *Haemophilus influenzae* type b carriage and immunity four years after receiving the *Haemophilus influenzae* oligosaccharide-CRM197 (HbOC) conjugate vaccine, *Pediat. Infect. Dis. J.* 12:478.

Belanger, M., Dubreuil, D., Harel, J., Girard, C., and Jacques, M., 1990, Role of lipopolysaccharides in adherence of *Actinobacillus pleuropneumoniae* to porcine tracheal rings, *Infect. Immun.* 58:3523.

Belanger, M., Rioux, S., Foiry, B., and Jacques, M., 1992, Affinity for porcine respiratory tract mucus is found in some isolates of *Actinobacillus pleuropneumoniae*, *FEMS Microbiol. Lett.* 97:119.

Belanger, M., Dubreuil, D., and Jacques, M., 1994, Proteins found within porcine respiratory tract secretions bind lipopolysaccharides of *Actinobacillus pleuropneumoniae*, *Infect. Immun.* 62:868.

Beynon, L.M., Moreau, M., Richards, J.C., and Perry, M.B., 1991a, Structure of the O-antigen of *Actinobacillus pleuropneumoniae*, *Carbohydr. Res.* 209:225.

Beynon, L.M., Perry, M.B., Richards, J.C., 1991b, Structure of the O-antigen of *Actinobacillus pleuropneumoniae* serotype 12 lipopolysaccharide, *Can. J. Chem.* 69:218.

Beynon, L.M., Griffith, D.W., Richards, J.C., and Perry, M.B., 1992, Characterization of the lipopolysaccharide O antigens of *Actinobacillus pleuropneumoniae* serotypes 9 and 11: antigenic relationships among serotypes 9, 11, and 1, *J. Bacteriol.* 174:5324.

Binger, E., Goury, V., Bennani, H., Lambert-Zechovsky, N., Aujard, Y., and Darbord, J.C., 1992, Bactericidal activity of beta-lactams and amikacin against *Haemophilus influenzae*: effect on endotoxin releas, *J. Antimicrob. Chemother.* 30:165.

Bowersock, T.L., Walker, R.D., Maddux, J.M., Fenner, D., and Moore, R.N., 1990, Hematological changes in calves exposed to a mixture of lipopolysaccharide and crude leukotoxin of *Pasteurella haemolytica*, *Can. J. Vet. Res.* 54:425.

Breider, M.A., Kumar, S., and Corstvet, R.E., 1990, Bovine pulmonary endothelial cell damage mediated by *Pasteurella haemolytica* pathogenic factors, *Infect. Immun.* 58:1671.

Breider, M.A., Kumar, S., and Corstvet, R.E., 1991a, Interaction of bovine neutrophils in *Pasteurella haemolytica* mediated damage to pulmonary endothelial cells, *Vet. Immun. Immunopath.* 27:337.

Breider, M.A., Kumar, S., and Corstvet, R.E., 1991b, Protective role of bovine neutrophils in *Pasteurella haemolytica*-mediated endothelial cell damage, Infect. Immun. 59:4570.

Brogden, K.A., Adlam, K.A., Lehmkuhl, H.D., Cutlip, R.C., Knights, J.M., and Engen, R.L., 1989, Effect of *Pasteurella haemolytica* (A1) capsular polysaccharide on sheep lung *in vivo* and on pulmonary surfactant *in vitro*, *Am. J. Vet. Res.* 50:555.

Brophy, L.N., Kroll, J.S., Ferguson, D.F., and Moxon, E.R., 1991, Capsulation gene loss and 'rescue' mutations during the Cap+ to Cap- transition in *Haemophilus influenzae* type b, *J. Gen. Microbiol.* 137:2571.

Burroughs, M., Cabellos, C., Prasad, S., and Tuomanen, E., 1992, Bacterial components and the pathophysiology of injury to the blood-brain barrier: does cell wall add to the effects of endotoxin in gram-negative meningitis?, *J. Infect. Dis.* 165:S82.

Burroughs, M., Prasad, S., Cabellos, C., Mendelman, P.M., and Tuomanen, E., 1993, The biologic activities of peptidoglycan in experimental *Haemophilus influenzae* meningitis, *J. Infect. Dis.* 167:464.

Byrd, W., and Kadis, S., 1989, Structures and sugar compositions of lipopolysaccharides isolated from seven *Actinobacillus pleuropneumoniae* serotypes, *Infect. Immun.* 57:3901.

Byrd, W., and Kadis, S., 1992, Preparation, characterization, and immunogenicity of conjugate vaccines directed against *Actinobacillus pleuropneumoniae* virulence determinants, *Infect. Immun.* 60:3042.

Byrd, W., Harmon B.G., and Kadis, S., 1992, Protective efficacy of conjugate vaccines against experimental challenge with porcine *Actinobacillus pleuropneumoniae*. *Vet. Immunol. Immunopathol.* 34:307.

Campagnari, A.A., Spinola, S.M., Lesse, A.J., Kwaik, Y.A., Mandell, R.E., and Apicella, M.A., 1990, Lipooligosaccharide epitopes shared among gram-negative non-enteric mucosa pathogens, Microb. Pathog. 8:353.

Campagnari, A.A., Wild, L.M., Griffiths, G.E., Karalus, R.J., Wirth, M.A., and Spinola, S.M., 1991, Role of lipooligosaccharides in experimental dermal lesions caused by *Haemophilus ducreyi*, *Infect. Immun.* 59:2601.

Campagnari, A.A., Karalus, R., Apicella, M., Melaugh, W., Lesse, A.J., and Gibson, B.W., 1994, Use of pyrocin to select a *Haemophilus ducreyi* variant defective in lipooligosaccharide biosynthesis, *Infect. Immun.* 62:2379.

Car, B.D., Slauson, D.O., Suyemoto, M.M., Dore, M., and Neilson, N.R., 1991, Expression and kinetics of induced procoagulant activity in bovine pulmonary alveolar macrophages, *Exp. Lung. Res.* 17:939.

Clementsen, P., Milman, N., Kilian, M., Fomsgaard, A., Back, L., and Norn, S., 1990, Endotoxin from *Haemophilus influenzae* enhances IgE-mediated and non-immunological histamine release, *Allergy* 45:10.

Confer, A.W., Durham, J.A., and Clarke, C.R., 1992, Comparison of antigens of *Pasteurella hameolytica* serotype 1 grown *in vitro* and *in vivo*, *Am. J. Vet. Res. 53:472.*

Confer, A.W., and Durham, J.A., 1992, Sequential development of antigens and toxins of *Pasteurella haemolytica* serotype 1 grown in cell culture medium, *Am. J. Vet. Res.* 53:646.

Cook, G., Wilson, R., Kroll, S., Todd, H., Garbett, N., Moxon, R., and Cole, P., 1989, Opsonic requirements and interaction of *Haemophilus influenzae* with human polymorphonuclear neutrophil leukocytes studied by luminol-enhanced chemiluminescence, *Micro. Pathog.* 7:101.

Cope, L.D, Yogev, R., Mertsola, J., Argyle, J.C., McCracken Jr., G.H., and Hansen, E.J., 1990, Effect of mutations in lipooligosaccharide biosynthesis genes on virulence of *Haemophilus influenzae* type b, *Infect. Immun.* 58:2343.

Cope, L.D., Yogev, R., Maertsola, J., Latimer, J.L., Hanson, M.S., McCracken Jr., G.H., and Hansen, E.J., 1991, Molecular cloning of a gene involved in lipooligosaccharide biosynthesis and virulence expression by *Haemophilus influenzae* type B, Mol. Micorbiol. 5:1113.

Corn, P.G., Anders, J., Takala, A.K., Kayhty, H., and Hoiseth, S.K., 1993, Genes involved in *Haemophilus influenzae* type b capsule expression are frequently amplified, *J. Infect. Dis.* 167:356.

Cundell, D.R., Taylor, G.W., Kanthakumar, K., Wilks, M., Tabaqchall, S., Dorey, E., Devalia, J.L., Roberts, D.E., Davies, R.J., R. Wilson, and Cole, P.J., 1993, Inhibition of human neutrophil migration in vitro by low-molecular-mass products 2of nontypeable *Haemophilus influenzae*, *Infect. Immun.* 61:2419.

Czuprynski, C.J., Noel, E.J., and Adlam, C., 1991, *Pasteurella haemolytica* A1 purified capsular polysaccharide do not stimulant interleukin-1 and tumor necrosis factor release by bovine monocyte and alveolar macrophages, *Vet. Immun. Immunopath.* 28:157.

Davies, R.L., Ali, Q., Parton, R., Coote, J.G., Gibbs, A., and Freer, J.H., 1991, Optimal conditions for the analysis of *Pasteurella haemolytica* lipopolysaccharide by sodium dodecyl sulphate-polyacrylamide gel electrophoresis, *FEMS Microbiol. Lett.* 69:23.

Davies, R.L., Parton, R., Coote, J.G., Gibbs, A., and Freer, J.H., 1992, Outer-membrane protein and lipopolysaccharide variation in *Pasteurella haemolytica* serotype A1 under different growth conditions, *J. Gen. Microbiol.* 138:909.

Dubreuil, J.D., Gilbert, L., and Jacques, M., 1992, Cell surface characteristics and virulence in mice of *Pasteurella multocida*, *Zentralblatt fur Bakteriol.* 276:366.

Fenwick, B.W., and Osburn, B.I., 1986a, Isolation and biologic characterization of two lipopolysaccharides and a capsular enriched polysaccharide fraction isolated from *Haemophilus pleuropneumoniae*, *Am. J. Vet. Med. Res. 47:1433.*

Fenwick, B.W., and Osburn, B.I., 1986b, Immune responses to the lipopolysaccharides and capsular polysaccharides of *Haemophilus pleuropneumoniae* in convalescent and immunized pigs, *Infect. Immun.* 54:575.

Fenwick, B.W., and Osburn, B.I., 1986c, Vaccine potential of *Haemophilus pleuropneumoniae* oligosaccharide tetanus-toxoid conjugates, *Infect. Immun.* 54:583.

Fenwick, B.W., Cullor, J.S., Osburn, B.I., and Olander, H., 1986a, Mechanisms involved in the protection provided by immunization against core lipopolysaccharides of *E. Coli* (J5) from lethal *Haemophilus pleuropneumoniae* infections in swine, *Infect. Immun.* 53:298-304.

Fenwick, B.W., Osburn, B.I., and Olander, H.J., 1986b, Resistance of C3H/HeJ Mice to the effects of *Haemophilus pleuropneumoniae*. *Infect. Immun.* 53:474.

Fenwick, B.W., 1990, Virulence attributes of the lipopolysaccharides of the HAP group Organisms, *Can. J. Vet. Res.* 54:S28.

Ficken, M.D., Barnes, H.J., Avakian, A.P., and Carver, D.K., 1992a, Vaccination of turkeys with cell-free culture filtrates of *Pasteurella multocida*: effects of dilution, iron chelation, and heterologous challenge, *Avian Dis.* 36:975.

Ficken, M.D., Barnes, H.J., Avakian, A.P., and Carver, D.K., 1992b, Vaccination of turkeys with cell-free culture filtrate of *Pasteurella multocida*: effect of ultrafiltration and endotoxin removal, *Avian Dis.* 36:423.

Frosch, M., Edwards, U., Bousset, K., Krausse, B., and Weisgerber, C., 1991, Evidence for a common molecular origin of the capsule gene loci in gram-negative bacteria expressing group II capsular polysaccharides, *Mol. Microbiol.* 5:1251.

Graham, L.L., Harris, R., Villiger, W., and Beverridge, T.J., 1991, Freeze-substitution of gram-negative eubacteria: general cell morphology and envelope profiles, *J. Bacteriol.* 173:1623.

Hansen, L.M., and Hirsh, D.C., 1989, Serum resistance is correlated with encapsulation of avian strains of *Pasteurella multocida*, *Vet. Microbiol.* 21:177.

Harmon, B.G., Glisson, J.R., Latimer, K.S., Steffens, W.L., and Nunnally, J.C., 1991, Resistance of *Pasteurella multocida* A:3,4 to phagocytosis by turkey macrophages and heterophils, *Am. J. Vet. Res.* 52:1507.

Hensel, A., Pabst, R., Bunka, S., and Petzoldt, K., 1994, Oral and aerosol immunization with viable or inactivated *Actinobacillus pleuropneumoniae* bacteria: Antibody response to capsular polysaccharide in bronchoalveolar lavage fluids (BALF) and sera of pigs, *Clin. Exp. Immunol.* 96:91.

Inzana, T.J., 1990, Capsules and virulence of the HAP group bacteria, *Can. J. Vet. Res.* 54:S22.

Inzana, T.J., Clark, G.F., and Todd, J., 1990, Detection of serotype-specific antibodies or capsular antigen of *Actinobacillus pleuropneumoniae* by a double-label radioimmunoassay, *J. Clin. Microbiol.* 28:312.

Inzana, T.J., Gogolewski, R.P., and Corbeil, L.B., 1992a, Phenotypic phase variation in *Haemophilus somnus* lipooligosaccharide during bovine pneumonia and after *in vitro* passage, *Infect. Immun.* 60:2943.

Inzana, T.J., Todd, J., Kock, C., and Nicolet, J., 1992b, Serotype specificity of immunological assays for the capsular polymer *Actinobacillus pleuropneumoniae* serotype 1 and 9, *Vet. Microbiol.* 31:351.

Inzana, T.J., and Todd, J., 1992, Immune response of cattle to *Haemophilus somnus* lipid A-protein conjugates vaccine and efficacy in a mouse abortion model, Am. J. Vet. Res. 53:175.

Inzana, T.J., Todd, J., and Veit, H.P., 1993, Safety, stability, and efficacy of noncapsulated mutants of *Actinobacillus pleuropneumoniae* for use in live vaccines, *Infect. Immun.* 61:1682.

Jacques, M., Lebrun, A., Foiry, B., Dargis, M., and Malouin, F., 1991, Effects of antibiotics on the growth and morphology of *Pasteurella multocida*, *J. Gen. Microbiol.* 137:2663.

Jacques, M., Kobisch, M., Belanger, M., and Dugal, F., 1993, Virulence of capsulated and noncapsulated isolates of *Pasteurella multocida* and their adherence to porcine respiratory tract cells and mucus, *Infect. Immun.* 61:4785.

Jacques, M., Belanger, M., Diarra, M.S., Dargis, M., and Malouin, F., 1994, Modulation of *Pasteurella multocida* capsular polysaccharide during growth under iron-restricted conditions and *in vivo*, *J. Gen. Microbiol.* 140:263.

Kroll, J.S., Loynds, B., Brophy, L.N., and Moxon, E.R., 1990, The bex locus in encapsulated *Haemophilus influenzae*: a chromosomal region involved in capsule polysaccharide export, *Mol. Microbiol.* 4:1853 (1990).

Kumar, S., Breider, M.A., Corstvet, R.E., and Maddux, J.L., 1991, Effect of *Pasteurella haemolytica* saline capsular extract of bovine pulmonary endothelial cells, *Am. J. Vet. Res.* 52:1774.

Kuratana, K., and Anderson, P., 1991, Host metabolites that phenotypically increase the resistance of *Haemophilus influenzae* type b to clearance mechanisms, *J. Infect. Dis.* 163:1073.

Lebrun, A., Caya, M., and Jacques, M., 1992, Effects of sub-MICs of antibiotics on cell surface characteristics and virulence of *Pasteurella multocida*, *Antimicrob. Agents Chemother.* 36:2093.

Lee, M.D., Glisson, J.R., and Wooley, R.E., 1992, Factors affecting endotoxin release from the cell surface of avian strains of *Pasteurella multocida*, *Vet. Microbiol.* 31:369.

Maheswaran, S.K., Weiss, D.J., Kannan, M.S., Townsend, E.L., Reddy, K.R., Whiteley, L.O., and Srikumaran, S., 1992, Effects of *Pasteurella haemolytica* A1 leukotoxin on bovine neutrophils: degranulation and generation of oxygen-derived free radicals, *Vet. Immunol. Immunopath.* 33:51.

Maheswaran, S.K., Kannan, M.S., Weiss, D.J., Reddy, K.R., Townsend, E.L., Yoo, H.S., Lee, B.W., and Whiteley, L.O., 1993, Enhancement of neutrophil-mediated injury to bovine pulmonary endothelial cells by *Pasteurella haemolytica* leukotoxin, *Infect. Immun.* 61:2618.

Mandrell, R.E., and Apicella, M.A., 1993, Lipooligosaccharides (LOS) of mucosal pathogens: molecular mimicry and host modification of LOS, *Immunobiology* 187:382.

McCluskey, J., Gibbs, H.A., and Davies, R.L., 1994, Variation in outer-membrane protein and lipopolysaccharide profiles of *Pasteurella haemolytica* isolates of serotype A1 and A2 obtained from pneumonic and healthy cattle, *J. Gen. Microbiol.* 140:807.

McDermott, C.M., and Fenwick, B.W., 1992, Mechanisms Responsible for Increases in Bovine Neutrophil Acyloxyacyl Hydrolase (AOAH) Activity During Inflammation, *Am. J. Vet. Res.* 53:803.

Melaugh, W., Phillips, N.J., Campagnari, A.A., Karalus, R., Gibson, B.W., 1992, Partial characterization of the major lipooligosaccharide from a strain of *Haemophilus ducreyi*, the causative agent of chancroid, a genital ulcer disease, J. Biol. Chem. 267:13434.

Mertsola, J., Cope, L.D., Saez Llorens, X., Ramilo, O., Kennedy, W., McCracken Jr., G.H., and Hansen, E.J., 1991, *In vivo* and *in vitro* expression of *Haemophilus influenzae* type b lipooligosaccharide epitopes, *J. Infect. Dis.* 164:555.

Nakai, T., Kawahara, K., Danbara, H., and Kume, K., 1991, The epitope structure in lipopolysaccharide recognized by the serotype 2-specific monoclonal antibody of *Actinobacillus pleuropneumoniae*, *FEMS Microbiol. Lett.* 82:143.

Nakai, T., Kawahara, K., Danbara, H., and Kume, K., 1992, Identification of the cross-reacting antigen among *Actinobacillus pleuropneumoniae* strains of serotype 1, 9, and 11 by use of monoclonal antibodies, *J. Vet. Med. Sci.* 54:707.

Patrick, C.C., Pelzel, S.E., Miller, E.E., Haanes-Fritz, E., Radolf, J.D., Gulig, P.A., McCracken Jr, G.H., and Hansen, E.J., 1989, Antigenic evidence for simultaneous expression of two different lipooligosaccharides by some strains of *Haemophilus influenzae* type b, *Infect. Immun.* 57:1971.

Paulsen, D.B., Mosier, D.A., Clinkenbeard, K.D., and Confer, A.W., 1989, Direct effects of *Pasteurella haemolytica* lipopolysaccharide on bovine pulmonary endothelial cells *in vitro*, *Am. J. Vet. Res.* 50:1633.

Paulsen, D.B., Confer, A.W., Clinkenbeard, K.D., and Mosier, D.A., 1990, *Pasteurella haemolytica* lipopolysaccharide-induced arachidonic acid release from and neutrophil adherence to bovine pulmonary artery endothelial cells, *Am. J. Vet. Res.* 51:1635.

Perry, M.B., 1990, Structural analysis of the lipopolysaccharide of *Actinobacillus (Haemophilus) pleuropneumoniae* serotype 10, *Biochem. Cell. Biol.* 68:808.

Read, R.C., Rutman, A.A., Jeffery, P.K., Lund V.J., Brain A.P., Moxon, E.R., Cole, P.J., and Wilson,R., 1992, Interaction of capsulate *Haemophilus influenzae* with human airway mucosa in vitro, *Infect. Immun.* 60:3244 (1992).

Richards, J.C., and Leitch, R.A., 1989, Elucidation of the structure of the *Pasteurella haemolytica* serotype T10 lipopolysaccharide O-antigen by n.m.r. spectroscopy, *Carbohydr. Res.* 186:275.

Rimler, R.B., 1990, Comparisons of *Pasteurella multocida* lipopolysaccharides by sodium dodecyl sulfate-polyacrylamide gel electrophoresis to determine relationship between group B and E hemorrhage septicemia strains and serologically related group A strains, *J. Clin. Microbiol.* 28:654.

Rimler, R.B., 1994, Presumptive identification of *Pasteurella multocida* serogroups A, D, and F by capsule depolymerisation with mucopolysaccharidases. *Vet. Rec.* 134:191.

Rosendal, S., and MacInnes, J.I., 1990, Characterization of an attenuated strain of *Actinobacillus pleuropneumoniae*, serotype 1, *Am. J. Vet. Res.* 51:711.

Rubin, L.G., and St. Geme, J.W., 1993, Role of lipooligosaccharide in virulence of the Brazilian purpuric fever clone of *Haemophilus influenzae* biogroup aegyptius for infant rats, *Infect. Immun.* 61:650.

Sato, S., Takamatsu, N., Okahashi, N., Matsunoshita, N., Inoue, M., Takehara, T., and Koga., T, 1992, Construction of mutants of *Actinobacillus actinomycetemcomitans* defective in serotype b-specific polysaccharide antigen by insertion of transponson Tn916, *J. Gen. Microbiol.* 138:1203.

Schweda, E.K.H., Sundstrom, A.C., Eriksson, L.M., Jonasson, J.A., and Lindberg, A.A., 1994, Structural studies of the cell envelope lipopolysaccharides from *Haemophilus ducreyi* strains ITM 2665 and ITM 4747, *J. Biol. Chem.* 269:12040.

Severn, W.B., and Richards, J.C., 1993, Characterization of the O-polysaccharide of *Pasteurella haemolytica* serotype A1, *Carbohydr. Res.* 240:277.

Sharma, S.A., Olchowy, T.W., Yang, Z., and Breider, M.A., 1992, Tumor necrosis factor alpha and interleukin 1 alpha enhance lipopolysaccharide-mediated bovine endothelium cell injury, *J. Leuk. Biol.* 51:579.

Slocombe, R.F., Mulks, M., Killingsworth, C.R., Derksen, F.J., and Robinson, N.E., 1990, Effect of *Pasteurella haemolytica*-derived endotoxin on pulmonary structure and function in calves, *Am. J. Vet. Res.* 51:433.

Spinola, S.M., Kwaik, Y.A., Lesse, A.J., Campagnari, A.A., and Apicella, M.A., 1990, Cloning and expression in *Escherichia coli* of a *Haemophilus influenzae* type b lipooligosaccharide synthesis gene(s) that encodes 2-keta-3-deoxyoctulosonic acid epitope, *Infect. Immun.* 58:1558.

St. Geme, J.W., and Falkow, S., 1991, Loss of capsule expression by *Haemophilus influenzae* type b results in enhanced adherence to and invasion of human cells, *Infect. Imun.* 59:1325.

Steffens, W.L., Byrd, W., and Kadis, S., 1990, Identification and localization of surface sialyated glycoconjugates in *Actinobacillus pleuropneumoniae* by direct enzyme-colloidal gold cytochemistry, *Vet. Microbiol.* 25:217.

Sutherland, A.D., Davies, R.C., and Murray, J., 1993, An experimental anti-idiotype vaccine mimicking lipopolysaccharide give protection against *Pasteurella multocida*, *FEMS Immunol. Med. Microbiol.* 7:105.

Swift, A.J., Moxon E.R., Zwahlen, A., and Winkelstein, J.A., 1991, Complement-mediated serum activities against genetically defined capsular transformants of *Haemophilus influenzae*, *Microb. Pathog.* 10:261.

Tsai, C.M., Gu, X.X., and Byrd, R.A., 1994, Quantification of polysaccharide in *Haemophilus influenzae* type b conjugate and polysaccharide vaccines by high-performance anion-exchange chromatography with pulsed amperometric detection, *Vaccine* 12:700.

Tsuji, M., and Matsumoto, M., 1989, Pathogenesis of fowl cholera: influence of encapsulation of the fate of *Pasteurella multocida* after intravenous inoculation into turkeys, *Avian Dis.* 33:238.

Utley, S.R., Bhat, U.R., Byrd, W., and Kadis, S., 1992, Characterization of lipopolysaccharides for four *Pasteurella haemolytica* serotype strains, evidence for presence of sialic acid in serotypes 1 and 5, *FEMS Microbiol. Lett.* 92:211.

Virji, M., Kayhty, H., Ferguson, D.J., Alexandrescu, C., and Moxon, E.R., 1991, Interaction of *Haemophilus influenzae* with cultured human endothelial cells, *Microb. Pathog.* 10:231.

Ward, C.K., and Inzana, T.J., 1993, Cloning of a DNA region putatively involved in the encapsulation of *Actinobacillus pleuropneumoniae*, *Proc. Conf. Res. Works Animal Dis.* 74:44.

Whiteley, L.O., Maheswaran, S.K., Weiss, D.J., and Ames, T.R., 1990, Immunohistochemical localization of *Pasteurella haemolytica* A1-derived endotoxin, leukotoxin and capsular polysaccharide in experimental bovine pasteurella pneumonia, *Vet. Path.* 27:150.

Wilson, C.F., Sutherland, A.D., Inglis, L., and Donachie, W., 1992, Characterization and biological activity of monoclonal antibodies specific for *Pasteurella haemolytica* A1 capsule and lipopolysaccharide, *Vet. Microbiol.* 31:161.

Zwahlen A., Kroll, J.S., Rubin, L.G., and Moxon, E.R., 1989, The molecular basis of pathogenicity in *Haemophilus influenzae*: comparative virulence of genetically-related capsular transformants and correlation with changes at the capsulation locus cap, *Microb. Pathog.* 7:225.

ADHESIN-RECEPTOR INTERACTIONS BY *HAEMOPHILUS INFLUENZAE* AND OTHER BACTERIA FROM THE HAP GROUP

L. van Alphen

Department of Medical Microbiology
Academic Medical Center, room 162L
Faculty of Medicine
Meibergdreef 15,
NL - 1105 AZ Amsterdam
The Netherlands

INTRODUCTION

Many microbes in nature can thrive in a variety of ecological niches, whereas others are restricted to specific microenvironments. *Haemophilus influenzae* f.i. is only able to colonize and infect humans (Turk and May, 1967). Host range, tissue tropism and target cell specificity demonstrated by a particular microbe are determined at least in part by a stereochemical fit between microbial adhesins and complementary receptor architectures on host surfaces. This is a dynamic process which is meant to direct the microbe to its final target. This target may be a niche on the mucosa during carriership, inflamed tissue after bacterial action on the epithelium, or the circulation when the bacteria have passed the epithelium. Also during systemic spread, bacteria have preference for certain tissues. In patients with meningitis *H. influenzae is* able to bind to and pass the blood brain barrier on its way to the cerebrospinal fluid and meningi (Moxon, 1992). Each step in these processes requires specific interactions between bacteria and host cells. Regulation of the synthesis of the bacterial adhesins allows the bacteria to bind to receptors available, thereby directing the bacteria to the next target tissue.

In this paper I focus on (i) the bacterial and host cell receptors involved in the specific interaction of bacteria from the HAP group, especially *H. influenzae* with respiratory epithelium, and (ii) the regulation of expression of the major adhesins, the fimbriae, of *H. influenzae*.

STEPS IN THE INTERACTION OF *H. INFLUENZAE* WITH NASOPHARYNGEAL MUCOSA IN HEALTH AND DISEASE

During asymptomatic colonization of the nasopharynx *H. influenzae* is able to adapt to the local environment, multiply, and avoid mucociliary clearance and local defense mechanisms of the host (Moxon, 1986; Murphy and Sethi, 1992). Therefore, this complex colonization process requires more than just adherence of the bacteria to the epithelium of the upper respiratory tract. Morphological studies on infected human organ culture models of nasal polyps and adenoids have revealed that *H. influenzae* is trapped in the mucus layer (Read *et al.*, 1991). The mucus structures to which *H. influenzae* binds have not been identified yet,

Haemophilus, Actinobacillus, and Pasteurella
Edited by W. Donachie *et al.*, Plenum Press, New York, 1995

and it is not clear to what *H. influenzae* specifically binds. After trapping in the mucus layer of organ cultures *H. influenzae* invariably causes functional and structural damage of the respiratory epithelium. Mucus production is even stimulated by the presence of extracellular products of *H. influenzae* (Adler *et al.*, 1986). Increased mucus production is one of the characteristics of infections in patients with chronic bronchitis whose bronchi are persistently colonized and infected by *H. influenzae*. Mucus production is the result of mucus gland hyperplasia in these patients. Trapping of the bacteria in the mucus prevents the bacteria from being killed by local defense mechanisms, since mucus is a diffusion barrier for solutes and is impermeable to cells involved in host defense.

The epithelium exposed to the mucus-entrapped *H. influenzae* becomes disorganized, ciliary beating stops, and ciliary epithelial cells are sloughed off (Read *et al.*, 1991; Loeb *et al.*, 1988; Wilson *et al.*, 1992). Then the bacteria adhere to the nonciliated basal cells and to the epithelial basement membrane (Read, *et al.*, 1991). Both these substrates are unavailable for adherence in intact mucosa. The adherent bacteria are not expelled by mechanical (coughing) and residual mucociliary clearance and remain as long as (immunological) defense mechanisms are not eradicating the bacteria.

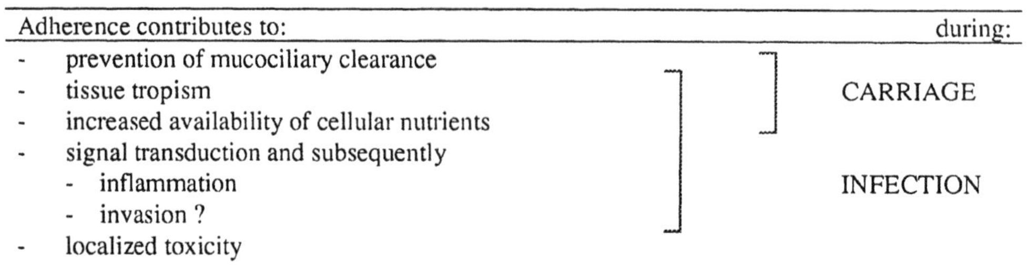

Figure 1. Processes affected by adherence (van Alphen *et al.*, 1991).

In health, dissemination of bacteria from the nasopharynx throughout the respiratory tract (sinuses, eustachian tubes, lower respiratory tract) occurs continuously and the bacteria are cleared efficiently without clinical symptoms by various host defense mechanisms. Growth of persistent *H. influenzae* results in colonization of the respiratory epithelium of the nasopharynx. When the interaction of *H. influenzae* with the epithelium of the respiratory tract results in an inflammatory tissue reaction, the asymptomatic carriage may convert into a respiratory tract infection. Inflammation causes subsequent further damage of the mucosa. Pertubation of the epithelial layer allows bacteria to reach the submucosa and eventually the bloodstream (Turk and May, 1967; Read *et al.*, 1991).

Our knowledge of adherence mechanisms is mainly restricted to those of the first steps in the infection, i.e. establishment to avoid mucociliary clearance, specific binding to epithelial cells and subsequent entrance into host tissues. Figure 1 summarizes the most important effects of adherence.

Table 1. Some characteristics of adhesins of bacteria of the HAP group

Adhesins of the HAP group:	Adherence to:	Ref.
Haemophilus influenzae:		
fimbriae	various cells and tissues	Loeb *et al.*, 1988; Wilson *et al.*, 1992; Alphen *et al.*, Sterk *et al.*, 1991; Alphen *et al.*, 1988; Brinton Jr *et al.*, 1989; Guerina *et al.*, 1981; Mason *et al.*, 1985; Pichichero *et al.*, 1982; Pichichero; 1984; Read *et al.* 1992
High Molecular Weight (HMW) non-fimbrial adhesins	some epithelial cells	St. Geme III and Falkow, 1991; Barenkamp *et al.*, 1992; St Geme III *et al.*, 1993; Noel *et al.*, 1994
?	mucus	Read *et al.*, 1991; Adler *et al.*, 1986; Loeb *et al.*, 1988; Wilson *et al.*, 1992
Haemophilus parainfluenzae:		
fimbriae		
clustering outer membrane adhesin	epithelial cells?	Kahn and Gromkova, 1981
(34 kDa protein)	oral streptococci, salivary pellicle, neuraminidase treated erythroyctes	Liljemark *et al.*, 1992; Lai *et al.*, 1990
Haemophilus parasuis:		
fimbriae	epithelial cells?	Munch *et al.*, 1992
Actinobacillus pleuropneumoniae:		
smooth lipopolysaccharide	porcine tracheal rings	Belanger *et al.*, 1994; Belanger *et al*, 1990.
(serotypes 2 and 7)	(10 and 11kb mucus proteins)	
fimbriae	polymorphonuclear leukocytes?	Inzana, 1991
?	haemagglutination	Jaques *et al*, 1991.
Actinobacillus actinomycetemcomitans		
fimbriae (subunit 54 kDa)	KB epithelial cells	Meyer and Fives-Taylor, 1994
amorphous surface material	KB epithelial cells	Meyer and Fives-Taylor, 1994
extracellular microvesicles	KB epithelial cells	Meyer and Fives-Taylor, 1994
Pasteurella multocida:		
fimbriae	epithelial cells?	Pijoan and Trigo, 1990

ADHESINS OF *H. INFLUENZAE* AND OTHER BACTERIA OF THE HAP GROUP

A variety of adhesins have been described for members of the group of HAP bacteria. The current data are summarized in Table 1. The adhesins include fimbriae, outer membrane proteins and lipopolysaccharide. The receptors for these adhesins are found on a variety of cell types, including epithelial cells, leukocytes, and on mucus, other bacteria and tooth pellicle. The receptor structures for these adhesins are unknown, except the receptor for the *H. influenzae* fimbriae and the receptor for lipopolysaccharide of *A. pleuropneumoniae*. The various fimbriae types, their receptors, and the nonfimbrial adhesins of *H. influenzae* are summarized in Table 2.

Table 2. Specificity of adhesins of *H. influenzae* (Loeb *et al.*, 1988; Noel *et al.*, 1994).

Adhesin	Specificity	Remarks
fimbriae:		
LKP	nasopharyngeal epithelial cells, AnWj positive erythrocytes	sialyllactosylceramide receptor
SNN	nasopharyngeal epithelial cells	
LNN	not identified	
VNN	not identified	
peritrichous pili	Hela cells, (Hep2)	non-haemagglutinating, on genital isolates (biotype IV)
nonfimbrial adhesins (outer membrane proteins):		
HMW1	Chang, HEp-2,kb ME-180 cell lines macrophages	FHA-like
HMW2	HEp-2, Hec-1B cells	FHA-like
not identified	mucus	

THE LKP FIMBRIAE GENE CLUSTER OF *H. INFLUENZAE*

The genes coding for several types of LKP fimbriae have been cloned. Minimally a DNA insert of 8kb is required to confer the ability to express morphologically and antigenically intact *H. influenzae* fimbriae upon *E. coli,* indicating that a large gene cluster is required for the biogenesis of *H. influenzae* fimbriae (van Ham *et al.*, 1989). Cloning and deletion analysis of the DNA encoding the LKP1 fimbriae of a nontypable *H. influenzae* also pointed to the presence of a gene cluster, since a fragment of 7kb was necessary for fimbriae expression in *E. coli* (Kar *et al.*, 1990). The major fimbrial subunits, composing the majority of the fimbrial protein, have been cloned from various encapsulated and nonencapsulated *H. influenzae* strains (St Geme III *et al.*, 1993; van Ham *et al.*, 1989; Kar *et al.*, 1990; Kilsdorf *et al.*, 1990; Langermann and Wright, 1990; Coleman *et al.*, 1991; Forney, *et al.*, 1991; Whitney and Farley, 1993). These major subunits are very similar to one another with homologies ranging from 68 - 100% at the amino acid level. The predicted amino acid sequence from the nucleotide sequence of the major subunit gene *hifA* correlates well with the amino acid sequence determined for the purified protein (Armes and Forney, 1990).

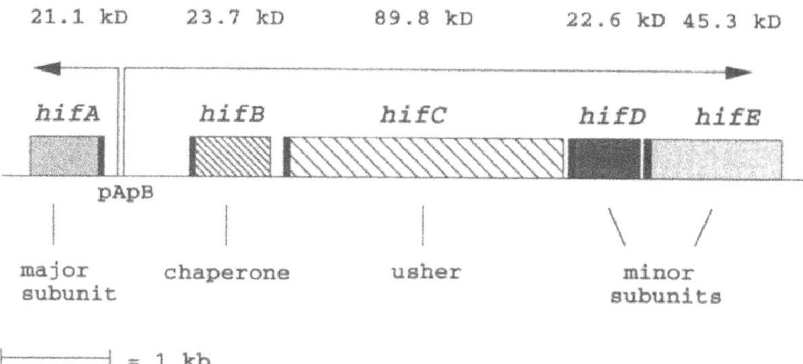

Figure 2. Physical and genetic map of the *H. influenzae* LKP4 fimbriae gene cluster. Filled boxes represent parts of genes coding for signal peptides. Closed arrows show the directions of transcription from the clustered fimbrial promotors (pApB).

The gene cluster contains 5 more genes *(hifB - hifE)* as deduced from the DNA sequence analysis of the complete gene cluster of *H. influenzae* type b LKP4 fimbriae (Figure 2) (van Ham *et al.*,). Of these, only fimbrial genes *hifB* and *hifC* have been sequenced from different strains (van Ham *et al.*, 1994; Smith *et al.*,unpublished; Watson *et al.*, 1994). The predicted proteins of the various strains are almost identical (98%). The homology of the different fimbrial genes in various strains indicates that the fimbrial gene clusters are strongly conserved. The small changes in *hifB* and *hifC* are probably the result of evolutionary drift, and the more pronounced differences between the *hifA* genes are likely the result of selection under immunological pressure. The lack of data about the genetic organization of the various gene clusters and the nucleotide sequence of the *hifD - hifE* genes in other strains makes a definitive conclusion regarding the conservation of *H. influenzae* fimbriae presently not possible.

The *hifB* gene shows homology to several bacterial chaperone proteins which transport bacterial subunits through the periplasm protecting them against proteolytic degradation and folding them into an assembly-competent conformation (Holmgren *et al.*, 1992). The HifB protein has a molecular weight of 24 kD, and the *hifC* gene codes for a 90 kD protein. The HifC protein has significant homology with a class of outer membrane proteins designated as molecular ushers. In the pap-pili system these proteins are involved in the export of fimbrial subunits in an ordered sequence to ensure a proper construction of the fimbriae (Dodson *et al.*, 1993). The *HifD* and *hifE* genes encode mature proteins of 23 and 45 kd, respectively. The HifD protein is highly homologous to HifA. HifE shows a high degree of similarity at the C-terminal end to HifA and overall similarities to large fimbrial subunits in other bacterial species, indicating that they are minor subunits. HifD has various characteristics of a lipoprotein (the Thr-Ala-Ser-Cys consensus sequence of amino acids 17 - 20; more than one Lys within the first 7 amino acids; turn-promoting residues immediately downstream of the peptidase II cleavage site) (van Ham *et al.*, 1994).

The fimbrial gene cluster of *H. influenzae* belongs to the family of *E. coli*-like fimbriae, which includes the well-characterized pyelonephritis-associated-pili (Pap-pili) and type 1 pili of *E. coli* (reviewed in refs. 60,61). The homology between the fimbrial genes of *H. influenzae* and the type 1 genes of *E. coli* is the strongest within this family of *E. coli*-like fimbriae (van Ham *et al.*, 1989; van Ham *et al.*, unpublished), although a *hifD*-like gene is lacking in the other gene clusters of the *E. coli* family. *HifA* has its own transcript and *hifB - hifE* are transcribed in the opposite direction into a polycystronic mRNA (van Ham *et al.*,1989; van Ham *et al.*, 1993). Mutational insertion analysis of *hifD* and *hifE* revealed that the products of these genes influenced the amount of fimbriae expressed on the cell surface, possibly by stabilizing the initiation complex required for fimbriae production or termination of the fimbrial synthesis (van Ham *et al.*,1994).

Adherence studies with fimbrial mutants in *H. influenzae* and *E. coli* clones expressing defined genes of the *H. influenzae* fimbriae gene cluster demonstrated that these adhesive properties reside in the major fimbrial subunit (van Ham *et al.*, submitted for publication). Analogous to other bacteria (St Geme III *et al.*, 1993), the minor subunits may be involved in binding to different substrates. Components of the extracellular matrix may be good candidates for fimbriae-mediated binding of *H. influenzae*, since *H. influenzae* was detected in association with the basement membrane in organ culture models (Read *et al.*, 1991).

SPECIFICITY OF LKP-4 FIMBRIAE-MEDIATED ADHERENCE OF *H. INFLUENZAE*

Over the years many groups have shown that fimbriae mediate adherence to oropharyngeal epithelial cells and erythrocytes (van Alphen *et al.*, 1988; Guerina *et al.*, 1981; Mason *et al.*, 1985; Pichichero *et al.*, 1982; Pichichero, 1984). In haemagglutination experiments the receptor for fimbriated *H. influenzae* on human erythrocytes was identified as the blood group AnWj antigen (van Alphen *et al.*, 1986; Poole and van Alphen, 1988) (Table 3). This receptor is present on human and primate erythrocytes and is absent on erythrocytes from a variety of other animal species.

Table 3. Properties of the blood group AnWj (Anton)-receptor for *H. influenzae* LKP-4 fimbriae

-	Present on almost all adult erythrocytes. These are AnWj-positive
	Absent on cord blood erythrocytes (negative for AnWj)
-	Full expression of fimbriae receptor and AnWj starts within two weeks after birth, and is completed within one day
-	Insensitive to proteolysis, sensitive to mild periodate oxidation
-	Absent in genetically and phenotypically AnWj$^-$ individuals (Lu-)
-	Bacterial haemagglutination is blocked by anti-AnWj antibodies

The erythrocyte receptor appeared different from the epithelial cell receptor, since antibodies to the two types of receptors do not cross-react and since the two receptors are expressed independently (van Alphen *et al.*, 1987). However, a defect in the expression of both receptors was observed in members of a family showing that both receptors are related (van Alphen *et al.*, 1990) (Table 4).

Table 4. Properties of the epithelial cell receptor for LKP-4 fimbriae of *H. influenzae*

-	Present at birth
-	Insensitive to proteolysis, sensitive to mild periodate oxidation
-	Absent in genotypically rare AnWj-individuals
-	Adherence is not blocked with anti-AnWj antibodies

Inhibition experiments with purified gangliosides revealed that ganglioside GM3 inhibited adherence of fimbriated *H. influenzae* to epithelial cells and haemagglutination in nM concentrations (van Alphen *et al.*, 1991). GM2 and more extensive gangliosides were up to 10 times more potent inhibitors. N-acetyl-sialyl-lactose and the macromolecular organization provided by the ceramide tail are essential. From these data it was concluded that the fimbriae of *H. influenzae* belong to the family of low affinity lactosylceramide binding fimbriae, and that this receptor structure is part of more complex molecules on epithelial cells and erythrocytes which are different (van Alphen *et al.*, 1991). Lactosylceramide receptors are common receptors (Table 5). They are thought to be involved as second adhesins for close contact (invasive bacteria, viruses) (Stromberg *et al.*, 1991; Karlsson *et al.*, 1992; Karlsson, 1989). The aminoacids in the receptor binding domain of the HifA protein subunit of the fimbriae have not been identified yet.

Table 5. Properties of lactosylceramide receptors

-	Common receptor for respiratory pathogens
-	Contain isoreceptors -> tissue specificity and tropism
-	Low affinity
-	Requires multivalent adhesins for efficient binding
-	Not found in glycoproteins
-	Internal structure of many glycolipids

NON-FIMBRIAL ADHESINS IN *H. INFLUENZAE:* COMPARISON WITH FHA OF *BORDETELLA PERTUSSIS*

Two nonfimbrial adhesins have been identified, which are only present in nontypable *H. influenzae* (Barenkamp and Leininger, 1992; Barenkamp and Bodor, 1990). These proteins, designated HMW1 and HMW2 are partly homologous, and antigenically related to the filamentous haemagglutinin, a colonization factor of *Bordetella pertussis* (Barenkamp and Leininger, 1992). Both HMW1 and HMW2 mediate binding to distinct epithelial cells (St Geme III *et al.*, 1993; Hultgren *et al.*, 1993).

ADHESINS AND TISSUE TROPISM

Both fimbrial and nonfimbrial adhesins have been suggested to mediate the tissue tropism of *H. influenzae* (Sterk *et al.*, 1991; Hultgren *et al.*, 1993).

Table 6. Dependence on fimbriae for the binding of *H. influenzae* to various respiratory cell types

Cell type	Binding by *H. influenzae*	
	fimbriated	nonfimbriated
Epithelium*	++	-
Fibroblasts in connective tissue, underlining epithelia and surrounding vessels	++	++
Lymphocytes, monocytes, tissue macrophages including type II pneumocytes, follicle cetre cells, dendritic reticulum cells	++	++

*for epithelial cell type see Table 7

Indeed, fimbriae and the nonfimbrial adhesins HMW1 and HMW2 differ in epithelial cell specificity. The specificity of the HMW adhesins differed with respect to various epithelial cell lines. Using tissue sections of the respiratory epithelium, fimbriae-mediated adherence to ciliated and pseudostratified columnar epithelial cells, but not to squamous and cubic cells was demonstrated (Sterk *et al.*, 1991) (Tables 6,7). In addition, fimbriae-mediated adherence to various tissues and cell types correlated with the ability of *H. influenzae* to colonize or to cause infections in the tissues or organs with these cell types (Sterk *et al.*, 1991).

Table 7. Fimbriae-dependent binding of *H. influenzae* to various epithelia.

Epithelium cell type	Tissue	*H. influenzae* binding	
		+ fimbriae	- fimbriae
squamous,cubic	pharynx, epiglottis	-	-
pseudostratified columnar	pharynx, epiglottis	++	-
ciliated columnar	rhinopharynx, nasopharynx, adenoid, bronchi	++	-
conjunctival	eye	-	-
glomerular, distal and proximal tubules	kidney	+/-	-

An apparently important functional characteristic of *H. influenzae* in the development of disease is variation in the expression of the fimbriae. During natural infection, nasopharyngeal isolates are often fimbriated, while their isogenic counterparts from systemic sites are, as a rule, nonfimbriated. The nonfimbriated isolates have often the genes required for fimbriae expression, since these bacteria can be enriched for fimbriation *in vitro* (Guerina *et al.*, 1981; Mason *et al.*, 1985; Pichichero *et al.*, 1982). These observations suggest that expression of fimbriae is beneficial during the initial stages of colonization and infection, but disadvantageous in establishing systemic disease. Fimbriae-mediated binding to erythrocytes and other cells in the circulation may enhance clearance of the bacteria by these cells.

The expression of the fimbriae of *H. influenzae* type b appeared subject to reversible phase variation in vitro in high frequency (10^{-4}) between two or three expression levels (van Ham *et al.*, 1993). This frequency is so high that in each site of colonization or infection the fimbriated or nonfimbriated variants occur next to each other, allowing selection of the variant best equipped for that particular environment. Phase variation is controlled at the transcriptional level of all fimbrial genes. The *hifA* and *hifB* - *hifE* promoter regions were found to be clustered through an almost complete divergent overlap. The nucleotide sequence between the -10 and -35 positions of the overlapping promoters consists of repetitive TA units. Variation in the number of TA units changes the normally strictly constrained spacing between the -35 and -10 sequences, thereby controlling the bidirectional transcription initiation. Nine repeats result in abolishment of the expression, 10 and 11 repeats provide the proper distance between the -10 and -35 sequences, and thus expression. Twelve repeats would be too much, but expression may be through an alternative -35 promoter site (van Ham *et al.*, 1993) (Figure 3).

HifB GTCTTTCCCACGTAAGCTAAAGCCAACCTACGGCAAAATTAATTCAAAATAAAATATGAAA

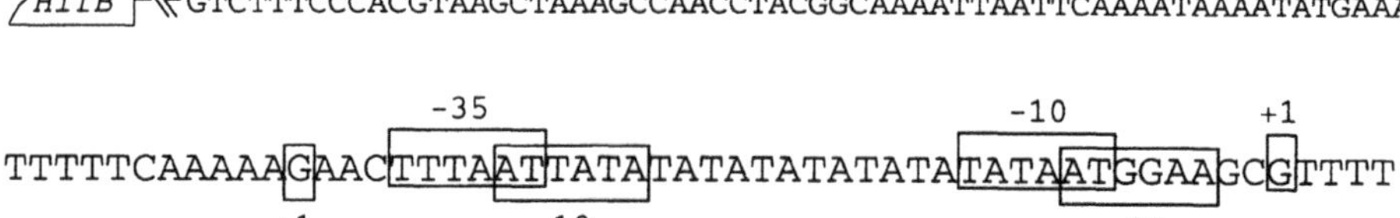

TTACTTTTTTTGGAAAAACAAATCTTGCTGTTTATTAAGGCTTTAGCATTTTAATAAACGGTTTATATTA

ATCTTTTGCCTTATTTATAAGGCACAAACCTCTTTATGGAGCAATTTATT HifA

Figure 3. The transcription of the two divergently directed fimbrial genes *hifA* and *hifB is* controlled by a variable sequence of reiterated TA units in their combined promotor region. In the case of 9 units, the putative -10 and -35 promotor sequences for both *hifA* and *hifB* are separated by 14 bases, which does not allow transcription of either *hifA* or *hifB*. With 10 or 11 TA units, the -10 and -35 motifs are separated by 16 or 18 bases, respectively, allowing transcription of both genes, resulting in fimbriae expression. A spacing of 16 bases results in the highest level of fimbriation.

In contrast to *H. influenzae* type b, blood isolates of *H. influenzae*, subspecies *aegyptius* causing Brazilian purpuric fever (BPF) (Brenner *et al.*, 1988), are generally fimbriated (Mayer *et al.*, 1989). This suggests that fimbriae confer a survival advantage to intravascular BPF bacteria, but may also point to a different mechanism of expression. The major fimbrial subunit of the BPF *H. influenzae,* subspecies *aegyptius,* shows high homology to the HifA subunit of *H. influenzae* (St Geme III *et al.,* 1993; Whitney and Farley, 1993). Interestingly, a fimbriated isolate of the BPF strains contains only 9 TA units in its promoter region, a number too low to allow fimbriae gene transcription in *H. influenzae* type b. Thus, the mechanism determining fimbriae expression in *H. influenzae,* subspecies *aegyptius,* seems to be different from that in *H. influenzae.*

CONCLUSIONS

1. Fimbriae and nonfimbrial adhesins are common on respiratory pathogens
2. Fimbriae of *H. influenzae* help to overcome electrostatic repulsion between (encapsulated) bacteria and host cells
3. Fimbriae-mediated adherence of *H. influenzae is* probably one step in a dynamic process, since:
   - Fimbriae-mediate low affinity adherence to ganglioside GM3-like receptors on epithelial cells and erythrocytes
   - the expression is subject to phase variation
4. Tissue-specific adherence correlates with tissue tropism of *H. influenzae*
5. Fimbriae genes are coded for by an 8kb DNA fragment, which is sufficient for expression of *H. influenzae* fimbriae in *E. coli*
5. The genes are part of a gene cluster; *hifA* codes for the fimbrial subunit
6. Phase variation of the fimbriae expression is caused by slipped mispair transcription in a region consisting of AT-repeats, which serves as a bidirectional promotor for *hifA* and *hifB*

REFERENCES

Adler, K.B., Henley, D.D. and Davis, G.S., 1986, Bacteria associated with obstructive pulmonary disease elaborate extracellular products that stimulate mucin secretion by explants of guinea pig airways. *Am. J. Pathol.* 125:501-514.

Armes, L.G. and Forney, L.J., 1990, The complete primary structure of pilin from *Haemophilus influenzae* type b strain Eagan. *J. Protein Chemistry* 9:45-52.

Barenkamp, S.J. and Bodor, F.F., 1990, Development of serum bactericidal activity following nontypable *Haemophilus influenzae* acute otitis media. *Pediatr. Infect. Dis. J.* 9:333-339.

Barenkamp, S.J. and Leininger, E., 1992, Cloning, expression, and DNA sequence analysis of genes encoding nontypable *Haemophilus influenzae* high-molecular-weight surface-exposed proteins related to filamentous haemagglutinin of *Bordetella pertussis. Infect. Immun.* 60:1302-1313.

Belanger, M. Dubreuil, D. and Jaques, M., 1994, Proteins found within respiratory tract secretions bind lipopolysaccharides of *Actinobacillus pleuropneumoniae. Infect. Immun.* 62:868-873.

Belanger, M., Dubreuil, D., Harel, J., Girard, C. and Jaques, M., 1990. Role of lipopolysaccharides in adherence of *Actinobacillus pleuropneumoniae* to porcine tracheal rings. *Infect. Immun.* 58:3523-3530.

Brenner, D.J., Mayer, L.W., Carlone, G.M., Harrison, L.H., Bibb, W.F., de Cunto Brandileone, M.C., Sottnek, F.O., Irino, K., Reeves, M.W., Swenson, J.M., Birkness, K.A., Weyant, R.S., Berkley, S.F., Woods, T.C., Steigerwalt, A.G., Grimont, P.A.D., McKinney, R.M., Fleming, D.W., Gheesling, L.L., Cooksey, R.C., Arko, R.J., Broome, C.V., and The Brazilian Purpuric Fever Study Group, 1988, Biochemical, genetic, and epidemiologic characterization of *Haemophilus influenzae* biogroup *aegyptius (Haemophilus aegyptius;* strains associated with Brazilian purpuric fever. *J. Clin. Microbiol.* 26:1524-1534.

Brinton Jr.,C.C., Carter, M.J., Derber, D.B., Kar, S., Kramarik, J.A., To, A.C., To, S.C., Wood, S.W., 1989, Design and development of pilus vaccines for *Haemophilus influenzae* diseases. *Pediatr. Infect. Dis. J.* 8:S54S61.

Coleman, T., Grass, S. and Munson Jr., R., 1991, Molecular cloning, expression, and sequence of the pilin gene from nontypable *Haemophilus influenzae* M37. *Infect. Immun.* 59:1716-1722.

Dodson, K.W., Jacob-Dubuisson, F., Striker, R.T. and Hultgren, S.J., 1993, Outer membrane PapC molecular usher discriminately recognizes periplasmic chaperone-pilus subunit complexes. *Proc. Natl. Acad. Sci USA* 90:3670-3674.

Forney, L.J., Marrs, C.F., Bektesh, S.L. and Gilsdorf, J.R., 1991, Comparison and analysis of the nucleotide sequences of pilin genes from *Haemophilus influenzae* type b strains Eagan and M43. *Infect. Immun.* 59:1991-1996.

Gilsdorf, J.R., McCrea, K. and Forney, L., 1990, Conserved and nonconserved epitopes among *Haemophilus influenzae* type b pili. *Infect. Immun.* 58:2252-2257.

Guerina, N.G., Langermann, S., Clegg, H.W., Kessler, T.W., Goldmann, D.A. and Gilsdorf, J.R., 1982, Adherence of piliated *Haemophilus influenzae* type b to human oropharyngeal cells. *J. Infect. Dis.* 146:564.

Holmgren, A., Kuehn, M.J., Branden, C.-I. and Hultgren, S.J., 1992, Conserved immunoglobulin-like features in a family of periplasmic pilus chaperones in bacteria. *EMBO J.* 11:1617-1622.

Hultgren, S.J., Abraham, S., Caparon, M., Falk, P., St. Geme III, J.W. and Normark, S., 1993, Pilus and nonpilus bacterial adhesins: assembly and function in cell recognition. *Cell* 73:887-901.

Inzana, T.J., 1991, Virulence properties of *Actinobacillus pleuropneumoniae*. *Microb. Pathog.* 11:305-316.

Jaques, M., Belanger, M., Roy, G. and Foiry, B., 1991, Adherence of *Actinobacillus pleuropneumoniae* to porcine tracheal epithelial cells and frozen lung sections. *Vet. Microbiol.* 27:133-143.

Kahn, M.E. and Gromkova, R., 1981, Occurrence of pili on and adhesive properties of *Haemophilus parainfluenzae*. *J. Bacteriol.* 145:1075-1078.

Kar, S., To, S.C.-M., and Brinton Jr., C.C., 1990, Cloning and expression in *E. coli* of LKP pilus genes from a nontypable *Haemophilus influenzae* strain. *Infect. Immun.* 58:903-908.

Karlsson, K.-A. J. Angstrom, J. Bergstrom, and Lanne, B., 1992, Microbial interactions with animal cell surface carbohydrates. *APMIS Suppl.* 100:71-83.

Karlsson, K.-A., 1989, Animal glycosphingolipids as membrane attachment sites for bacteria. *Ann. Rev. Biochem.* 58:309-350.

Lai, C.H., Bloomquist, C.G. and Liljemark, W.F., 1990, Purification and characterization of an outer membrane protein adhesin from *Haemophilus parainfluenzae* HP-28. *Infect. Immun.* 58:3833-3839.

Langermann, S. and Wright, A., 1990, Molecular analysis of the *Haemophilus influenzae* type b pilin gene. *Mol. Microbiol.* 4:221-230.

Liljemark, W.F., Bloomquist, C.G. and Lai, C.H., 1992, Clustering of an outer membrane adhesin of *Haemophilus parainfluenzae*. *Infect. Immun.* 60:687-689.

Loeb, M.R., Connor, E. and Penney, P.A., 1988, A comparison of the adherence of fimbriated and nonfimbriated *Haemophilus influenzae* type b to human adenoids in organ culture. *Infect. Immun.* 56:484-489.

Mason Jr, E.O., Kaplan, S.L., Wiedermann, B.L., Norrod, E.P. and Stenback.W.A., 1985, Frequency and properties of naturally occurring adherent piliated strains of *Haemophilus influenzae* type b. Infect. Immun. 49:98-103.

Mayer, L.W., Bibb, W.F. and Birkness, K.A., 1989, Distinguishing clonal characteristics of the Brazilian purpuric fever-producing strain. *Pediatr. Infect. Dis. J.* 8:241-243.

Meyer, D.H. and Fives-Taylor, P.M., 1994, Characteristics of adherence of *Actinobacillus actinomycetemcomitans* to epithelial cells. *Infect. Immun.* 62:928-935.

Moxon, E.R., 1992, Molecular basis of invasive *Haemophilus influenzae* type b disease. *J Infect Dis* 165:S77-S81.

Moxon, E.R., 1986, The carrier state: *Haemophilus influenzae*. *J. Antimicrob. Chemother.* 18:17-24.

Munch, S. Grund, S. and Kruger, M., 1992, Fimbriae and membranes on *Haemophilus parasuis*. *Zentralbl. Veterinarmed. [B~.* 39:59-64.

Murphy, T.F. and Sethi, S., 1992, Bacterial infection in chronic obstructive pulmonary disease. *Am. Rev. Respir. Dis.* 146:1067-1083.

Noel, G.J., Barenkamp, S.J., St. Geme III, J.W., Haining, W.N. and Mosser, D.M., 1994, High-molecular-weight surface-exposed proteins of *Haemophilus influenzae* mediate binding to macrophages, *J. Infect. Dis.* 169:425-429.

Pichichero, M.E., 1984, Adherence of *Haemophilus influenzae* to human buccal and pharyngeal epithelial cells: relationship to piliation. *J. Med. Microbiol.* 18:107-116.

Pichichero, M.E., Anderson, P., Loeb, M. and Smith, D.H., 1982, Do pili play a role in pathogenicity of *Haemophilus influenzae* type b? *Lancet* ii:960-962.

Pijoan, C. and Trigo, F., 1990, Bacterial adhesion to mucosal surfaces with special reference to *Pasteurella multocida* isolates from atrophic rhinitis. *Can. J. Vet. Res.* 54:S16-S21.

Poole, J. and van Alphen, L., 1988, *Haemophilus influenzae* receptor and the AnWj antigen. *Transfusion* 28:289.

Read, R.C., Rutman, A.A., Jeffery, P.K., Lund, V.J., Brain, A.P.R., Moxon, E.R., Cole, P.J. and Wilson, R., 1992, Interaction of capsulate *Haemophilus influenzae* with human airway mucosa in vitro. *Infect. Immun.* 60:3244-3252.

Read, R.C., Wilson, R., Rutman, A., Lund, V., Todd, H.C., Brain, A.P.R., Jeffrey, P.K. and Cole, P.J., 1991, Interaction of nontypable *Haemophilus influenzae* with human respiratory mucosa in vitro. J. *Infect. Dis.* 163:549-558.

Smith, A.L., Forney, L. and Chanyangam, M. GenBank accession number X66606, unpublished.

St. Geme III, J.W. and Falkow, S., 1991, Loss of capsule expression by *Haemophilus influenzae* type b results in enhanced adherence to and invasion of human cells. *Infect. Immun.* 59:1325-1333.

St. Geme III, J.W., Falkow, S. and Barenkamp., S.J., 1993, High-molecular-weight proteins of nontypable *Haemophilus influenzae* mediate attachment to human epithelial cells. *Proc. Natl. Acad. Sci. USA* 90:2875-2879.

Sterk, L.T.M., van Alphen, L., Geelen-van den Broek, L., Houthoff, H.J. and Dankert, J., 1991, Differential binding of *Haemophilus influenzae* to human tissues by fimbriae. *J. Med. Microbiol.* 35:129-138.

Stromberg, N., Nyholm, P.-G., Pascher, I. and Normark, S., 1991, Saccharide orientation at the cell surface affects glycolipid receptor function. *Proc. Natl. Acad. Sci. USA* 88:9340-9344.

Turk, D.C. and May, J.R., 1967, *Haemophilus influenzae*, its clinical significance. The English University Press Ltd, London.

van Alphen, L. Geelen-van den Broek, L. Blaas, L., van Ham, M. and Dankert, J., 1991, Blocking of fimbriae mediated adherence of *Haemophilus influenzae* by sialyl gangliosides. *Infect. Immun.* 59:4473-4477.

van Alphen, L., Dankert, J. and Jansen, H.M. Virulence factors in the colonization and persistence of bacteria in the airways. *Am. Rev. Resp. Dis.* in press.

van Alphen, L., Levene, C., Geelen-van den Broek, L., Poole, J., Bennett, M. and Dankert, J., 1990, Combined inheritance of the epithelial and erythrocyte receptor for *Haemophilus influenzae*. *Infect. Immun.* 58:3807-3809.

van Alphen, L., Poole, J. and Overbecke, M., 1986, The Anton blood group antigen is the erythrocyte receptor for *Haemophilus influenzae*. *FEMS Microbiol. Lett.* 37:69-71.

van Alphen, L., Poole, J., Geelen, L. and Zanen, H.C., 1987, The erythrocyte and epithelial cell receptor for *Haemophilus influenzae* are expressed independently. *Infect. Immun.* 55:2355-2358.

van Alphen, L., van den Berghe, N. and Geelen-van den Broek, L., 1988, Interaction of *Haemophilus influenzae* with human erythrocytes and oropharyngeal epithelial cells is mediated by a common fimbrial epitope. *Infect. Immun.* 56:1800-1806.

van Ham, S.M., Mooi, F.R., Sindhunata, M.G., Maris, W.R. and van Alphen, L., 1989, Cloning and expression in *E. coli* of *Haemophilus influenzae* fimbrial genes establishes adherence to oropharyngeal epithelial cells. *EMBO J.* 11:35353540.

van Ham, S.M., van Alphen, L., Mooi, F.R. and van Putten, J.P.M., 1993, Fimbrial phase variation in *Haemophilus influenzae is* trancriptionally regulated by changes in the *HifA* promotor region. *Cell* 73:1187-1196.

Watson, W.J., Gilsdorf, J.R., Tucci, M.A., McCrea, K.W., Forney, L.J. and Marrs, C.F., 1994, Identification of a gene essential for piliation in *Haemophilus influenzae* type b with homology to the pilus assembly platform genes of Gram-negative bacteria. *Infect. Immun.* 62:468-475.

Whitney, A.M. and Farley, M.M., 1993, Cloning and sequence analysis of the structural pilin gene of Brazilian purpuric fever-associated *Haemophilus influenzae* biogroup *aegyptius*. *Infect. Immun.* 61:1559-1562.

Wilson, R. Read, R. and Cole, P., 1992, Interaction of *Haemophilus influenzae* with mucus, cilia, and respiratory epithelium. *J. Infect. Dis.* 165:S100-S102.

EXOTOXINS OF *ACTINOBACILLUS PLEUROPNEUMONIAE*

J. Frey

Institute for Veterinary Bacteriology
Laenggasstrasse 122
CH-3012 Berne
Switzerland

INTRODUCTION

Actinobacillus pleuropneumoniae (previously *Haemophilus pleuropneumoniae*) is the etiological agent of swine pleuropneumonia, a severe, contagious disease of swine causing important economic losses in industrialized swine production worldwide. The disease is either acute and often fatal or chronic localized and necrotizing. It seems to be mainly transmitted via the respiratory route from pig to pig (Nicolet and Konig, 1966; Nicolet, 1968; Nicolet *et al.*, 1969; Nicolet, 1992; Shope, 1964). Two different biovars, the ß-NAD dependent biovar 1 and the ß-NAD independent biovar 2 (Niven and O'Reilly, 1990), and twelve serotypes of *A. pleuropneumoniae* have been described (Nicolet, 1988; Nielsen, 1988). The chemical composition of the capsular polysaccharides (K-antigens) and lipopolysaccharides (O-antigens) of all serotypes is known (Perry *et al.*, 1990; Beynon *et al.*, 1993). Among the twelve serotypes, significant differences in virulence have been observed. In particular, serotypes 1 and 5 and to some extent also 9 and 11 are involved in severe outbreaks with high mortality and severe pulmonary lesions. Serotypes 2,4,6,7,8 and 12 are less virulent, although some of these serotypes are frequently isolated. In spite of the fact that serotypes 2,4,6,7,8 and 12 cause virtually the same lesions as serotypes 1, 5, 9 and 11, they cause lower mortality (Fales *et al.*, 1989; Rapp *et al.*, 1985; Brandreth and Smith, 1985; Martineau *et al.*, 1984; Nicolet, 1992; Rosendal *et al.*, 1985). In addition some serotypes, in particular serotype 3, seem to be of very low virulence and of no epidemiological importance in many countries. Serotyping is therefore one of the most important and most frequently used tools in epidemiology and sanitation programmes to control porcine pleuropneumonia (Nicolet, 1992; Nielsen, 1990).

Several factors seem to be involved in pathogenicity of *A. pleuropneumoniae*, including capsular polysaccharides (Jacques *et al.*, 1988; Jensen and Bertram, 1986; Inzana *et al.*, 1988; Inzana *et al.*, 1993) lipopolysaccharides (Fenwick and Osburn, 1986; Udeze *et al.*, 1987; Idris *et al.*, 1993), membrane proteins (Gerlach *et al.*, 1992; Deneer and Potter, 1989; Gonzalez *et al.*, 1990; Gerlach *et al.*, 1993; Rapp and Ross, 1988; Thwaits and Kadis, 1993), adhesion factors (Jacques *et al.*, 1991; Dom *et al.*, 1994a; Utrera and Pijoan, 1991) and exotoxins (Frey and Nicolet, 1988b; Frey *et al.*, 1993b; Frey *et al.*, 1994a; Frey, 1994; Kume *et al.*, 1986; Kamp and van Leengoed, 1989; Nakai *et al.*, 1984). It is most likely that the degree of virulence of the different *A. pleuropneumoniae* serotypes is, to a large extent, associated with the exotoxins expressed by the different strains (Frey, 1994; Frey *et al.*, 1994a; Devenish and Rosendal, 1991a; Rosendal *et al.*, 1980). Exotoxins of *A. pleuropneumoniae* have been postulated to be directly involved in generating clinical signs, following the discovery that cell free supernatant could produce lung lesions similar to those caused by the natural infection (Rosendal *et al.*, 1980). The purpose of this paper is to review the biochemical and genetic data on known exotoxins of *A. pleuropneumoniae*, their impact on virulence, their antigenic behavior and their potential to induce protective immunity.

BIOCHEMICAL AND GENETIC PROPERTIES OF *A.PLEUROPNEUMONIAE* RTX-TOXINS APXI, APXII AND APXIII

Characteristic for all *A. pleuropneumoniae* strains is a secreted hemolytic activity which was postulated to play a central role in pathogenesis. Hemolysin of *A. pleuropneumoniae* was purified first as a protein with a molecular mass of approximately 105 kDa. Biochemical and biophysical analysis revealed it to be a pore forming protein of the family of RTX-toxins (Lalonde *et al.*, 1989; Frey and Nicolet, 1988b; Frey and Nicolet, 1988a) with hemolytic and neutrophil toxic activity (Kamp and van Leengoed, 1989; van Leengoed *et al.*, 1989; Rosendal *et al.*, 1988) suggesting it to be an important virulence factor. The hemolysins or cytotoxins of *A. pleuropneumoniae* have therefore been a subject of intensive research during the last few years. Three protein toxins, all belonging to the family of RTX-toxins which are widely spread among Gram-negative pathogenic bacteria (Strathdee and Lo, 1989; Welch, 1991) have been found in *A. pleuropneumoniae*. Initial use of different names for these toxins and the fact that they strongly resemble each other created confusion and misinterpretation of genetic, biochemical and immunological results. These toxins are therefore now called *Actinobacillus* ***pleuropneumoniae*** RTX-toxins (**Apx** toxins) and include the strongly hemolytic and cytotoxic **ApxI**, the weakly hemolytic and cytotoxic **ApxII** and the non hemolytic but cytotoxic **ApxIII** (Frey *et al.*, 1993b). From the current genetic knowledge on Apx-toxins, it can be deduced that most of the earlier work on cellular interactions with exotoxins of *A. pleuropneumoniae* was carried out with preparations containing more than one toxin. The present review therefore does not include a detailed description on these interactions since more detailed work on the activity of the different individual toxins using single recombinant toxins will have to be done before any conclusions can be drawn.

ApxI

ApxI, a protein with an apparent molecular mass of 105 kDa, is strongly hemolytic (Frey and Nicolet, 1988a; Frey and Nicolet, 1988b) and strongly cytotoxic to phagocytic cells (Kamp *et al.*, 1991). It is produced and secreted by reference strains of serotypes 1, 5a, 5b, 9, 10 and 11 (Frey and Nicolet, 1990; Frey *et al.*, 1992; Kamp *et al.*, 1991; Frey *et al.*, 1993b). The deduced amino acid sequence of the *apxIA* gene predicts a protein with a calculated molecular mass of 110.1 kDa, an amphipathic N-terminus, three strongly hydrophobic domains and a domain with 13 glycine rich nonapeptides (Frey *et al.*, 1991b). ApxIA is very similar to the *E. coli* hemolysin HlyA, and much less similar to the other *A. pleuropneumoniae* Apx toxins and to *Pasteurella haemolytica* leukotoxin LktA. ApxI strongly binds calcium, which is required for haemolytic activity (Devenish and Rosendal, 1991b; Frey *et al.*, 1991b). Calcium binding by RTX toxins is depending on the glycine rich nonapeptide repeats. This was demonstrated for the *E.coli* HlyA which has 13 glycine rich repeats. Mutant HlyA with fewer repeats than wild type HlyA had a reduced Ca^{2+} binding capacity and a lower haemolytic activity; the haemolytic activity of the mutant could be restored to some extent in buffer containing high concentrations of Ca^{2+} (Ludwig *et al.*, 1988). ApxI is encoded by the *apxI* operon which contains the four genes: the posttranslational activator gene *apxIC*, the toxin structural gene *apxIA* and the secretion genes *apxIB* and *apxID*. These genes are arranged in the order *apxICABD* (Gygi *et al.*, 1992; Frey *et al.*, 1994b; Jansen *et al.*, 1993b), which is common to most RTX operons (Hughes *et al.*, 1992; Welch, 1991; Felmlee *et al.*, 1985). The entire *apxI* operon is present in reference strains of serotypes 1, 5a, 5b, 9, 10 and 11 (figure 1) (Frey *et al.*, 1993a; Jansen *et al.*, 1992b; Jansen *et al.*, 1993b; Frey *et al.*, 1993b). These serotypes are known to produce and secrete ApxI (Kamp *et al.*, 1991; Frey *et al.*, 1992; Frey and Nicolet, 1990). The nucleotide sequences of the entire *apxI* operon in the serotype 1 reference strain and the *apxICA* genes from serotypes 5, 9, 10 and 11 have been determined. In serotypes 5a, 5b and 10, *apxI* shows variations located at the 3' end of the gene (Nagai *et al.*, 1993; Frey *et al.*, 1993a; Jansen *et al.*, 1993b; Frey *et al.*, 1994b). These variations must lead to a slightly different ApxI protein in serotypes 5a, 5b and 10 compared to the serologically related serotypes 1, 9 and 11, but no immunological differences have been found yet (van den Bosch *et al.*, 1990). The sequence differences, however, are important in the design of oligonucleotide primers for the PCR detection of *apxIA* genes (Frey *et al.*, 1993a). The reference strains for serotypes 2, 4, 6, 7, 8 and 12, which do not produce ApxI, contain a partial operon with the genes *apxIBD* and a small segment from the 3' end of *apxIA* (figure 1). Only the serotype 3 reference strain is completely devoid of genes of the *apxI* operon (figure 1) (Frey *et al.*, 1993a; Jansen *et al.*, 1992b; Jansen *et al.*, 1993b; Frey *et al.*, 1993b).

The *apxI* operon is transcribed as two RNA species, a major one containing *apxICA* and a minor one containing *apxICABD*. The level of *apxICA* RNA is substantially higher in the type strain *A. pleuropneumoniae* 4074^T grown in the presence of 1.5 mM free Ca^{2+} ions than in the presence of < 10 µM free Ca^{2+} (Gygi *et al.*, 1992), supporting the view that the expression of *apxI* determinant is Ca^{2+} regulated (Frey and Nicolet, 1988b). The transcription start point is the same under Ca^{2+} induced and non-induced conditions and is located 133 bp upstream from the translation start codon of *apxIC* (Frey *et al.*, 1994b). The potential promoter sequence of the *apxI* operon contains an near optimal Pribnow box at the correct distance upstream from the transcription start point and a corresponding -35 box which is preceded by a -44 box (CAAAAT) (Frey *et al.*, 1994b) which has a potential for DNA bending and activation of transcription initiation.

ApxII

ApxII, a protein with an apparent molecular mass between 103-105 kDa, is weakly hemolytic and weakly cytotoxic and is produced by all serotype reference strains except serotype 10 (Frey and Nicolet, 1988b; Frey *et al.*, 1991a; Frey *et al.*, 1992; Kamp *et al.*, 1991; Smits *et al.*, 1991; Anderson *et al.*, 1991). Serotype 1,2,4,5,6,7,8,9,11 and 12 secrete ApxII into the growth medium, while in serotype 3 ApxII is mainly located in the cytoplasm (Frey *et al.*, 1994a). Since the ApxII protein is difficult to distinguish from ApxI on SDS-polyacrylamide and since several serotypes produce both toxins, many studies on *A. pleuropneumoniae* hemolysins before 1992 did not distinguish between ApxI and ApxII and most involved both toxins. The amino acid sequence deduced from *apxIIA* predicts a structural protein (ApxIIA) with a molecular mass of 102.5 kDa, with very similar features as ApxI, but with only 8 glycine rich nonapeptide repeats. The fewer glycine rich repeats explain in part the requirement for much higher concentrations of free Ca^{2+} and the lower hemolytic activity of ApxII compared to ApxI (Frey and Nicolet, 1988b; Frey *et al.*, 1991b), since the number of glycine rich repeats in these toxins seems to be involved in the binding of Ca^{2+}. The amino acid sequence of ApxIIA strongly resembles that of *P. haemolytica* leukotoxin LktA and to a much lesser extent other Apx toxins and *E.coli* HlyA (Jansen *et al.*, 1992a; Chang *et al.*, 1989; Frey, 1994; Frey *et al.*, 1994b). Genetic studies have shown that the *apxII* operon contains the activator gene *apxIIC*, and the structural gene *apxIIA*, but no secretion genes *BD*. A short segment of a 5' end of a *B* gene is located downstream of *apxIIA* (Chang *et al.*, 1989; Jansen *et al.*, 1992a; Smits *et al.*, 1991). The *apxIICA* genes are present in all serotype reference strains except 10 (figure 1) and share a high degree of similarity (Jansen *et al.*, 1992a; Frey *et al.*, 1993a; Frey *et al.*, 1993b) (figure 1). The *apxIICA* genes have been completely sequenced from serotypes 5, 7 and 9 (Chang *et al.*, 1989; Smits *et al.*, 1991) and partially sequenced from serotype 7 (Anderson *et al.*, 1991). The lack of *BD* genes in the *apxII* operon is striking. It seems that secretion of ApxII occurs via the gene products of the *apxIBD* genes, which are present in all serotypes except in serotype 3. Serotype 3, which contains *apxIICA* but lacks *apxIBD* genes secretes only faint amounts of ApxII (Frey *et al.*, 1994a) and hence is also much less hemolytic than the other ApxII producing serotypes, in spite of the presence of the secretion genes of the *apxIII* operon (figure 3). ApxII in serotype 3 reference strain, however, can be secreted to a level similar to the other ApxII producing serotypes when the *apxIBD* genes are provided *in trans* on a broad host range expression vector (our unpublished results). If cloned in *E. coli*, in contrast, ApxII is secreted via the *apxIIIBD* gene products as well (Macdonald and Rycroft, 1993).

By the construction of hybrid genes between *apxIIA* and the structural gene of *P. haemolytica* leukotoxin (*lktA*), the cytotoxic potential of ApxII was mapped to the C-terminal half of the protein. This region is physically separable from the region specifying erythrocyte lysis (McWhinney *et al.*, 1992). In addition it was observed that when the *apxIICA* genes were expressed and secreted in *E.coli* by *apxIBD*, by *hlyBD* or by *lktBD* gene products, two forms of ApxIIA, a cytotoxic/hemolytic and a cytotoxic/non-hemolytic form was found in the cytoplasm, while the extracellular ApxIIA was mainly cytotoxic/non-hemolytic (Tu *et al.*, 1994).

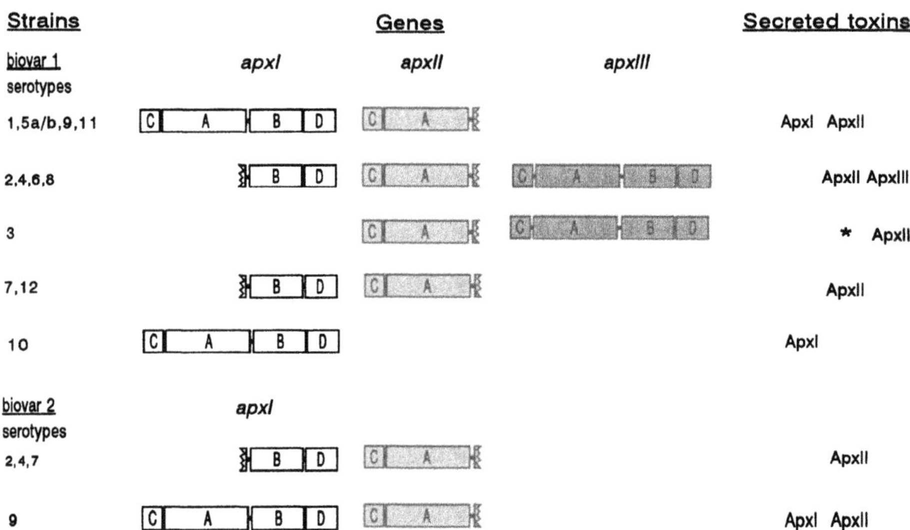

Figure 1. Genes of the three different *apx* operons and secreted Apx toxins in the different *A. pleuropneumoniae* biovar 1 serotype reference strains and different biovar 2 serotypes. Broken boxes represent truncated 3'-terminal *apxIA* or 5'-terminal *apxIIB* gene fragments. Data were taken from our hybridization experiments and from various published data referenced in the text. * = Serotype 3 reference strain S1421 releases faint amounts of ApxII to the culture medium which could be due to cell lysis, non-specific release, or partial secretion by the ApxIIIBD proteins. Some of the serotype 3 field strains analyzed released substantially more ApxII to the medium than the reference strain, but still less than the other serotypes.

ApxIII

ApxIII, a protein with an apparent molecular mass of 120 kDa, is not hemolytic, but is strongly cytotoxic for alveolar macrophages and neutrophils. It is produced and secreted by the reference strains of serotypes 2, 4, 3, 6 and 8 (Kamp *et al.*, 1991; Macdonald and Rycroft, 1992; Rycroft *et al.*, 1991; van den Bosch *et al.*, 1992). The amino acid sequence deduced from the gene *apxIIIA* predicts for ApxIIIA a molecular mass of 112.8 kDa and shows the typical domains of RTX-toxins with 13 glycine rich nonapeptides (Chang *et al.*, 1993a; Jansen *et al.*, 1993a). ApxIIIA is more similar to ApxIA (54% identical amino acids) and HlyA (54%) than to ApxIIA (49%) or LktA (47%). The *apxIII* operon has the typical genes *apxIIICABD* analogous to the *apxI* operon (Chang *et al.*, 1993b; Jansen *et al.*, 1993a; Macdonald and Rycroft, 1992). The whole operon is present in serotypes reference strains 2, 3, 4, 6 and 8 (Jansen *et al.*, 1992b; Frey *et al.*, 1993b; Frey *et al.*, 1994a) (figure 1). The nucleotide sequence of the whole *apxIII* operon from serotype 2 (Chang *et al.*, 1993a) and the *apxIIICA* genes of serotype 8 are reported. The C and A genes from these serotypes are very similar except for minor differences at the 3' end of the *apxIIIA* gene.

Distribution of Apx Toxins in *A. pleuropneumoniae* field strains

Genetic and biochemical analysis of over 200 biovar 1 field strains of all serotypes isolated from different geographical areas show that they contain the same *apx* genes and express the same Apx toxins as their respective serotype reference strains (J. Frey and J. F. van den Bosch unpublished results). Only two exceptions have been found. These strains, one of which was isolated from a healthy animal, lack *apx* genes compared to their respective reference strains. Several field strains which initially showed an aberrant *apx* gene profile were found to be improperly typed. The *apx* gene profiles, as shown in figure 1, are therefore specific for a given serotype, and can serve for DNA probe or PCR based toxin typing complementary to serotyping in *A. pleuropneumoniae* diagnostics and epidemiology.

Biovar 2 strains do not express the same Apx toxins as their corresponding biovar 1 strains. Analysis of biovar 2 strains of serotypes 2, 4, 7 and 9 by DNA:DNA hybridization and immunoblot analysis show that none of them expresses ApxIII and no genes of the *apxIII* operon can be detected. They, however, express the same ApxI and ApxII toxins and contain the same *apxI* and *apxII* genes as their biovar 1 analogs (figure 1) (J. Frey and J. F. van den Bosch unpublished results). A serotype 2 biovar 2 strain which has been shown to produce ApxII, but not ApxIII kills pulmonary alveolar macrophages less rapidly and multiplies slower in presence of the macrophages than serotype 2 biovar 1 strains, indicating a role for ApxIII killing pulmonary alveolar macrophages (Dom *et al.*, 1994b). These results may explain why certain biovar 2 strains, in particular serotype 2 biovar 2 strains (Dom and Haesebrouck, 1992), are less virulent than their biovar 1 analogues.

Genes for ApxI, ApxII and ApxIII have also been found in several other *Actinobacillus* species including *Actinobacillus suis*, *Actinobacillus lignieresii*, *Actinobacillus equuli* and *Actinobacillus rossi* (Burrows and Lo, 1992) (M. Beck, thesis, University of Berne, 1994). The hemolysin genes *ashCA* from *A. suis* have been sequenced and shown to be virtually identical to *apxIICA* from *A. pleuropneumoniae* (Burrows and Lo, 1992). It is interesting to note that although *A. lignieresii* strain NCTC4189 contains the complete *apxICABD* operon it is not expressed (J. Frey unpublished results).

IMPACT OF *A. PLEUROPNEUMONIAE* RTX-TOXINS ON VIRULENCE

In *A. pleuropneumoniae* outbreaks as well as in experimental infections of swine and mice, several studies show that certain serotypes of *A. pleuropneumoniae* are particularly virulent, while others are much less virulent (Rosendal *et al.*, 1985; Vaillancourt *et al.*, 1990; Martineau *et al.*, 1984; Komal and Mittal, 1990; Rosendal and Mitchell, 1983; Nicolet, 1970). Comparison of these data with the types of Apx toxins which are produced by the different serotypes reveal that those serotypes which produce ApxI, and to a lesser extent those serotypes which produce two different Apx toxins, are particularly virulent (Kamp *et al.*, 1991; Frey and Nicolet, 1990; Frey and Nicolet, 1988b; Frey *et al.*, 1992; Frey, 1994). These studies, although founded on multiple observations, have the disadvantage that they are based on comparisons between different serotypes and strains which imply that other factors than the Apx toxins could be the cause of differences in virulence. Mutants in *apx* genes and genetic complementation with cloned toxin genes would give a more straightforward answer to what extent the virulence of *A. pleuropneumoniae* is dependent on Apx toxins.

A hemolysin-deficient serotype 2 mutant which still produces ApxIII, but no ApxII was shown to cause the same lung lesions in infected pigs as the wild type (w.t.) strain. The clinical signs in pigs infected with the mutant seemed to be less severe compared to the w.t. strain (Rycroft *et al.*, 1991). These data indicate that ApxIII is sufficient for virulence in serotype 2, but that ApxII also is involved to some extent. A hemolysin-deficient serotype 5 mutant, obtained by chemical mutagenesis, which secretes no ApxI and no ApxII was shown to be non-pathogenic in pigs and mice even at much higher doses than the LD_{50} of the parent strain. These experiments indicate that the hemolytic toxins ApxI and ApxII are essential virulence factors of *A. pleuropneumoniae* serotype 5 (Inzana *et al.*, 1991).

Direct evidence of the role of ApxI and ApxII was recently obtained by using the non-hemolytic serotype 5 mutant (Inzana *et al.*, 1991) which was shown to have lost its entire *apxI* operon, but had retained the *apxIICA* genes and hence produced ApxII intracellularly but did not secrete it. Complementation of this serotype 5 mutant with *apxIBD* genes cloned on a broad host range plasmid results in the recombinant strain JF1060 which secretes ApxII. This recombinant strain is able to induce pleuropneumonia in pigs, but higher doses compared to the w.t. strain are required. This strain is avirulent in a mouse infection model. Complementation of the same mutant with *apxICABD* genes (strain JF1064) results in secretion of both ApxI and ApxII and completely restores virulence in pigs and mice. These results show that either ApxI or ApxII is an essential virulence factor of *A. pleuropneumoniae* and that the toxin must be secreted in order to cause disease. The experiments also show that the toxin ApxII alone is sufficient (together with the other non-RTX virulence factors) to produce pleuropneumonia in pigs, but that in combination with ApxI, the degree of virulence of the strain strongly increases thus confirming previous assumptions that ApxI is involved in particularly high virulence (D. Reimer, J. Frey, R. Jansen and T. Inzana, manuscript in preparation). Using isogenic strains to clarify the impact of Apx toxins on virulence, has the advantage that interference from other potential virulence factors can be excluded.

THE ANTIGENIC ROLE OF EXOTOXINS APXI, APXII AND APXIII AND THEIR USE IN VACCINES

The hemolytic *A. pleuropneumoniae* ApxI and ApxII toxins were shown by various methods to be major immunogenic proteins in pigs naturally or experimentally infected with any of the 12 serotypes (Devenish *et al.*, 1989; Devenish *et al.*, 1990a; Devenish *et al.*, 1990b; Frey and Nicolet, 1991). Since ApxI and ApxII could not be separated biochemically and there was only little evidence for the existence of two different toxins (Frey and Nicolet, 1990) it was initially believed that ApxI and ApxII were immunologically cross reacting (Frey and Nicolet, 1991), or that all serotypes secreted the same toxin (Devenish *et al.*, 1989). Strong immunological reactions to the 105 kDa ApxI+ApxII proteins were detected with sera of pigs from herds with a history of porcine pleuropneumonia, but also with sera from animals of herds known to be free of porcine pleuropneumonia (Devenish *et al.*, 1990b; Frey and Nicolet,

1991). These sera also reacted with 105 kDa proteins of *A. suis*, *A. rossii*, which showed immunological cross reactions with polyclonal antiserum against purified ApxI+ApxII. Other bacterial species from the family of *Pasteurellaceae* were therefore postulated to produce cytotoxins closely related to ApxI or ApxII (Frey and Nicolet, 1991; Devenish *et al.*, 1989), results which have now been confirmed by genetic data showing that *A. suis* and *A. rossii* do contain the genes for the ApxII toxin (M. Beck, thesis, University of Berne, 1994).

Natural infection with less virulent *A. pleuropneumoniae* strains was shown to induce cross protective immunity against virulent serotypes (Nielsen, 1979). The knowledge of the exotoxic and strong immunogenetic character of Apx toxins in *A. pleuropneumoniae* infected animals prompted researchers to use these toxins to induce protective immunity against porcine pleuropneumonia or *A. pleuropneumoniae* infection (Fedorka Cray *et al.*, 1993). Vaccination with a preparation of ApxI+ApxII from *A. pleuropneumoniae* serotype 1 reference strain protects against mortality in pigs infected with a serotype 1 field strain, but lung lesions can still be detected in immunized pigs (Devenish *et al.*, 1990a). Using a mouse model, purified ApxI+ApxII from the serotype 1 reference strain gives partial protection against mortality in serotype 1 and 5 infected animals, while ApxI+ApxII in combination with a membrane protein fraction gives nearly full protection. The membrane proteins alone induce only partial protection (Bhatia *et al.*, 1991).

The combination of a crude hemolysin preparation from serotype 1 including ApxI and ApxII with a cell extract gives good protection against infection with *A. pleuropneumoniae* serotype 1 and serotype 5 (Beaudet *et al.*, 1994).

The impact of ApxI+ApxII on induction of protective immunity in pigs was recently shown by the use of the avirulent non-hemolytic mutant of *A. pleuropneumoniae* serotype 5 (described above), which does not secrete ApxI or ApxII. This mutant does not give any protection in a challenge in pigs with the homologous serotype 5 w.t. strain (Inzana *et al.*, 1991). In contrast, an avirulent non-capsulated mutant of the same serotype 5 strain, which was shown to induce antibodies against ApxI and ApxII, efficiently protected pigs against clinical disease following challenge with the virulent homologous serotype 5 strain (Inzana *et al.*, 1993). These experiments show that ApxI and ApxII or at least one of the two Apx toxins, are essential (but supposedly not sufficient) to induce protective immunity against porcine pleuropneumonia caused by serotype 5.

Serotype specific protection was obtained using recombinant ApxII and recombinant 60 kDa outer membrane protein (transferrin binding protein), cloned from *A. pleuropneumoniae* serotype 7. Using this vaccine, protection against *A. pleuropneumoniae* serotype 7, but not against serotype 1 challenge was obtained (Rossi Campos *et al.*, 1992). The lack of protection against serotype 1 can be explained at least in part by the lack of the antigen ApxI in this vaccine preparation.

Taking advantage of the knowledge of the distribution of Apx toxins in the different *A. pleuropneumoniae* serotypes, a biochemically defined subunit vaccine which contains ApxI, ApxII, ApxIII and an outer membrane protein of 42 kDa, which is common to all serotypes, gives most efficient protection against all *A. pleuropneumoniae* serotypes (van den Bosch *et al.*, 1992; Kobisch and van den Bosch, 1992). Using the toxins or the outer membrane protein as immunogens alone it was shown that some protection against mortality after *A. pleuropneumoniae* serotype 1 or serotype 5 infection was obtained, but protection against the development of typical lung pathology was obtained only by the use of both Apx toxins and the outer membrane protein together (van den Bosch *et al.*, 1990) (J.F. van den Bosch, personal communication).

OTHER POTENTIAL *A. PLEUROPNEUMONIAE* EXOTOXINS

Besides the RTX-toxins, several different exotoxins have been claimed in the early literature on *A. pleuropneumoniae*, but few have been confirmed.

The CAMP phenomenon (first described by Christie *et al.*, 1994), a synergistic hemolysis of sheep erythrocytes by the CAMP factor in the diffusion zone of the sphingomyelinase of *Staphylococcus aureus*, is characteristic for *A. pleuropneumoniae* (Kilian, 1976). Recent evidence obtained by recombinant DNA technology shows that the CAMP reaction of *A. pleuropneumoniae* can be caused by all three Apx toxins, since ApxI, ApxII and also the non hemolytic ApxIII expressed in *E.coli* produce a CAMP effect (J. Frey, unpublished results). Moreover, the non-hemolytic serotype 5 mutant, which secretes no ApxI, shows no CAMP reaction, while the w.t. strain is CAMP positive and the mutant, complemented with cloned genes that restore ApxII secretion or with cloned genes restoring ApxI + ApxII production and secretion, is also CAMP positive (J. Frey, unpublished results). These data confirm earlier observations made by neutralization tests implying that the CAMP phenomenon could be related to an Apx toxin (Devenish *et al.*, 1992).

An apparent hemolysin gene named *hlyX* or *cfp* was cloned from *A. pleuropneumoniae* serotype 1 (Lian *et al.*, 1989; Frey *et al.*, 1989). This gene expresses a 29.5 kDa protein in *E. coli* which confers a hemolytic effect and a CAMP phenomenon. Sequence analysis of the *hlyX* (*cfp*) gene revealed that it has strong homology to the *E. coli fnr* gene (fumarate-nitrate-reductatse regulation) (MacInnes *et al.*, 1990) and can complement *E.coli fnr* mutants (Soltes and MacInnes, 1994). These data imply that *hlyX* or (*cpf*) is not directly a hemolysin or cohemolysin gene, but a regulatory gene which also can induce hemolytic and co-hemolytic activity (Soltes and MacInnes, 1994). It is not known whether HlyX or (Cfp) plays a role in regulation of hemolytic activities in *A. pleuropneumoniae*.

A heat stable hemolytic cytocidal and anti-phagocytic substance of carbohydrate nature was identified in *A. pleuropneumoniae* serotype 2 culture supernatant under defined culture conditions (Kume *et al.*, 1986), but has not been analyzed any further.

Secreted proteases from *A. pleuropneumoniae* serotype 1 with different molecular masses have been reported which can degrade porcine IgA and porcine, human and bovine hemoglobin (Negreteabascal *et al.*, 1994). IgA protease activity is also produced by other pathogenic *Haemophilus* (*Actinobacillus*) species and may also be involved in the virulence of *A. pleuropneumoniae* (Kilian *et al.*, 1979).

CONCLUDING REMARKS

The RTX-toxins of *A. pleuropneumoniae* ApxI, ApxII and ApxIII are major virulence factors. Genetic studies suggest that at least one of the Apx toxins is required for *A. pleuropneumoniae* to be virulent. They are so far the best analyzed exotoxins of this species. In general, the presence of ApxI is associated with a high level of virulence, while the presence of ApxII alone seems to confer moderate virulence. Strains secreting two different toxins are significantly more virulent than strains producing only one, indicating a possible synergistic effect between the toxins. It should be noted that *A. pleuropneumoniae* is the only bacterial species known so far which can produce two different RTX toxins. The presence of *apxI*, *apxII* and *apxIII* genes is specific to a given serotype. Typing of *apx* genes in *A. pleuropneumoniae* or analysis of expression of the different Apx toxins by means of specific antibodies can therefore be used as valuable tools to complement serotyping of *A. pleuropneumoniae* in diagnostics and epidemiology. In addition, the knowledge of distribution of the different Apx toxins in the different serotypes and the potential of these antigens to induce, together with other constituents, protective immunity against porcine pleuropneumonia is important in the design of effective vaccines.

ACKNOWLEDGMENTS

I am most grateful to J. Nicolet, Institute for Veterinary Bacteriology University of Berne, for initiating the project on *A. pleuropneumoniae* virulence factors and for his continuous interest and support. I also thank C. Hughes, Department of Pathology, Cambridge University Cambridge U.K. and R. Welch, Department of Medical Microbiology, University of Wisconsin, Madison, WI, U.S.A. for helpful advice on RTX-toxins, J. MacInnes, Department of Veterinary Microbiology, University of Guelph, Canada, J.F. van den Bosch and R.P.A.M. Segers, Intervet International B.V., Boxmeer, The Netherlands, for help and critical reading of this manuscript. This project was supported by the Swiss National Science Foundation grant # 3100.39123.93.

REFERENCES

Anderson, C., Potter, A.A. and Gerlach, G.F., 1991, Isolation and molecular characterization of spontaneously occurring cytolysin-negative mutants of *Actinobacillus pleuropneumoniae* serotype 7, *Infect.Immun.* 59:4110-4116.

Beaudet, R., Mcsween, G., Boulay, G., Rousseau, P., Bisaillon, J.G., Descoteaux, J.P. and Ruppanner, R., 1994, Protection of mice and swine against infection with *Actinobacillus pleuropneumoniae* by vaccination, *Vet.Microbiol.* 39:71-81.

Beck, M., 1994, Thesis, University of Berne.

Beynon, L.M., Richards, J.C. and Perry, M.B., 1993, Characterization of the *Actinobacillus pleuropneumoniae* serotype K11/01 capsular antigen, *Eur.J.Biochem.* 214:209-214.

Bhatia, B., Mittal, K.R. and Frey, J., 1991, Factors involved in immunity against *Actinobacillus pleuropneumoniae* in mice, *Vet.Microbiol.* 29:147-158.

Brandreth, S.R. and Smith, I.M., 1985, Prevalence of pig herds affected by pleuropneumonia associated with *Haemophilus pleuropneumoniae* in eastern England, *Vet.Rec.* 117:143-147.

Burrows, L.L. and Lo, R.Y., 1992, Molecular characterization of an RTX toxin determinant from *Actinobacillus suis*, *Infect.Immun.* 60:2166-2173.

Chang, Y.F., Young, R. and Struck, D.K., 1989, Cloning and characterization of a hemolysin gene from *Actinobacillus (Haemophilus) pleuropneumoniae*, *DNA* 8:635-647.

Chang, Y.F., Shi, J.R., Ma, D.P., Shin, S.J. and Lein, D.H., 1993b, Molecular Analysis of the *Actinobacillus pleuropneumoniae* RTX Toxin-III Gene Cluster, *DNA Cell Biol.* 12:351-362.

Christie, R., Atkins, N.E., and Munch-Peterson, E., 1994, A note on a lytic phenomenon shown by group B *Streptococci*. *Aust. J. Exp. Biol. Med. Sci.* 22:197-200.

Deneer, H.G. and Potter, A.A., 1989, Identification of a maltose-inducible major outer membrane protein in *Actinobacillus (Haemophilus) pleuropneumoniae*, *Microb. Patholog.* 6:425-432.

Devenish, J., Rosendal, S., Johnson, R. and Hubler, S., 1989, Immunoserological comparison of 104-kilodalton proteins associated with hemolysis and cytolysis in *Actinobacillus pleuropneumoniae*, *Actinobacillus suis*, *Pasteurella haemolytica*, and *Escherichia coli*, *Infect.Immun.* 57:3210-3213.

Devenish, J., Rosendal, S. and Bosse, J.T., 1990a, Humoral antibody response and protective immunity in swine following immunization with the 104-kilodalton hemolysin of *Actinobacillus pleuropneumoniae*, *Infect.Immun.* 58:3829-3832.

Devenish, J., Rosendal, S., Bosse, J.T., Wilkie, B.N. and Johnson, R., 1990b, Prevalence of seroreactors to the 104-kilodalton hemolysin of *Actinobacillus pleuropneumoniae* in swine herds, *J.Clin.Microbiol.* 28:789-791.

Devenish, J., Brown, J.E. and Rosendal, S., 1992, Association of the RTX proteins of *Actinobacillus pleuropneumoniae* with hemolytic, CAMP, and neutrophil-cytotoxic activities, *Infect.Immun.* 60:2139-2142.

Devenish, J. and Rosendal, S., 1991a, Immunological characterization of breakdown peptides of the 104 kilodalton hemolysin of *Actinobacillus pleuropneumoniae* serotype 1, *Vet.Microbiol.* 29:85-93.

Devenish, J. and Rosendal, S., 1991b, Calcium binds to and is required for biological activity of the 104-kilodalton hemolysin produced by *Actinobacillus pleuropneumoniae* serotype 1, *Can.J.Microbiol.* 37:317-321.

Dom, P. and Haesebrouck, F., 1992, Comparative virulence of NAD-dependent and NAD-independent *Actinobacillus pleuropneumoniae* strains, *Zbl.Veterinarmed.* 39:303-306.

Dom, P., Haesebrouck, F., Ducatelle, R. and Charlier, G., 1994a, In vivo association of *Actinobacillus pleuropneumoniae* serotype 2 with the respiratory epithelium of pigs, *Infect.Immun.* 62:1262-1267.

Dom, P., Haesebrouck, F., Kamp, E.M. and Smits, M.A., 1994b, NAD-independent *Actinobacillus pleuropneumoniae* strains: production of RTX toxins and interactions with porcine phagocytes, *Vet.Microbiol.* 39:205-218.

Fales, W.H., Morehouse, L.G., Mittal, K.R., Bean Knudsen, C., Nelson, S.L., Kintner, L.D., Turk, J.R., Turk, M.A., Brown, T.P. and Shaw, D.P., 1989, Antimicrobial susceptibility and serotypes of *Actinobacillus (Haemophilus) pleuropneumoniae* recovered from Missouri swine, *J.Vet.Diagn.Invest.* 1:16-19.

Fedorka Cray, P.J., Stine, D.L., Greenwald, J.M., Gray, J.T., Huether, M.J. and Anderson, G.A., 1993, The importance of secreted virulence factors in *Actinobacillus pleuropneumoniae* bacterin preparation: A comparison, *Vet.Microbiol.* 37:85-100.

Felmlee, T., Pellett, S. and Welch, R.A., 1985, Nucleotide sequence of an *Escherichia coli* chromosomal hemolysin, *J.Bacteriol.* 163:94-105.

Fenwick, B.W. and Osburn, B.I., 1986, Immune responses to the lipopolysaccharides and capsular polysaccharides of *Haemophilus pleuropneumoniae* in convalescent and immunized pigs, *Infect.Immun.* 54:575-582.

Frey, J. and Nicolet, J., 1988a, Purification and partial characterization of a hemolysin produced by *Actinobacillus pleuropneumoniae* type strain 4074, *FEMS Microbiol.Lett.* 55:41-46.

Frey, J. and Nicolet, J., 1988b, Regulation of hemolysin expression in *Actinobacillus pleuropneumoniae* serotype 1 by Ca^{2+}, *Infect.Immun.* 56:2570-2575.

Frey, J., Perrin, J. and Nicolet, J., 1989, Cloning and expression of a cohemolysin, the CAMP factor of *Actinobacillus pleuropneumoniae*, *Infect.Immun.* 57:2050-2056.

Frey, J. and Nicolet, J., 1990, Hemolysin patterns of *Actinobacillus pleuropneumoniae*, *J.Clin.Microbiol.* 28:232-236.

Frey, J. and Nicolet, J., 1991, Immunological properties of *Actinobacillus pleuropneumoniae* hemolysin I, *Vet.Microbiol.* 28:61-73.

Frey, J., Deillon, J.B., Gygi, D. and Nicolet, J., 1991a, Identification and partial characterization of the hemolysin (HlyII) of *Actinobacillus pleuropneumoniae* serotype 2, *Vet.Microbiol.* 28:303-312.

Frey, J., Meier, R., Gygi, D. and Nicolet, J., 1991b, Nucleotide sequence of the hemolysin I gene from *Actinobacillus pleuropneumoniae*, *Infect.Immun.* 59:3026-3032.

Frey, J., van den Bosch, H., Segers, R. and Nicolet, J., 1992, Identification of a second hemolysin (HlyII) in *Actinobacillus pleuropneumoniae* serotype 1 and expression of the gene in *Escherichia coli*, *Infect.Immun.* 60:1671-1676.

Frey, J., Beck, M., Stucki, U. and Nicolet, J., 1993a, Analysis of hemolysin operons in *Actinobacillus pleuropneumoniae*, *Gene* 123:51-58.

Frey, J., Bosse, J.T., Chang, Y.F., Cullen, J.M., Fenwick, B., Gerlach, G.F., Gygi, D., Haesebrouck, F., Inzana, T.J., Jansen, R., Kamp, E.M., Macdonald, J., MacInnes, J.I., Mittal, K.R., Nicolet, J., Rycroft, A.N., Segers, R.P.A.M., Smits, M.A., Stenback, E., Struck, D.K., Vandenbosch, J.F., Willson, P.J. and Young, R., 1993b, *Actinobacillus pleuropneumoniae* RTX-toxins: uniform designation of haemolysins, cytolysins, pleurotoxin and their genes, *J.Gen.Microbiol.* 139:1723-1728.

Frey, J., 1994, RTX-toxins in *Actinobacillus pleuropneumoniae* and their potential role in virulence, in: "Molecular Mechanisms of Bacterial Virulence," C.I. Kado and J.H. Crosa, eds., pp.325-340. Kluwer Academic Publishers, Dordrecht, Boston, London.

Frey, J., Beck, M. and Nicolet, J., 1994a, RTX-toxins of *Actinobacillus pleuropneumoniae*, in: "Bacterial Protein Toxins," J. Freer, R. Aitken, J.E. Alouf, B. Boulnois, P. Falmagne, F. Fehrenbach, C. Montecucco, Y. Piemont, R. Rappouli, T. Wadstrom and B. Witholt, eds., pp.322-332. Gustav Fischer, Stuttgart, Jena, New York.

Frey, J., Haldimann, A., Nicolet, J., Boffini, A. and Prentki, P., 1994b, Sequence analysis and transcription of the *apxI* operon (hemolysin I) from *Actinobacillus pleuropneumoniae*, *Gene* 142:97-102.

Gerlach, G.F., Anderson, C., Potter, A.A., Klashinsky, S. and Willson, P.J., 1992, Cloning and expression of a transferrin-binding protein from *Actinobacillus pleuropneumoniae*, *Infect.Immun.* 60:892-898.

Gerlach, G.F., Anderson, C., Klashinsky, S., Rossicampos, A., Potter, A.A. and Willson, P.J., 1993, Molecular Characterization of a protective outer membrane lipoprotein (OmlA) from *Actinobacillus pleuropneumoniae* Serotype-1, *Infect.Immun.* 61:565-572.

Gonzalez, G.C., Caamano, D.L. and Schryvers, A.B., 1990, Identification and characterization of a porcine-specific transferrin receptor in *Actinobacillus pleuropneumoniae*, *Mol.Microbiol.* 4:1173-1179.

Gygi, D., Nicolet, J., Hughes, C. and Frey, J., 1992, Functional analysis of the Ca^{2+}-regulated hemolysin I operon of *Actinobacillus pleuropneumoniae* serotype 1, *Infect.Immun.* 60:3059-3064.

Hughes, C., Stanley, P. and Koronakis, V., 1992, *E.coli* hemolysin interactions with prokaryotic and eukaryotic cell membranes, *Bioessays* 14:519-525.

Idris, U.E.A., Harmon, B.G., Udeze, F.A. and Kadis, S., 1993, Pulmonary lesions in mice inoculated with *Actinobacillus pleuropneumoniae* hemolysin and lipopolysaccharide, *Vet.Pathol.* 30:234-241.

Inzana, T.J., Ma, J., Workman, T., Gogolewski, R.P. and Anderson, P., 1988, Virulence properties and protective efficacy of the capsular polymer of *Haemophilus (Actinobacillus) pleuropneumoniae* serotypes, *Infect.Immun.* 56:1880-1889.

Inzana, T.J., Todd, J., Ma, J.N. and Veit, H., 1991, Characterization of a non-hemolytic mutant of *Actinobacillus pleuropneumoniae* serotype 5: role of the 110 kilodalton hemolysin in virulence and immunoprotection, *Microb. Patholog.* 10:281-296.

Inzana, T.J., Todd, J. and Veit, H.P., 1993, Safety, stability, and efficacy of noncapsulated mutants of *Actinobacillus pleuropneumoniae* for use in live vaccines, *Infect.Immun.* 61:1682-1686.

Jacques, M., Foiry, B., Higgins, R. and Mittal, K.R., 1988, Electron microscopic examination of capsular material from various serotypes of *Actinobacillus pleuropneumoniae*, *J.Bacteriol.* 170:3314-3318.

Jacques, M., Belanger, M., Roy, G. and Foiry, B., 1991, Adherence of *Actinobacillus pleuropneumoniae* to porcine tracheal epithelial cells and frozen lung sections, *Vet.Microbiol.* 27:133-143.

Jansen, R., Briaire, J., Kamp, E.M. and Smits, M.A., 1992a, Comparison of the cytolysin II genetic determinants of *Actinobacillus pleuropneumoniae* serotypes, *Infect.Immun.* 60:630-636.

Jansen, R., Briaire, J., Kamp, E.M. and Smits, M.A., 1992b, The cytolysin genes of *Actinobacillus pleuropneumoniae*, *Proc. Int. Pig. Vet. Soc. Congr.* 12:197.

Jansen, R., Briaire, J., Kamp, E.M., Gielkens, A.L.J. and Smits, M.A., 1993a, Cloning and Characterization of the *Actinobacillus pleuropneumoniae* RTX-toxin III (ApxIII) gene, *Infect.Immun.* 61:947-954.

Jansen, R., Briaire, J., Kamp, E.M., Gielkens, A.L.J. and Smits, M.A., 1993b, Structural Analysis of the *Actinobacillus pleuropneumoniae* RTX-toxin-I (ApxI) operon, *Infect.Immun.* 61:3688-3695.

Jensen, A.E. and Bertram, T.A., 1986, Morphological and biochemical comparison of virulent and avirulent isolates of *Haemophilus pleuropneumoniae* serotype 5, *Infect.Immun.* 51:419-424.

Kamp, E.M. and van Leengoed, L.A., 1989, Serotype-related differences in production and type of heat-labile hemolysin and heat-labile cytotoxin of *Actinobacillus (Haemophilus) pleuropneumoniae*, *J.Clin.Microbiol.* 27:1187-1191.

Kamp, E.M., Popma, J.K., Anakotta, J. and Smits, M.A., 1991, Identification of hemolytic and cytotoxic proteins of *Actinobacillus pleuropneumoniae* by use of monoclonal antibodies, *Infect.Immun* 59:3079-3085.

Kilian, M., 1976, The haemolytic activity of *Haemophilus* species, *Acta Pathol.Microbiol.Scand.(Sect.B)* 84:339-341.

Kilian, M., Mestecky, J. and Schrohenloher, R.E., 1979, Pathogenic species of the genus *Haemophilus* produce immunglobulin A1 protease, *Infect.Immun.* 26:143-149.

Kobisch, M. and van den Bosch, J.F., 1992, Efficacy of an *Actinobacillus pleuropneumoniae* subunit vaccine, *Proc. Int. Pig. Vet. Soc. Congr.* 12:216.

Komal, J.P. and Mittal, K.R., 1990, Grouping of *Actinobacillus pleuropneumoniae* strains of serotypes 1 through 12 on the basis of their virulence in mice, *Vet.Microbiol.* 25:229-240.

Kume, K., Nakai, T. and Sawata, A., 1986, Interaction between heat-stable hemolytic substance from *Haemophilus pleuropneumoniae* and porcine pulmonary macrophages in vitro, *Infect.Immun.* 51:563-570.

Lalonde, G., McDonald, T.V., Gardner, P. and O'Hanley, P.D., 1989, Identification of a hemolysin from *Actinobacillus pleuropneumoniae* and characterization of its channel properties in planar phospholipid bilayers, *J.Biol.Chem.* 264:13559-13564.

Lian, C.J., Rosendal, S. and MacInnes, J.I., 1989, Molecular cloning and characterization of a hemolysin gene from *Actinobacillus (Haemophilus) pleuropneumoniae*, *Infect.Immun.* 57:3377-3382.

Ludwig, A., Jarchau, T., Benz, R., and Goebel, W., 1988, The repeat domain of *Escherichia coli* haemolysin (HlyA) is responsible for its Ca^{2+}-dependent binding to erythrocytes. *Mol. Gen. Genet.* 214:553-561.

Macdonald, J. and Rycroft, A.N., 1992, Molecular cloning and expression of ptxA, the gene encoding the 120-kilodalton cytotoxin of *Actinobacillus pleuropneumoniae* serotype 2, *Infect.Immun.* 60:2726-2732.

Macdonald, J. and Rycroft, A.N., 1993, *Actinobacillus pleuropneumoniae* haemolysin-II is secreted from *Escherichia coli* by *A. pleuropneumoniae* pleurotoxin secretion gene products, *FEMS Microbiol.Lett.* 109:317-322.

MacInnes, J.I., Kim, J.E., Lian, C.J. and Soltes, G.A., 1990, *Actinobacillus pleuropneumoniae hlyX* gene homology with the *fnr* gene of *Escherichia coli*, *J.Bacteriol.* 172:4587-4592.

Martineau, G.P., Desrosiers, R., Charette, R. and Moore, C., 1984, Control measures and economical aspects of swine pleuropneumonia in Quebec, *Proc.Am.Assoc.Swine Pract.* 97-111.

McWhinney, D.R., Chang, Y.F., Young, R. and Struck, D.K., 1992, Separable domains define target cell specificities of an RTX hemolysin from *Actinobacillus pleuropneumoniae*, *J.Bacteriol.* 174:291-297.

Nagai, S., Yagihashi, T. and Ishihama, A., 1993, DNA sequence analysis of an allelic variant of the *Actinobacillus pleuropneumoniae* RTX toxin-I (ApxIA) from serotype-10, *Microb. Patholog.* 15:485-495.

Nakai, T., Sawata, A. and Kume, K., 1984, Pathogenicity of *Haemophilus pleuropneumoniae* for laboratory animals and possible role of its hemolysin for production of pleuropneumonia, *Nippon Juigaku Zasshi* 46:851-858.

Negreteabascal, E., Tenorio, V.R., Serrano, J.J., Garcia, C. and Delagarza, M., 1994, Secreted proteases from *Actinobacillus pleuropneumoniae* serotype 1 degrade porcine gelatin, hemoglobin and immunoglobulin A, *Can.J.Vet.Res.* 58:83-86.

Nicolet, J. and Konig, H., 1966, (On *Haemophilus* pleuropneumonia in swine. Bacteriologic, patho-anatomic and histological findings. Preliminary report), *Pathol.Microbiol.* 29:301-306.

Nicolet, J., 1968, Sur l'hémophilose du porc. Identification d'un agent fréquent: *Haemophilus parahaemolyticus*, *Pathol.Microbiol.* 31:215-225.

Nicolet, J., Konig, H. and School, E., 1969, (On *Haemophilus pleuropneumonia* in swine. II. A contagious disease of scientific value), *Schweiz.Arch.Tierheilkd.* 111:166-174.

Nicolet, J., 1970, "Aspects microbiologiques de la pleuropneumoniae contagieuse du porc," Habilitation, University of Berne, Berne, Switzerland.

Nicolet, J., 1988, Taxonomy and Serological Identification of *Actinobacillus pleuropneumomiae*, *Can.Vet.J.* 29:578-580.

Nicolet, J., 1992, *Actinobacillus pleuropneumoniae*, in: "Diseases of Swine," A.D. Leman, B.E. Straw, W.L. Mengeling, S. D'Allaire and D.J. Taylor, eds., pp.401-408. Iowa State University Press, Ames, Iowa U.S.A.

Nielsen, R., 1979, *Haemophilus parahaemolyticus* serotypes: pathogenicity and cross immunity, *Nord.Vet.Med.* 31:407-413.

Nielsen, R., 1988, Seroepidemiology of *Actinobacillus pleuropneumoniae*, *Can.Vet.J.* 29:580-582.

Nielsen, R., 1990, New diagnostic techniques: a review of the HAP group of bacteria, *Can.J.Vet.Res.* 54 Suppl:S68-S72.

Niven, D.F. and O'Reilly, T., 1990, Significance of V-factor dependency in the taxonomy of *Haemophilus* species and related organisms, *Int.J.Syst.Bact.* 40:1-4.

Perry, M.B., Altman, E., Brisson, J.R., Beynon, L.M. and Richards, J.C., 1990, Structural characteristics of the antigenic capsular polysaccharides and lipopolysaccharides involved in the serological classification of *Actinobacillus (Haemophilus) pleuropneumoniae* strains, *Serodiagn.Immunother.Infect.Disease* 4:299-308.

Rapp, V.J., Ross, R.F. and Erickson, B.Z., 1985, Serotyping of *Haemophilus pleuropneumoniae* by rapid slide agglutination and indirect fluorescent antibody tests in swine, *Amer.J.Vet.Res.* 46:185-192.

Rapp, V.J. and Ross, R.F., 1988, Immunogenicity of outer membrane components of *Haemophilus (Actinobacillus) pleuropneumoniae*, *Can.Vet.J.* 29:585-587.

Rosendal, S., Mitchell, W.R., Weber, M., Wilson, M.R. and Zaman, M.R., 1980, *Hemophilus pleuropneumonia*. Lung lesions induced by sonicated bacteria and sterile culture supernatant, *Proc. Int. Pig. Vet. Soc. Congr.* 5:221.

Rosendal, S. and Mitchell, W.R., 1983, Epidemiology of *Haemophilus pleuropneumoniae* infection in pigs: a survey of Ontario pork producers, 1981, *Can.J.Comp.Med.* 47:1-5.

Rosendal, S., Boyd, D.A. and Gilbride, K.A., 1985, Comparative virulence of porcine *Haemophilus* bacteria, *Can.J.Comp.Med.* 49:68-74.

Rosendal, S., Devenish, J., MacInnes, J.I., Lumsden, J.H., Watson, S. and Xun, H., 1988, Evaluation of heat-sensitive, neutrophil-toxic, and hemolytic activity of *Haemophilus (Actinobacillus) pleuropneumoniae*, *Amer.J.Vet.Res.* 49:1053-1058.

Rossi Campos, A., Anderson, C., Gerlach, G.F., Klashinsky, S., Potter, A.A. and Willson, P.J., 1992, Immunization of pigs against *Actinobacillus pleuropneumoniae* with two recombinant protein preparations, *Vaccine* 10:512-518.

Rycroft, A.N., Williams, D., McCandlish, I.A. and Taylor, D.J., 1991, Experimental reproduction of acute lesions of porcine pleuropneumonia with a haemolysin-deficient mutant of *Actinobacillus pleuropneumoniae*, *Vet.Rec.* 129:441-443.

Shope, R.E., 1964, Porcine contagious pleuropneumonia. Experimental transmission, etiology and pathology, *J.Exp.Med.* 119:357-368.

Smits, M.A., Briaire, J., Jansen, R., Smith, H.E., Kamp, E.M. and Gielkens, A.L., 1991, Cytolysins of *Actinobacillus pleuropneumoniae* serotype 9, *Infect.Immun.* 59:4497-4504.

Soltes, G.A. and MacInnes, J.I., 1994, Regulation of gene expression by the HlyX protein of *Actinobacillus pleuropneumoniae*, *Microbiology-UK.* 140:839-845.

Strathdee, C.A. and Lo, R.Y., 1989, Cloning, nucleotide sequence, and characterization of genes encoding the secretion function of the *Pasteurella haemolytica* leukotoxin determinant, *J.Bacteriol.* 171:916-928.

Thwaits, R.N. and Kadis, S., 1993, Purification of surface-exposed integral outer membrane proteins of *Actinobacillus pleuropneumoniae* and their role in opsonophagocytosis, *Am.J.Vet.Res.* 54:1462-1470.

Tu, A.H.T., Hausler, C., Young, R. and Struck, D.K., 1994, Differential expression of the cytotoxic and hemolytic activities of the ApxIIA toxin from *Actinobacillus pleuropneumoniae*, *Infect.Immun.* 62:2119-2121.

Udeze, F.A., Latimer, K.S. and Kadis, S., 1987, Role of *Haemophilus pleuropneumoniae* lipopolysaccharide endotoxin in the pathogenesis of porcine *Haemophilus* pleuropneumonia, *Am.J.Vet.Res.* 48:768-773.

Utrera, V. and Pijoan, C., 1991, Fimbriae in *A. pleuropneumoniae* strains isolated from pig respiratory tracts, *Vet.Rec.* 128:357-358.

Vaillancourt, J.P., Martineau, G.P., Lariviere, S., Higgins, R. and Mittal, K.R., 1990, Seroprevalence of *Actinobacillus (Haemophilus) pleuropneumoniae* serotype-1 infection in swine herds in Quebec, *J.Am.Vet.Med.Assn.* 196:301-306.

van den Bosch, J.F., Pennings, A.M.M.A., Cuiijpers, M.E.C.M., Pubben, A.N.B., van Vugt, F.G.A. and van der Linden, M.F.I., 1990, Heterologous protection induced by an *A. pleuropneumoniae* subunit vaccine, *Proc. Int. Pig. Vet. Soc. Congr.* 11:11.

van den Bosch, J.F., Jongenelen, I.M.C.A., Pubben, A.N.B., van Vugt, F.G.A. and Segers, R.P.A.M., 1992, Protection induced by a trivalent *A. pleuropneumoniae* subunit vaccine, *Proc. Int. Pig. Vet. Soc. Congr.* 12:194.

van Leengoed, L.A., Kamp, E.M. and Pol, J.M., 1989, Toxicity of *Haemophilus pleuropneumoniae* to porcine lung macrophages, *Vet.Microbiol.* 19:337-349.

Welch, R.A., 1991, Pore-forming cytolysins of Gram-negative bacteria, *Mol.Microbiol.* 5:521-528.

RECEPTOR-MEDIATED IRON ACQUISITION FROM TRANSFERRIN IN THE *PASTEURELLACEAE*

S.D. Kirby, J.A. Ogunnariwo and A.B. Schryvers

Department of Microbiology and Infectious Diseases,
Faculty of Medicine, University of Calgary, Calgary,
Alberta, Canada

INTRODUCTION

Iron is an essential growth factor for microorganisms. Its unique redox qualities makes it an important cofactor in many biological processes, including such fundamental functions as respiratory electron transport, energy metabolism and DNA biosynthesis. As such, a continuous supply of iron is crucial in the establishment and maintenance of a microorganism in its environment. Despite its abundance in nature, free Fe^{3+} is not readily available in aqueous solutions due to the presence of chelating anions and to the formation of insoluble ferric complexes at neutral pH in the presence of oxygen. In the vertebrate host, most iron is intracellular, in the form of ferritin or heme-compounds, making it unavailable to invading pathogens in the extracellular milieu (Otto *et al.*, 1992). Although small amounts of intracellular iron are released in the host in the form of hemoglobin, heme or other complexes as a result of cell lysis, these are readily eliminated by an efficient host hepatocyte scavenging mechanism. In the extracellular compartment, the small amount of iron that is available is complexed by the host glycoproteins, transferrin and lactoferrin.

Transferrin and lactoferrin are bilobed, monomeric glycoproteins of approximately 80 kDa, which bind Fe^{3+} with high affinity (Ka $\simeq 10^{-24}$ M) at a 2:1 molar stoichiometry (Baker *et al.*, 1987). One ferric ion, as well as one anionic cofactor, are bound per lobe. The N-terminal and C-terminal lobes of transferrins, which are similar in both amino acid sequence and three dimensional structure, are connected by an interdomain bridge that varies in length with different transferrins and lactoferrins. It has been hypothesized that the bilobed structure of transferrin and lactoferrin has resulted from an evolutionary event in which an ancestral transferrin gene was duplicated and fused (Funk and MacGillivray, 1989). In addition to the similarities between the lobes of transferrins, there is considerable amino acid homology among transferrins and lactoferrins from different species, reflecting conservation in the structure of these iron-binding glycoproteins (Baker and Lindley, 1992). In contrast, there is considerable variation in the number, structure and location of the N-linked oligosaccharide side chains of transferrins and lactoferrins from different species (Brock, 1985).

Lactoferrin is found mainly on mucosal surfaces, in mucosal secretions and at sites of infection where it is released from intracellular stores in neutrophils. While transferrin is found mainly in serum, it is also present in significant amounts (3% of total protein) in bronchoalveolar lavage fluid (Krek-Staples *et al.*, 1992), implicating it as a potential source of iron in the lung to the many species of the *Pasteurellaceae* which are respiratory pathogens. The iron sequestering effects of transferrin and lactoferrin lowers the concentration of free

Haemophilus, Actinobacillus, and Pasteurella
Edited by W. Donachie *et al.*, Plenum Press, New York, 1995

iron to levels which are unable to support microbial growth. This establishes a form of innate host immunity which acts as a barrier to invasive microorganisms at potential sites of infection. Indeed, a number of *in vivo* experiments have demonstrated that artificial supplementation of iron during host colonization can exacerbate the course of disease, as well as lower the effective titre of bacteria required to initiate infection; an effect that is believed to be due to transferrin and lactoferrin saturation leading to systemic hyperferremia (Weinberg, 1978). Although transferrin lowers the concentration of free serum iron, it is in sufficiently high concentrations and levels of saturation in most extracellular compartments (Bezkorovainy, 1987) to provide an adequate supply of iron for microbial growth. Since transferrin is the only source of extracellular iron present in significant quantities in the vascular fluids of humans and other vertebrate hosts, successful bacterial pathogens have evolved high-affinity iron uptake mechanisms capable of obtaining iron from this protein.

By far, the best studied iron acquisition mechanism is siderophore-mediated iron uptake, whereby low molecular weight molecules with extremely high affinity for iron compete with insoluble and complexed iron in the environment (Neilands, 1981). These molecules, which are expressed under conditions of iron deprivation, solubilize iron, making it available for cell surface iron-regulated outer membrane protein receptors specific for the iron-siderophore complex. This complex is internalized and transported to the cytosol where the iron can be removed by either degrading the ligand or reducing the ferric ion associated with the protein and recycling the ligand (Crosa, 1989).

RECEPTOR-MEDIATED IRON ACQUISITION

A number of bacteria have been identified which are apparently incapable of producing siderophores, yet are able to use host transferrin as a source of iron for growth. Many of these species belong to the *Pasteurellaceae* and are pathogens of significant veterinary, medical and economic importance (Table 1). These species have been shown to possess a bacterial mechanism of iron acquisition involving direct binding of host transferrin to receptor proteins expressed on the bacterial surface. In most species two receptor proteins have been identified, designated transferrin binding proteins (Tbps) 1 and 2.

Table 1. Prevalence of receptor-mediated iron acquisition from transferrin in the *Pasteurellaceae*.

Host	Bacterial Species	Disease	Reference
Human	*Haemophilus influenzae* (type b)	Meningitis	(Schryvers, 1988; Morton and Williams, 1989)
	Haemophilus influenzae (nontypable)	Otitis media, Pneumonia	(Schryvers, 1988; Morton and Williams, 1989)
Bovine	*Pasteurella haemolytica*	'Shipping fever'	(Ogunnariwo and Schryvers, 1990)
	Haemophilus somnus	TEME	(Ogunnariwo et al., 1990)
	Pasteurella multocida	Haemorrhagic septicemia	(Ogunnariwo et al., 1991)
Ovine	*Haemophilus agnii*	Septicemia	(Yu and Schryvers, 1994)
Equine	*Actinobacilus equiili*	Septicemia	(Yu et al., 1992)
Porcine	*Actinobacillus pleuropneumoniae*	Pleuropneumonia	(Gonalez et al., 1990; Morton and Williams 1989)
	Haemophilus suis	Glasser's disease	(Morton and Williams, 1989)
Avian	*Haemophilus paragallinarum*	Infectious coryza	(Ogunnariwo and Schryvers, 1992)
	Haemophilus avium	Sinusitis	(Ogunnariwo and Schryvers, 1992)

Specificity

Unlike mammalian transferrin receptors which exhibit the ability to bind a broad spectrum of vertebrate transferrin proteins, bacterial transferrin receptors demonstrate a marked host transferrin specificity (Schryvers and Gonzalez, 1990). In some cases the interaction is quite specific, in that only transferrins from closely related host species are recognized by the bacterial receptor proteins. In human pathogens, such as *H. influenzae,* receptors have been identified which were capable of binding transferrins from species within the hominoid lineage (human, chimpanzee, gorilla and orangutan), but not to transferrins from more distantly related primates (Macaque, Rhesus or Spider monkeys) (Gray-Owen and Schryvers, 1993). Similarly, receptors in the bovine pathogen, *H. somnus,* show strict specificity for bovine transferrin and are incapable of binding to transferrin from related ruminant species (sheep and goats) (Yu *et al.,* 1992). On the contrary, receptors from the ruminant pathogen *P. haemolytica* exhibit a less specific interaction, binding bovine, caprine and ovine transferrins (Yu *et al.*, 1992; Yu and Schryvers, 1994). In virtually all instances, the binding specificity encompasses a range of host species that are phylogenetically related, which is likely a reflection of the sequence and structural similarities of transferrins amongst these species. However, it has been reported that type b *H. influenzae* is capable of acquiring iron from bovine, rabbit and human transferrin, but not from mouse, dog, horse, rat, guinea-pig or ovotransferrin (Morton and Williams, 1989; Morton and Williams, 1990). In the absence of any strict correlation between utilization, binding assays and affinity isolation of Tbps, it is not clear whether these results reflect interactions with the transferrin receptor proteins or involve other bacterial components. The availability of isogenic receptor mutants in this species (Gray-Owen *et al.*, 1994) should faciltate resolution of any outstanding questions pertaining to specificity of the receptor-mediated pathway.

Composition

The presence of transferrin receptors was first demonstrated by binding of HRP-conjugated transferrin by intact cells grown under iron-deficient conditions or by crude membranes prepared from iron-deficient cells (Schryvers and Morris, 1988b). The loss of binding after protease treatment of intact cells indicated that the receptors were accessible at the surface and were protein in nature. Exposure of electroblotted proteins to HRP-conjugated transferrin tentatively identified a single receptor protein (Tbp2), but true identification of the receptor proteins awaited the development of affinity isolation techniques (Schryvers and Morris, 1988a). The original method involved prebinding of biotinylated transferrin to membranes, isolating the receptor-ligand complex by immobilization with streptavidin-agarose and eluting the receptor proteins in SDS-PAGE sample buffer. This procedure has not been without its drawbacks, which has prompted development of several modifications to the original procedure (Schryvers and Lee, 1993).

The affinity isolation technique has allowed the identification of Tbps 1 and 2 from *H. influenzae* (Schryvers, 1989), *H. somnus* (Yu and Schryvers, 1994) , *P. haemolytica* (Yu and Schryvers, 1994), *P. multocida* (Yu and Schryvers, 1994), *A. pleuropneumoniae* (Gonzalez *et al.*, 1990) and *H. paragallinarum* (Ogunnariwo and Schryvers, 1992). Only one receptor protein has been isolated from *H. agnii* (Yu and Schryvers, 1994). In previous reports using biotinylated bovine transferrin (bTf), only Tbp1 could be isolated from *P. haemolytica* (Ogunnariwo and Schryvers, 1990) and only Tbp2 was isolated from *P. multocida* (Ogunnariwo *et al.*, 1991). Thus the failure to isolate both receptor proteins from *H. agnii* may be a limitation of the affinity isolation procedures rather than an indication of the absence of one of the receptor proteins. This interpretation is supported by the observation that only the method utilizing biotinylated bTf, and not bTf directly immobilized to the resin, was capable of isolating the receptor protein from *H. agnii.*

Recently, isogenic mutants in the genes encoding the transferrin receptor proteins have been prepared in *H. influenzae* (Gray-Owen *et al.*, 1994), *Neisseria meningitidis* (Irwin *et al.*, 1993) and *Neisseria gonorrhoeae* (Cornelissen *et al.*, 1992; Anderson *et al.*, 1994), allowing assessment of their role in iron acquisition. These studies have revealed that both proteins play an important role in efficient binding and utilization of transferrin. Tbp1-deficient mutants in *H. influenzae* (Gray-Owen *et al.*, 1994), *N. gonorrhoeae* (Cornelissen *et al.*, 1992) and *N. meningitidis* (Irwin *et al.*, 1993) are all able to weakly bind human transferrin (hTf) in solid phase binding assays, but they are incapable of growing on media supplemented with transferrin as a sole iron source. These results clearly demonstrate that Tbp1 is absolutely required for acquisition of iron from transferrin. The precise role of Tbp2 is less clear. In *N. gonorrhoeae*, Tbp2-deficient strains are capable of binding transferrin at wild type levels, whereas *N. meningitidis* and *H. influenzae* strains have a much lower binding capacity. As well, *N. gonorrhoeae* and *H. influenzae* Tbp2-deficient mutants are capable of growing on media supplemented with transferrin as a sole iron source, although the growth of *H. influenzae* is markedly reduced. In contrast, *N. meningitidis* Tbp2-deficient mutants are incapable of growth on media supplemented solely with transferrin as an iron source. As would be expected, Tbp1/Tbp2-deficient double mutants were incapable of binding or utilization of transferrin in all species. It is clear that further studies will be necessary to determine whether the apparent differences in the role of Tbp2 in these species is a function of the intrinsic properties of the receptor proteins themselves, the nature of the cell surface or due to differences in experimental approaches.

Structure/Pathway Model

Recent advances in cloning of the genes encoding the Tbps and other pathway components has facilitated the development of a schematic model of the iron acquisition pathway (Figure 1). The predicted protein sequence of Tbp2 contains a region homologous to the consensus lipoprotein signal peptidase sequence, suggesting that this protein may be lipidated and anchored to the outer membrane by N-terminal fatty acid residues. The lipidation of Tbp2 has been substantiated by experiments with *A. pleuropneumoniae* (Gerlach *et al.*, 1992b). By virtue of its homology with TonB-dependent outer membrane proteins in *E. coli*, Tbp1 is predicted to be an integral transmembrane protein which may act as a gated pore in the outer membrane (Rutz *et al.*, 1992). Similar to other TonB-dependent outer membrane proteins in *E. coli*, it is hypothesized that Tbp1 functions through the use of a TonB-dependant energy coupling system (Klebba *et al.*, 1993; Postle, 1993). Naturally, this implies that these bacteria must also utilize TonB homologues, similar to that of *E. coli*, as a method of ATP-driven energy transduction between the inner and outer membranes. Recently, a TonB homologue has been identified and cloned in *H. influenzae*, as well as the coding regions for two other open reading frames which share sequence similarity to the *E. coli* ExbB and ExbD, proteins believed to be important in the stabilization and function of TonB (Jarosik *et al.*, 1994). Although the exact nature of iron removal from transferrin has not been characterized, it is likely that the binding of transferrin to Tbp1 and Tbp2, along with energy coupling through TonB, induces conformational changes in both the receptor and ligand. This would result in the release of Fe^{3+} from transferrin and transfer through Tbp1 across the outer membrane.

Since transferrin is not internalized by these pathogens, a mechanism is required for the transport of Fe^{3+} through the periplasm to the transport proteins on the cytoplasmic membrane. A 37 kDa iron-regulated periplasmic protein (and its corresponding gene) has been isolated and characterized in *N. gonorrhoeae* and *N. meningitidis* (Chen *et al.*, 1993). This protein is believed to be the high affinity periplasmic iron transport protein required for receptor-mediated iron acquisition from transferrin This protein, originally designated MIRP (major iron regulated protein) for the large amount expressed under iron limited conditions, has been found to be transiently associated with transferrin bound Fe^{55} and reversibly binds Fe^{3+} with a 1:1 stoichiometry (Chen *et al.*, 1993). This protein has been renamed Fbp (ferric binding protein) and has since been identified in many commensal and pathogenic isolates of the *Neisseriaceae* (Genco *et al.*, 1994). Its presence is anticipated in the equivalent pathways of the *Pasteurellaceae* and evidence for its existence is supported by the identification of 40 kDa and 35 kDa iron repressible periplasmic proteins in *H. influenzae* (Harkness *et al.*, 1992) and *P. haemolytica* (Lainson *et al.*, 1991), respectively. Recently, through a combination of PCR amplification and library screening, our laboratory has cloned the gene for the 40 kDa protein in *H. influenzae*. Sequence analysis has confirmed that one of these clones, designated pJ1, contains an *fbp* homologue. Western blotting with polyclonal antisera to the gonococcal Fbp demonstrates cross-reactivity with the 40 kDa protein expressed in iron starved periplasmic fractions from the original type b *H. influenzae* strain, as well as whole cell lysates of pJ1 expressed in *E. coli* (Kirby, S., unpublished observations). Taken in context, this provides evidence that Fbp is a common intermediate which functions as a periplasmic shuttle in receptor-mediated iron uptake from transferrin.

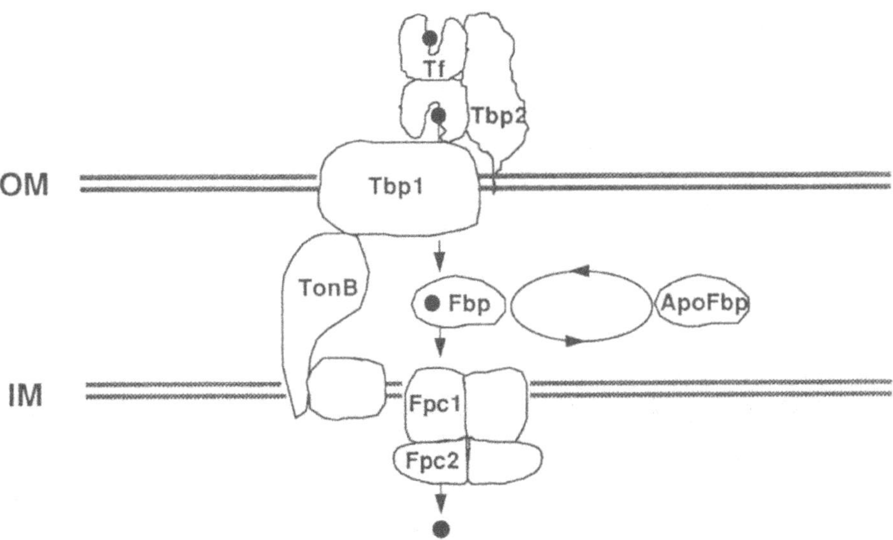

Figure 1. Schematic model of the transferrin receptor-mediated iron acquisition pathway. Interaction of transferrin with the Tbp1/2 complex and energy transduction via TonB leads to a conformational change in the transferrin molecule and the release of ferric cations. These ions are transported through Tbp1 where they become associated with Fbp. Fbp transports one ferric cation across the periplasmic space to an inner membrane permease complex consisting of Fpc1 and Fpc2, which subsequently translocates Fe^{3+} across the inner membrane into the cytoplasm. Apo-Fbp is released making it is available for another round of iron transfer. The transfer of iron across the inner membrane may concomitantly or subsequently occur with the reduction of the ferric ion to the ferrous state, making it available for biological use. Tf - transferrin, Tbp1 - transferrin binding protein 1, Tbp2 - transferrin binding protein 2, Fbp - ferric binding protein, Fpc1 - Fbp permease component 1, Fpc2 - Fbp permease component 2, OM - outer membrane, IM - inner membrane. The black circles represent iron.

Recently, a review of extracellular solute-binding proteins classified the gonococcal Fbp as a member of 'cluster 1' periplasmic binding proteins (Tam and Saier Jr., 1993). This group includes a number of periplasmic proteins with between 330 and 438 amino acid residues which are specific for a diverse range of substrates, including oligosaccharides, glycerol phosphate and iron. This group can be further divided into two size clusters, the larger carbohydrate binding proteins (396-438 residues) and the smaller iron binding proteins (330-338 residues). The iron binding proteins, which to date include the gonococcal Fbp and *Serratia marcescens* SfuA proteins, show a greater degree of homology to each other than to the carbohydrate binding proteins. All proteins in the cluster 1 group with less than 400 residues share a common consensus sequence, TGIKV, which is localized in the N-terminal region of the protein. Our putatively identified *H. influenzae* Fbp also possesses this consensus sequence and shows an intermediate degree of homology to the gonococcal Fbp and *S. marcescens* SfuA proteins. It is anticipated that similar proteins will be identified in other species of the *Pasteurellaceae*.

These periplasmic proteins are components of ATP binding cassette-type (ABC-type) transporter systems which are characterized by the presence of a high affinity extracytoplasmic solute binding protein and a permease complex consisting of: (a) one or two transmembrane proteins spanning the cytoplasmic membrane five or six times each, and (b) one or two cytoplasmic facing inner membrane ATPases (Tam and Saier Jr., 1993). Although permease components have not as yet been identified in either the *Neisseriaceae* or *Pasteurellaceae* transferrin iron uptake pathways, they have been identified in the *S. marcescens* iron uptake system (Angerer *et al.*, 1990). The genes for these proteins are organized in an operon structure with *sfuA* preceding the permease components *sfuB* and *C*. Preliminary investigation of an open reading frame identified immediately downstream of the putative *H. influenzae fbp* gene has revealed an amino acid encoding sequence with a large degree of homology to the *S. marcescens* SfuB (Kirby, S., unpublished observations). Given the similarity in structure and function of Fbp and SfuA, it may be that the permease components (illustrated as Fpc (Fbp permease components) 1 and 2 in Fig. 1) and their corresponding genes for the transferrin iron uptake systems in *H. influenzae, N. gonorrhoeae* and other members of the *Pasteurellaceae* and *Neisseriaceae* are organized in a similar manner to that of the *sfuABC* system in *S. marcescens.*

Regulation

The expression of transferrin receptor proteins and their corresponding binding activity has been shown to be enhanced in a number of *Pasteurellaceae* species under iron limiting conditions of growth (Ogunnariwo and Schryvers, 1992; Ogunnariwo *et al.*, 1991; Gonzalez *et al.*, 1990; Ogunnariwo and Schryvers, 1990; Ogunnariwo *et al.*, 1990), implicating iron as an important regulatory signal for the genes encoding these proteins. Cloning of the genes from several of these pathogens including *H. influenzae* (Gray-Owen *et al.*, 1994), *H. somnus* (Rioux, C., unpublished observations), *P. haemolytica* (Ogunnariwo, J., unpublished observations) and *A. pleuropneumoniae* (Gonzalez *et al.*, 1994) has provided some insights into these potential regulatory systems. The genes in these species are organized in an operon structure with the *tbpB* gene, encoding Tbp2, preceding the *tbpA* gene, encoding Tbp1, with an intergenic region ranging from 13 to 85 base pairs in the different species. There are no evident regulatory elements immediately upstream of the *tbpA* gene. The region upstream of *tbpB* not only contains appropriately located putative promoter and ribosome binding sites, but also contains potential binding sites for the iron-binding repressor protein, Fur (Gray-Owen *et al.*, 1994; Gonzalez *et al.*, 1994). The presence of a portion of a gene encoding a Fur homologue in these species (Karkhoff-Schweizer, R., unpublished observations) suggests that Fur-mediated regulation is present in these species. Fur functions by binding to highly conserved regions upstream of iron regulated genes and blocking transcription (Hennecke, 1990). Under conditions of iron starvation, Fur is incapable of binding these regions due to the scarcity of its cofactor, Fe^{2+}, which allows transcription of iron regulated genes. Although direct evidence for regulation of the *tbp* genes by Fur in these species has not yet been provided, it seems reasonable to propose that this mechanism is present (Figure 2).

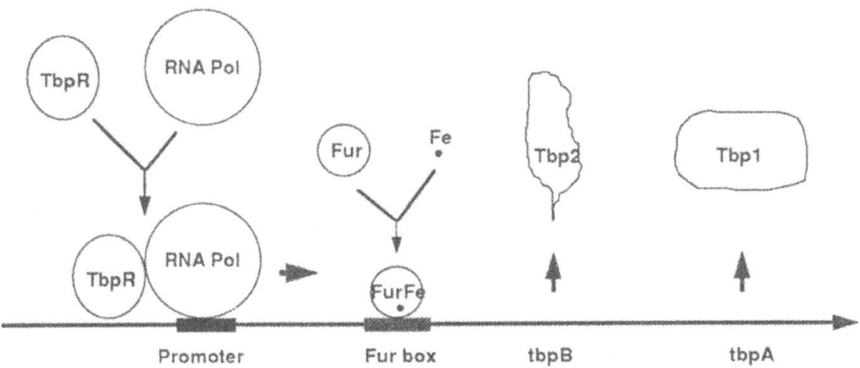

Figure 2. Predicted model for the regulation of transferrin receptor genes. Transcription of the *tbpB* and *tbpA* genes results in the production of Tbp2 and Tbp1. The association of Fe with Fur results in this complex binding to a Fur box, which inhibits transcription by RNA polymerase (RNA Pol). The binding of RNA polymerase to the *tbpB* promoter requires the presence of TbpR, which may possibly bind to a region upstream of the promoter.

Studies demonstrating that excess ferric nitrate, ferric citrate or ferric pyrophosphate in culture media was incapable of repressing expression of transferrin binding proteins in hemin-limited cultures of *H. influenzae* (Morton *et al.*, 1993), may indicate that hemin, and not iron, is involved in the regulation of *tbp* genes in this species. However, it is also possible that in this strain these sources of iron are not effective in raising intracellular iron to levels sufficient for Fur-mediated regulation. It is evident that direct evidence for Fur-mediated and/or hemin-mediated regulation in this species is required.

Recent studies have identified a gene in *H. influenzae* that encodes a protein with homologous to OxyR of *E. coli* (Henneck, 1990), a positive regulator responsive to oxygen concentrations in the medium (Maciver and Hansen, 1994). This protein, designated TbpR, is required for expression of transferrin binding in *H. influenzae*. The environmental signals that modulate TbpR activity and the mechanism by which TbpR acts have not be determined, but by analogy with OxyR, binding to regions upstream of the *tbpB* promoter are anticipated. It also needs to be established whether a homologue of TbpR is present in other species of the *Pasteurellaceae*. However, since the *tbp* genes in these species have likely evolved from a common ancestral gene, the proposed model for regulation of the *tbp* genes in this manner is reasonable (Fig. 2).

Finally, there is both direct and indirect evidence for the expression of the *tbp* genes *in vivo* during both human and veterinary infection (Williams and Griffiths, 1992). It has been demonstrated that type b *H. influenzae* recovered from the intraperitoneal cavities of infected infant rats are capable of binding human transferrin (hTf) and express both Tbps1 and 2 without subculturing (Holland *et al.*, 1992). As indirect evidence, sera collected from infected infant rats and human patients recovering from type b *H. influenzae* meningitis contain antibodies to the Tbp receptor proteins (Holland *et al.*, 1992). Similarly, convalescent sera from animals recovering from natural or experimental infections due to *A. pleuropneumoniae* or *P. haemolytica* react with a number of iron regulated outer membrane proteins, some of which have been identified as Tbps (Gerlach *et al.*, 1992b; Niven *et al.*, 1989; Donachie and Gilmour, 1988; Deneer and Potter, 1989; Sutherland *et al.*, 1990). Interestingly, fresh clinical isolates of strains of type b *H. influenzae* have been reported to show constitutive expression of Tbp receptors which did not revert back to iron regulated expression until after prolonged culturing in iron enriched media (Holland *et al.*, 1992). This was in contrast to *H. influenzae* strain Eagan which retained its iron regulated expression even after animal passage. This raises some interesting questions about the mechanism of these *in vivo* regulatory switches and the differences in expression between different strains of *H. influenzae*.

Vaccine Potential

Due to their essential role *in vivo* and their requisite accessibility at the bacterial surface, the Tbp proteins have been considered potentially ideal vaccine candidates. Studies with native Tbps (Schryvers *et al.*, unpublished observations) and with recombinant Tbp2s from *A. pleuropneumoniae* (Rossi-Campos *et al.*, 1992) have demonstrated their potential efficacy as vaccine antigens. However, there is no direct evidence that Tbp1 by itself is an effective vaccine antigen. Although Tbp2 can induce a protective immune response, variability in the Tbp2 proteins may necessitate the inclusion of a number of representative Tbp2s in order to provide sufficiently broad coverage in this species (Rossi-Campos *et al.*, 1992; Gerlach *et al.*, 1992a). Preliminary immunological and genetic analyses have indicated that there may be limited variability in the Tbp proteins from *P. haemolytica* (Ogunnariwo and Schryvers, 1994), even including isolates from different ruminant hosts. However, studies directly evaluating efficacy as a vaccine antigen will be necessary to determine whether this indicates that broad cross-protection is possible.

The use of intact native or recombinant transferrin receptor proteins as vaccine antigens may take advantage of the surface accessibility of these antigens, but may not fully exploit the functional attributes of these proteins. It is hoped that once we have a greater understanding of the structure of these proteins and their interaction with transferrin, we may be able to design broad spectrum vaccine agents based on conserved functional domains. This is based on the hypothesis that the transferrin-receptor interaction is conserved amongst different strains and species and that the conserved regions could serve as a basis for subunit vaccines. The demonstration of common antigenic domains in electroblotted Tbp2s from different species (Stevenson *et al.*, 1992) is encouraging and additional cross-reactivity may be present in conformational epitopes lost during electroblotting.

Receptor-Ligand Interaction

Experiments in which the N-linked oligosaccharide side chains of transferrin have been removed showed no changes in binding affinity (Padda and Schryvers, 1990). Therefore, the carbohydrate components of the transferrin molecule are not required for binding to the bacterial receptors and the ligand-receptor union is strictly mediated by protein-protein interactions. As a first step towards identifing the regions of transferrin involved in binding to bacterial receptor proteins, binding and affinity isolation experiments were performed with proteolytically-derived subfragments of transferrin. These results indicate that, in all species tested, regions on the C-lobe are involved in binding to Tbp1 (Table 2). In contrast, the regions involved in binding to Tbp2 from different species varied as to which lobe they were located on (Table 2). Tbp2s from *H. influenzae, A. pleuropneumoniae, H. avium* and *P. haemolytica* recognized regions on the C-lobe whereas Tbp2s from *H. somnus, H. agnii* and *P. multocida* recognized regions on the N-lobe. Both the C-lobe and N-lobe of ovotransferrin are involved in binding to Tbp2 from *H. paragallinarum* (Alcantara, 1994).

Table 2. Binding of transferrin subfragments to receptor proteins in *HAP* species.

Bacterial Species	Source*	Tbp1	Tbp2	Reference
Haemophilus influenzae	hTf	C-lobe	C-lobe	(Alcantara *et al.*, 1993)
Pasteurella haemolytica	bTf	C-lobe	C-lobe	(Yu and Schryvers, 1994)
Haemophilus somnus	bTf	C-lobe	N-lobe	(Yu and Schryvers, 1994)
Pasteurella multocida	bTf	C-lobe	N-lobe	(Yu and Schryvers, 1994)
Haemophilus agnii	bTf	NA	N-lobe	(Yu and Schryvers, 1994)
Actinobacillus pleuropneumoniae	pTf	C-lobe	C-lobe	(Gonzalez *et al.*, 1994)
Haemophilus paragallinarum	cTf	C-lobe	C & N-lobe	(Alcantara, 1994)
Haemophilus avium	cTf	C-lobe	C-lobe	(Alcantara, 1994)

* Source of transferrin for preparation of subfragments: hTf - human transferrin, bTf - bovine transferrin, pTf - porcine transferrin, cTf - chicken (ovo) transferrin. NA - data not available as Tbp1 has not been isolated from this species.

Additional insights into the transferrin-receptor interaction could be obtained from growth studies performed with the isolated transferrin subfragments (Alcantara, 1994). The inability of the individual lobes, alone or in combination, to support the growth of *H. paragallinarum* or *N. meningitidis* (which is similar to *H. influenzae* in its binding pattern) suggests that there may be additional interactions involved in iron acquisition from transferrin that are not detected by the binding or affinity isolation methods. One possible interpretation of these results is that regions on both lobes are involved in the transferrin-receptor interaction in different species (Figure 3) and variability which is observed is attributable to differences in the avidity of this interaction (Table 2). Further detailed structural studies will be necessary to determine whether the interaction as depicted in Figure 3 truly represents the receptor-ligand interaction in different species or whether there are major differences. This interpretation is consistent with an anticipated common mechanism of iron acquisition and common evolutionary origin of transferrin mediated iron uptake in this family.

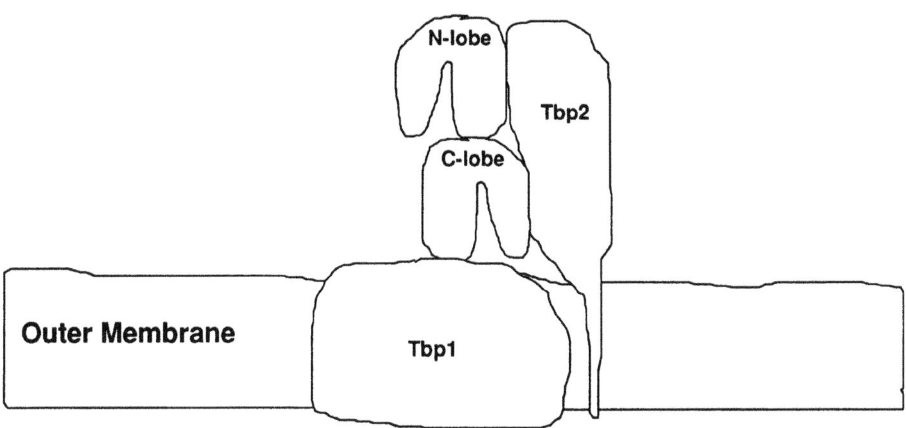

Figure 3. Proposed binding model of the interaction between the N and C lobes of transferrin and the Tbp1 and Tbp 2 transferrin receptor proteins on the bacterial outer membrane surface.

The approach of preparing proteolytic subfragments for defining the regions of interaction has a number of limitations that have hampered attempts at refining the localization of binding domains. In order to obtain more precise information, an approach involving the preparation of hybrid transferrin genes and production and testing of the resultant recombinant chimeric transferrins has been initiated. A limited set of hybrid bovine/human transferrin genes have been prepared by the PCR-based splicing by overlap extension (SOEing) method and these respective recombinant chimeric transferrins have been expressed in a baculovirus expression system (Kabani *et al.*, 1994). The chimeric recombinant transferrins were capable of binding to bacterial receptors in the fashion predicted by the prior studies with proteolytically-derived subfragments. These results demonstrate the validity of this approach and indicates that further studies directed at precisely localizing the determinants responsible for receptor binding are feasible.

CONCLUSIONS

Research in receptor-mediated iron acquisition from transferrin has gone through an interesting phase of rapid progress. This pathway, which is believed to be conserved and shared in many members of the *Pasteurellaceae* and *Neisseriaceae,* is becoming well understood as individual components are gradually being identified. Identification and characterization of both Tbp1 and Tbp2 have been achieved amongst most of the important pathogenic species of the *Pasteurellaceae.* Only in *H. agnii* has the Tbp2 protein not been isolated, likely due to a limitation of the affinity isolation procedure (Yu and Schryvers, 1994). At this point in time *Haemophilus influenzae* appears to be the species best suited for delineating the transferrin receptor-mediated iron acquisiton pathway and determining whether the proposed model (Fig. 1) is valid. The genes encoding the Tbp1 and Tbp2 receptor proteins (Gray-Owen *et al.*, 1994), Fbp and its associated inner membrane transport proteins (Kirby, S., unpublished observations) and TonB with its associated proteins (Jarosik *et al.*, 1994) have all been cloned. The availability of isogenic mutants in many of these loci and the ease of gene replacement in this species indicate that preliminary experiments to substantiate the model are close at hand. A detailed understanding of the iron acquisition process, however, will not be as readily attained and may ultimately require comparative study of the pathway in different species.

The precise nature of the functional receptor in the bacterial membrane and the respective roles of Tbp1 and Tbp2 are not fully understood. There is substantial evidence that both receptor proteins participate in and are required for effective binding to the ligand. Individual preparations of Tbp1 and Tbp2 from *P. haemolytica,* obtained by differential elution from a bovine transferrin affinity column without any apparent cross-contamination, were each capable of independantly binding to bovine transferrin (Ogunnariwo, unpublished observations). Similarly, recombinant Tbp2 (Gerlach *et al.*, 1992b, Gonzalez *et al.*, 1994) or recombinant Tbp1 (Gonzalez *et al.*, 1994) expressed from the cloned *tbpB* and *tbpA* genes of *A. pleuropneumoniae* were capable of binding to porcine transferrin. The analysis of isogenic mutants in *H. influenzae* deficient in either the *tbpA* or *tbpB* gene demonstrated that the presence of either Tbp1 or Tbp2 resulted in transferrin binding, albeit at a much reduced level (Gray-Owen *et al.*, 1994). Although it appears evident that both receptor proteins participate in binding to the ligand, it is not clear whether both proteins are essential for iron acquisition. Studies with an *H. influenzae* mutant deficient in Tbp1 demonstrate that this receptor protein is required for acquisition from human transferrin *in vitro* (Gray-Owen *et al.*, 1994). In contrast, the isogenic mutant deficient in Tbp2 was capable of growth on human transferrin, although to a reduced extent. This raises questions as to whether the requirement for Tbp2 may be a function of the nature of the cell surface and/or the environmental conditions. Clearly analysis of isogenic mutants in other *Pasteurellaceae* species would be of value in addressing this issue, particularly for analysis of growth under relevant *in vivo* conditions.

An area which is less well characterized is the regulation of the operons and genes expressing the components of this iron uptake pathway. Although it has been confirmed that the pathway is iron regulated, the mechanisms of this regulation are currently under investigation. The presence of a putative Fur box upstream of *tbpB* in *H. influenzae* (Gray-Owen *et al.*, 1994) and *A. pleuropneumoniae* (Gonzalez *et al.*, 1994), as well as the demonstration of Fur homologues in several species of the *Pasteurellaceae* (Karkhoff-Schweizer, R., unpublished observations) suggests a similar mechanism of iron regulated expression to that of *E. coli* and the many other species in which Fur has been identified . The recent isolation of an OxyR homologue (Tbp R), in *H. influenzae,* which is required for transferrin binding, suggests that other environmental signals have considerable impact on regulating gene expression (Maciver and Hansen, 1994). However, the mechanisms of regulation may be more complicated, as has been demonstrated in experiments with hemin limitation and *in vivo* isolates which revert to a constitutive transferrin binding phenotype until subcultured in iron enriched media (Morton *et al.*, 1993; Holland *et al.*, 1992). Clearly, more work is required for a complete understanding of these phenomena.

The prospects for developing effective medical and veterinary vaccines based on transferrin receptors look promising for a number of reasons. Their essential role *in vivo* and their requisite accessibility at the bacterial surface indicate that they are potentially ideal targets. Preliminary experiments indicate that they are antigenic and capable of inducing an effective immune response (Schryvers *et al.*, unpublished observations; Rossi-Campos *et al.*, 1992). Since pathogens affecting the same host seem to recognize and bind specifically their host transferrin, it is conceivable that their transferrin receptors may contain conserved, surface exposed regions which could serve as prospective vaccine candidates. The demonstration of considerable cross-reactivity amongst the Tbps of the human pathogens *H. influenzae, N. meningitidis* and *N. gonorrhoeae* (Stevenson *et al.*, 1992) supports this proposal. Similarly, antiTbp1/2 antisera from cattle isolates of *P. haemolytica* serotype A is capable of cross reacting with analogous surface proteins of other *P. haemolytica* isolates in cattle, sheep and goats (Ogunnariwo and Schryvers, 1994). These studies need to be extended to cover other species of the *Pasteurellaceae* and to specifically identify conserved surface epitopes. If in fact the transferrin-Tbp1/2 interaction is conserved, then antigenic domains based on epitopes essential for this function may prove useful for the design of subunit vaccines. It is possible that, in some cases, mixed vaccines will be required as a result of the heterogeneity in the Tbps of certian bacterial species. Further progress in this area will require a more detailed knowledge of the structure of Tbp molecules, of the transferrin/bacterial receptor interaction and of the antigenic domains of the receptor proteins.

REFERENCES

Alcantara, J., Yu, R-H., and Schryvers A.B., 1993, The region of human transferrin involved in binding to bacterial transferrin receptors is localized in the C-lobe, *Mol. Microbiol.* 8:1135-1143.

Alcantara, J., 1994, Identifying the binding domains of transferrin to its bacterial transferrin receptor. Thesis, University of Calgary, Calgary, Alberta, Canada.

Anderson, J.A., Sparling, P.F., and Cornelissen, C.N., 1994, Gonococcal transferrin-binding protein 2 facilitates but is not essential for transferrin utilization, *J. Bact.* 176:3162-3170.

Angerer, A., Gaisser, S., and Braun, V., 1990, Nucleotide sequences of the *sfuA*, *sfuB*, and *sfuC* genes of *Serratia marcescens* suggest a periplasmic binding protein dependant iron transport mechanism, *J. Bact.* 172:572-578.

Baker, E.N., Rumball, S.V., and Anderson, B.F., 1987, Transferrins: insights into structure and function from studies on lactoferrin, *Trends Biochem. Sci.* 12:350- 355.

Baker, E.N., and Lindley, P.F., 1992, New perspectives on the structure and function of transferrins, *J. Inorg. Biochem.* 47:147-160.

Bezkorovainy, A., 1987, Iron proteins, *in*:"Iron and Infection, Molecular, Physiological and Clinical Aspects," Bullen, J.J., and Griffiths, E., eds., John Wiley & Sons, Chichester.

Brock, J.H., 1985, Transferrins, *in*:"Metalloproteins (Part II)," Harrison, P.M., ed., MacMillan, London.

Chen, C-Y., Berish, S.A., Morse, S.A., and Mietzner T.A., 1993, The ferric iron-binding protein of pathogenic *Neisseria* spp. functions as a periplasmic transport in iron-acquisition from human transferrin, *Mol. Microbiol.* 10:311-318.

Cornelissen, C.N., Biswas, G.D., Tsai, J., Paruchuri, D.K., Thompson, S.A., and Sparling, P.F., 1992, Gonococcal transferrin-binding protein 1 is required for transferrin utilization and is homologous to TonB-dependant outer membrane receptors, *J. Bact*. 174:5788-5797.

Crosa, J.H., 1989, Genetics and molecular biology of siderophore-mediated iron transport in bacteria, *Microbiol. Lett.* 53:517-530.

Deneer, H.G., and Potter, A.A., 1989, Iron-repressible outer-membrane proteins of *Pasteurella haemolytica, J. Gen. Microbiol.* 135:435-443.

Donachie, W., and Gilmour, N.J.L., 1988, Sheep antibody response to cell wall antigens expressed in vivo by *Pasteurella haemolytica* serotype A2, *FEMS Microbiol. Lett*. 56:271-276.

Funk, W.D., and MacGillivray, R.T.A., 1989, Molecular biology of transferrins, *in*,"Iron Transport and Storage," Ponka, P., Schulman, H.M., and Woodworth, R.C., eds., CRC Press, Boston.

Genco, C.A., Berish, S.A., Chen, C-Y., Morse, S., and Trees, D.L., 1994, Genetic diversity of the iron-binding protein (Fbp) gene of the pathogenic and commensal *Neisseria, FEMS Microbiol. Lett.* 116:123-130.

Gerlach, G.F., Klashinsky, S., Anderson, C., Potter, A.A., and Willson, P.J., 1992a, Characterization of two genes encoding distinct transferrin-binding proteins in different *Actinobacillus pleuropneumoniae* isolates, *Infect. Immun.* 60:3253-3261.

Gerlach, G.F., Anderson, C., Potter, A.A., Klashinsky, S., and Willson, P.J., 1992b, Cloning and expression of a transferrin-binding protein from *Actinobacillus pleuropneumoniae*, *Infect. Immun.* 60:892-898.

Gonzalez, G.C., Caamano, D.L., and Schryvers, A.B., 1990, Identification and characterization of a porcine-specific transferrin receptor in *Actinobacillus pleuropneumoniae*, *Mol. Microbiol.* 4:1173-1179.

Gonzalez, G.C., Yu, R-H., Rosteck, P., and Schryvers, A.B., 1994, Cloning, expression and analysis of the transferrin binding protein genes from *Actinobacillus pleuropneumoniae,* Manuscript in preparation.

Gray-Owen, S.D., Loosemore, S., and Schryvers, A.B., 1994, Identification and characterization of genes encoding the human transferrin binding proteins from *Haemophilus influenzae*, *Infect. Immun.* Submitted for publication.

Gray-Owen, S.D., and Schryvers, A.B., 1993, The interaction of primate transferrins with receptors on bacteria pathogenic to humans, *Microb. Pathog.* 14:389-398.

Harkness, R.E., Chong, P., and Klein, M.H., 1992, Identification of two iron-repressed periplasmic proteins in *Haemophilus influenzae*, *J. Bact*. 174:2425-2430.

Hennecke, H., 1990, Regulation of bacterial gene expression by metal-protein complexes, *Mol. Micro.* 4:1621-1628.

Holland, J., Langford, P.R., Towner, K.J., and Williams, P., 1992, Evidence for *in vivo* expression of transferrin-binding proteins in *Haemophilus influenzae* type b, *Infect. Immun.* 60:2986-2991.

Irwin, S.W., Averill, N., Cheng, C.Y., and Schryvers, A.B., 1993, Preparation and analysis of isogenic mutants in the transferrin receptor protein genes, *tbp1* and *tbp2*, from *Neisseria meningitidis*, *Mol. Microbiol.* 8:1125-1133.

Jarosik, G.P., Sanders, J.D., Cope, L.D., Muller-Eberhard, U., and Hansen, E.J., 1994, A functional *tonB* gene is required for both utilization of heme and virulence expression by *Haemophilus influenzae* type b, *Infect. Immun.* 62:2470-2477.

Kabani, A.M., Button, L.L., and Schryvers, A.B., 1994, Identification of binding domains for bacterial transferrin receptors using chimeric transferrins, Manuscript in preparation.

Klebba, P.E., Rutz, J.M., Liu, J., and Murphy, C.K., 1993, Mechanisms of TonB-catalyzed iron transport through the enteric bacterial cell envelope, *J. Bioenerg. Biomemb.* 25:603-611.

Krek-Staples, J.A., Kew, R.R., and Webster, R.O., 1992, Ceruloplasmin and transferrin levels are altered in serum and brochoalveolar lavage fluid of patients with adult respiratory distress syndrome, *Am. Rev. Respir. Dis.* 145:1009-1015.

Lainson, F.A., Harkins, D.C., Wilson, C.F., Sutherland, A.D., Murray, J.E., Donachie, W. and Baird, G.D. 1991, Identification and localization of an iron-regulated 35 kDa protein of *Pasteurella haemolytica* serotype A2, *J. Gen. Microbiol.* 137:219-226.

Maciver, I., and Hansen, E.J., 1994, Identification of a positive regulatory factor involved in expression of transferrin-binding activity by *Haemophilus influenzae*, *Mol. Microbiol.* Submitted for publication.

Morton, D.J., Musser, J.M., and Stull, T.L., 1993, Expression of the *Haemophilus influenzae* transferrin receptor is repressible by hemin but not elemental iron alone, *Infect. Immun.* 61:4033-4037.

Morton, D.J., and Williams, P., 1989, Utilization of transferrin-bound iron by *Haemophilus* species of human and porcine origins, *FEMS Microbiol. Lett.* 65:123-128.

Morton, D.J., and Williams, P., 1990, Siderophore-independent acquisition of transferrin-bound iron by *Haemophilus influenzae* type b, *J. Gen. Microbiol.* 136:927-933.

Neilands, J.B., 1981, Microbial iron compounds, *Ann. Rev. Microbiol.* 50:715-731.

Niven, D.F., Donga, J., and Archibald, F.S., 1989, Responses of *Haemophilus pleuropneumoniae* to iron restriction: changes in the outer membrane protein profile and the removal of iron from porcine transferrin, *Mol. Microbiol.* 3:1083-1089.

Ogunnariwo, J.A., Cheng, C.Y., Ford, J.A., and Schryvers, A.B., 1990, Response of *Haemophilus somnus* to iron limitation: Expression and identification of a bovine-specific transferrin receptor, *Microb. Pathog.* 9:397-406.

Ogunnariwo, J.A., Alcantara, J., and Schryvers, A.B., 1991, Evidence for non-siderophore-mediated acquisition of transferrin-bound iron by *Pasteurella multocida, Microbial Pathogenesis.* 11:47-56.

Ogunnariwo, J.A., and Schryvers, A.B., 1990, Iron acquisition in *Pasteurella haemolytica*:: expression and identification of a bovine-specific transferrin receptor, *Infect. Immun.* 58:2091-2097.

Ogunnariwo, J.A., and Schryvers, A.B., 1992, Correlation between the ability of *Haemophilus paragallinarum* to acquire ovotransferrin-bound iron and the expression of ovotransferrin-specific receptors, *Avian Dis.* 36:655-663.

Ogunnariwo, J.A., and Schryvers, A.B., 1994, Antigenic and genetic analysis of transferrin receptors in ruminant isolates of *Pasteurella haemolytica*, Manuscript in preparation.

Otto, B.R., Verweij-Van Vught, A.M.J.J., and Maclaren, D.M., 1992, Transferrins and heme-compounds as iron sources for pathogenic bacteria, *CRC Crit. Rev. Microbiol.* 18:217-233.

Padda, J.S., and Schryvers, A.B., 1990, N-linked oligosaccharides of human transferrin are not required for binding to bacterial transferrin receptors, *Infect. Immun.* 58:2972-2976.

Postle, K., 1993, TonB protein and energy transduction between membranes, *J. Bioenerg. Biomemb.* 25:591-601.

Rossi-Campos, A., Anderson, C., Gerlach, G-F., Klashinsky, S., Potter, A.A., and Willson, P.J., 1992, Immunization of pigs against *Actinobacillus pleuropneumoniae* with two recombinant protein preparations, *Vaccine.* 10:512-518.

Rutz, J.M., Lui, J., Lyons, J.A., Goranson, J., Armstrong, S.K., McIntosh, M.A., Feix, J.B., and Klebba, P.E., 1992, Formation of gated channel by a ligand-specific transport protein in the bacterial outer membrane, *Science.* 258:471-475.

Schryvers, A.B., 1988, Characterization of the human transferrin and lactoferrin receptors in *Haemophilus influenzae*, *Mol. Microbiol.* 2:467-472.

Schryvers, A.B., 1989, Identification of the transferrin and lactoferrin binding proteins in *Haemophilus influenzae*, *J. Med. Microbiol.* 29.:121-130.

Schryvers, A.B., and Gonzalez, G.C., 1990, Receptors for transferrin in pathogenic bacteria are specific for the host's protein, *Can. J. Microbiol.* 36:145-147.

Schryvers, A.B., and Lee, B.C., 1993., Analysis of bacterial receptors for host iron binding proteins, *J. Microbiol. Methods.* 18:255-266.

Schryvers, A.B., and Morris, L.J., 1988a, Identification and characterization of the transferrin receptor from *Neisseria meningitidis*, *Mol. Microbiol.* 2:281-288.

Schryvers, A.B., and Morris, L.J., 1988b, Identification and characterization of the human lactoferrin-binding protein from *Neisseria meningitidis*, *Infect. Immun.* 56:1144- 1149.

Stevenson, P., Williams, P., and Griffiths, E., 1992, Common antigenic domains in transferrin binding protein 2 of *Neisseria meningitidis, Neisseria gonorrhoeae* and *Haemophilus influenzae* type b, *Infect. Immun.* 60:2391-2396.

Sutherland, A.D., Jones, G.E., and Poxton, I.R., 1990, The susceptibility of *in vivo* grown *Pasteurella haemolytica* to ovine defence mechanisms *in vitro*, *FEMS Microbiol. Immunol.* 64:269-278.

Tam, R., and Saier Jr., MH., 1993, Structural, Functional and Evolutionary Relationships among Extracellular Solute-Binding Receptors of Bacteria, *Microbiol. Rev.* 57:320-346.

Weinberg, E.D., 1978, Iron and infection, *Microbiol. Rev.* 42:45-66.

Williams, P., and Griffiths, E., 1992, Bacterial transferrin receptors - structure, function and contribution to virulence, *Med. Microbiol. Immunol.* 181:301-322.

Yu, R-H., Gray-Owen, S.D., Ogunnariwo, J., and Schryvers, A.B., 1992, Interaction of ruminant transferrin receptors in bovine isolates of *Pasteurella haemolytica* and *Haemophilus somnus*, *Infect. Immun.* 60:2992-2994.

Yu, R-H., and Schryvers, A.B., 1994, Transferrin receptors on ruminant pathogens vary in their interaction with the C-lobe and N-lobe of ruminant transferrins, *Can. J. Microbiol.* 40:532-540.

MOLECULAR STUDIES OF ANTIGENS IN HAP ORGANISMS

R.Y.C. Lo

Department of Microbiology
University of Guelph
Guelph, Ontario
Canada N1G 2W1

INTRODUCTION

The use of recombinant DNA technology to study the HAP organisms has greatly improved our ability to characterize the many potential virulence factors of these bacteria. Instead of relying on conventional identification and purification methods based on biochemical, biophysical or immunological techniques, the genes coding for the virulence factors could be isolated for more detailed studies. After isolation and characterization of the genes, the recombinant antigens could be expressed for detailed biochemical and immunological studies. Further, large scale production of the recombinant antigens is also possible by high level expression in a heterologous system. These recombinant antigens could be used in vaccine trial and challenge experiments to investigate their protective capabilities. As a result of these molecular approaches, we now have a better understanding of some of the virulence factors of the HAP organisms. Some of these characterized antigens are being used as potential candidates for the development of an efficacious vaccine against these opportunistic pathogens in animals. In addition to studying the virulence antigens of the bacteria, the accumulation of molecular data is beginning to give us a picture into the biology, genetics and evolution of these bacteria.

APPROACHES TO ISOLATE GENES OF INTEREST

Over the past several years, our laboratory has been working on a molecular approach to isolate genes from *Pasteurella haemolytica* A1 coding for its secreted/soluble antigens. We have successfully isolated and characterized the genes coding for the leukotoxin (Lo *et al.*, 1985; Lo *et al.*, 1987; Strathdee and Lo, 1989), the sialoglycoprotease (Abdullah *et al.*, 1991), a serotype-1 specific outer membrane protein (Gonzalez-Rayos *et al.*, 1986; Lo *et al.*,1991), several lipoproteins (Cooney and Lo, 1993) and the transferrin binding proteins (Woo and Lo, 1993). In addition, we have also initiated a research project to characterize the genes and the enzymes involved in the biosynthesis of the *P. haemolytica* A1 lipopolysaccharide (Potter and Lo, 1993; manuscript in preparation). Using our experience with *P. haemolytica* A1 as the example, I will illustrate some of the approaches which have been used for these studies. For some of these antigens, the cloned genes provide the first insight into the molecular size, biochemical, immunological properties and biological functions of the proteins.

I. Isolation by immunological screening

In this approach, an antiserum specific against the antigens of *P. haemolytica* A1 was used to screen a clone bank in an *Escherichia coli* host for the expression of the *P. haemolytica* A1 antigens. Clones which react positively in the screen could be isolated, and the recombinant plasmids characterized for the antigens they encode.

In our early experiments, we started with a rabbit polyclonal serum raised against all of the soluble antigens of *P. haemolytica* A1 which were present in the culture supernatant (Lo and Cameron, 1986). The culture supernatant contains molecules secreted by the bacterium and surface molecules which were slougthed off during cell division. A plasmid clone bank of *P. haemolytica* A1 DNA in the vector pBR322 was established in *E. coli* and the colonies screened with this antiserum. In these experiments, the colonies were grown up on hydrophobic membrane filters, lysed with choloroform, and the filters incubated with the polyclonal antiserum (Lo and Cameron, 1986). From the positively reacting colonies, we isolated a collection of recombinant plasmids coding for different secreted/surface antigens of *P. haemolytica* A1. From this initial collection, we have identified the genes coding for the leukotoxin (Lo *et al.*, 1985; Lo *et al.*, 1987; Strathdee and Lo, 1989), the sialoglycoprotease (Abdullah *et al.*, 1991), a serotype-1 specific outer membrane protein (Gonzalez-Rayos *et al.*, 1986; Lo *et al.*, 1991) and three lipoproteins (Cooney and Lo, 1993). These genes were identified either by the biological activity of the encoded protein (the leukotoxin and the sialoglycoprotease) or by the immunuological properties of the proteins (the serotype-1 specific outer membrane protein and the lipoproteins). Subsequent studies on these cloned gense will be described in a later section. Of course, there are still other clones isolated in this collection which await characterization.

Recently, we have returned to this screening approach in a more comprehensive study. Instead of using the rabbit antiserum raised against the soluble antigens of *P. haemolytica* A1, we used sera from calves which have been vaccinated with Presponse. Presponse is a commercially available vaccine based on the antigens in *P. haemolytica* A1 culture supernatant and has been shown to have greater than 70% efficacy in field trials (Shewen and Wilkie, 1988; Shewen *et al.*, 1988). By using the calve sera in the screening procedure, we hope to identify antigens which have induced an immune response in the animals after vaccination with Presponse. Further, instead of mass screening of the colonies from the *P. haemolytica* A1 clone bank, we decided to examine each recombinant clone from the library separately. To carry out this procedure, the *P. haemolytica* A1 plasmid clone bank was transformed into *E. coli*, individual clones were plated out and cultured separately. Total protein materials were prepared from individual clones, separated by SDS-PAGE and Western immunoblotted with the calf serum. This is a very labor intensive series of work. Out of over 1650 clones examined, we have identified thirty-five recombinant clones producing *P. haemolytica* A1 antigens recognized by the calve sera. The Western immunoblot for ten representative recombinant clones producing *P. haemolytica* A1 antigens are shown in Figure 1. Each of the recombinant plasmid from the clones will be further characterized for the antigens they encode. It is hoped that most, if not all, of the *P. haemolytica* A1 antigens which are recognized by the Presponse vaccinated animals will be identified in this comprehensive analysis.

In addition to using antiserum against all of the soluble antigens of *P. haemolytica* A1, we have also focused on a particular surface molecule, the lipopolysaccahride (LPS). A purified LPS preparation of *P. haemolytica* A1 was used to immunize rabbits in the production of an antiserum against only the LPS molecule. This anti-LPS serum was used to screen the *E. coli* clones *enmass* as described above. We have mapped the recombinant plasmids from eight independently isolated positive clones and identified an overlapping DNA region which encode a function that alters the LPS phenotype of the *E. coli* host. Detailed characterization of this DNA has been carried out, and the results will be described in a later section.

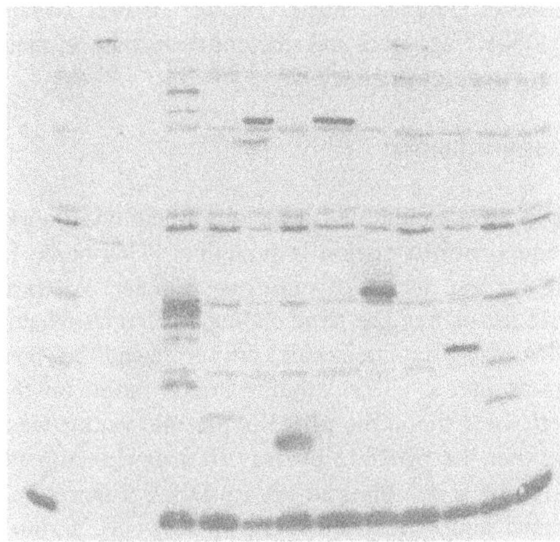

Figure 1. Western immunoblot of antigens expressed in *E. coli* clones carrying recombinant plasmids of *P. haemolytica* A1. Lane E: *E. coli* carrying pBR322 control; lanes 1 to 10 are representative *E. coli* clones expressing *P. haemolytica* A1 antigens as shown by the additional bands compared to lane E. These are total protein materials from 15 µl of cell suspension. Lane P: *P. haemolytica* A1 soluble proteins from 1 ml of culture supernatant precipitated with 6% trichloroacetic acid. Lane M: prestained molecular size standards, ranging from 200 kDa at the top to 20 kDa at the bottom. The first antibody used is pooled sera from calves vaccinated with Presponse.

II. Isolation by complementation

It may also be possible to isolate genes from *P. haemolytica* A1 by the complementation of well defined mutants in highly conserved biosynthetic pathways in a heterologous host such as *E. coli* or *Salmonella typhimurium*. One such well defined system in *S. typhimurium* is the *galE* gene (Adhya, 1987). *galE* codes for the enzyme UDP-galactose-4-epimerase which is involved in the production of UDP-galactose, one of the precursor molecules in LPS biosynthesis (Adhya, 1987). This function appears to be highly conserved, as noticed by the high homology between a number of characterized *galE* genes from *E. coli* (Lemaire and Muller-Hill, 1986), *S. typhimurium* (Houng *et al.*, 1990), *Neisseria gonorrhoeae* (Robertson *et al.*, 1993), *Neisseria meningitidis* (Jennings *et al.*, 1993) and *Haemophilus influenzae* (Maskell *et al.*, 1992). The *P. haemolytica* A1 plasmid clone bank was introduced into a *S. typhimurium galE* mutant and selected for growth on galactose as the carbon source. By this selection, we have isolated recombinant clones which restored GalE function in the *S. typhimurium* mutant. Upon futher characterization, the cloned DNA was shown to code for the *P. haemolytica* A1 *galE* gene.

III. Isolation by oligonucleotide hybridization

If the protein antigen of interest could be purified such that an N-terminal amino acid sequence could be obtained, then one could attempt screening for the gene (or the 5' portion of the gene) by oligonucleotide hybridization. For example, we have purified several secreted proteins of *P. haemolytica* A1 and obtained an N-terminal amino acid sequence for them (Abdullah, 1991). Oligonucleotides have been prepared based on a preferred codon usage table for *P. haemolytica* A1 (Lo, 1992). These oligonucleotides have been used, though with limited success, to hybridize with *E. coli* colonies carrying the *P. haemolytica* A1 plasmid bank. We have isolated several putative positive clones which hybridized specificially with the oligonucleotides. However, whether these are the correct clones await confirmation by cloning of the specific DNA fragments and nucleotide sequence analysis to see if they indeed encode for the proper amino acid sequences.

IV. Isolation by PCR amplification

PCR is a very powerful technique for the amplification of a particular DNA fragment if sufficient nucleotide sequence information is available. This could be done even if sequence information is available for the design of only one primer. We have used this approach to amplify a DNA fragment for part of the gene coding for the transferrin binding protein (Tbp1) of *P. haemolytica* A1. An oligonucleotide primer based on the N-terminal amino acid sequence of Tbp1 was prepared. Two other primers, based on the left and right junction sequences flanking the *Bam*HI site in the pBR322 plasmid vector was also prepared. The Tbp primer was used with either the pBR322 primers (left or right) in PCR to amplify a product from the entire *P. haemolytica* A1 plasmid library DNA (Figure 2). A 750 bp product was produced, purified, cloned and sequenced. The first twenty amino acids deduced from the nucleotide sequence correspond exactly to the twenty amino acids of the N-terminal of the purified Tbp1. This demonstrates that the PCR product is part of the *tbpA* gene. Using this 750 bp fragment as a probe, we have screened the *P. haemolytica* A1 genome and isolated DNA fragments that code for *tbpA* as well as the upstream *tbpB* gene. Further characterization of these genes will be described later.

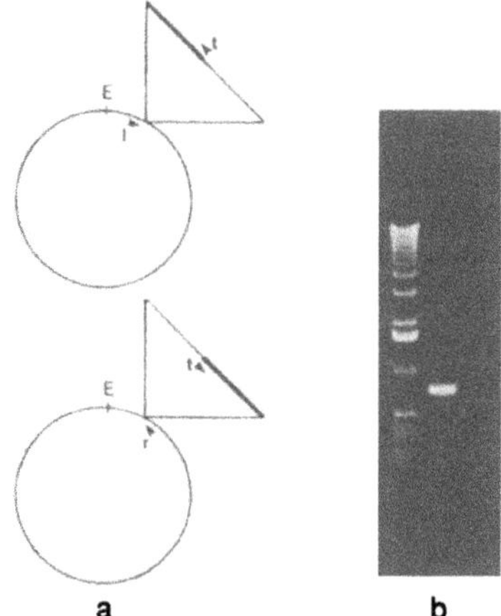

Figure 2. Amplification of *tbpA* sequence from the plasmid clone bank DNA using one primer based on the N-terminal amino acid sequence of Tbp1. a) Diagram showing the rationale for the amplification approach. By using either pBR322 primer left (l) or right (r) in combination with the tbp primer (t), a PCR product (the thick line) corresponding to the 5' of *tbpA* could be amplified depending on the orientation the insert DNA is cloned into pBR322. E - the *Eco*RI site on pBR322. b) An agarose gel showing a 750 bp PCR product corresponding to *tbpA*.

Of course a more direct approach could be carried out if sequence information is available for the design of both primers. As an example, we have used PCR to amplify a DNA fragment from *A. pleuropneumoniae* CM5 corresponding to the *rfaC* gene of *E. coli* and *S. typhimurium*. The *rfaC* gene codes for the enzyme heptosyltransferase 1, which is responsible for the addition of the first heptose residue to KDO I (Schnaitman and Klena, 1993). KDO I is the penultimate KDO moiety of the inner core of LPS on which all subsequent sugar moieties are attached. Hence, it is very likely that all Gram-negative bacteria possess a homologue of *rfaC* with similar function since most of the inner core are highly conserved. Based on the conserved nucleotide sequence of *E. coli* (Chen and Coleman Jr., 1993) and *S. typhimurium rfaC* (Sirisena *et al.*, 1992), two primers were designed and used to amplify from genomic DNA of *P. haemolytica* A1 and a number of *Actinobacillus* spp. The initial data showed that a product of similar size was amplified from the DNA of *P. haemolytica* A1, *A. pleuropneumoniae* CM5 and Shope. The product from *A. pleuropneumoniae* CM5 hybridized under high stringency conditions to the *E. coli rfaC* product. Verification that this is a *rfaC* homologue awaits the cloning and sequence determination of this product. Eventually, the *rfaC* gene could be used in complementation studies to see if it could replace the homologue in *E. coli* (or *S. typhimurium*) *rfaC* mutants.

All of the above approaches have been used in our laboratory, with varying degree of success, in the isolation of recombinant plasmids coding for the antigens of *P. haemolytica* A1. The following are brief descriptions of some of the continuing studies that follows after cloning of the genes.

STUDIES WITH THE LEUKOTOXIN GENES

The leukotoxin (Lkt) has been implicated as a major virulence factor of *P. haemolytica* A1 (Cho and Jericho, 1986; Sutherland and Donachie, 1986). After successful cloning and sequence analysis of the *lkt* genes, it was found to be highly homologous with the α-hemolysin genes (*hly*) of *E. coli* (Strathdee and Lo, 1987; Strathdee and Lo, 1989). These observations result in an accelerated understanding of the *P. haemolytica* A1 leukotoxin based on the characteristics of the *E. coli* α-hemolysin (Cavalieri *et al.*, 1984). The *lkt* and *hly* determinants subsequently form the basis for the identification of a family of widely disseminated toxins in Gram-negative bacteria. This family has been named the RTX toxins (Lo, 1990; Welch, 1991; Coote, 1992; Braun and Focareta, 1991; Welch *et al.*, 1992).

Genetic manipulation of the *lkt* genes allow high level expression of the leukotoxin in *E. coli* free from other *P. haemolytica* A1 antigens (Strathdee, 1989). The recombinant leukotoxin was recovered from the *E. coli* culture supernatant utilizing the functional homology and complementation between the *lkt* and *hly* secretion systems (Strathdee and Lo, 1989; Chang *et al.*, 1989; Koronakis *et al.*, 1992). In a vaccine trial and challenge experiment using the heterologously produced leukotoxin, it was shown that supplementation of leukotoxin to the culture supernatant vaccine Presponse greatly improved its efficacy (Conlon *et al.*, 1991). This is consistent with the notion that the leukotoxin is an important virulence factor and that it should be included in the formulation of an efficacious vaccine against *P. haemolytica* A1. Continuing work is being carried out for large scale production and incorporation of recombinant leukotoxin into Presponse.

Shortly after the cloning, characterization and discovery of homology between *lkt* and *hly*, similar related toxin determinants were reported to be present in hemolytic *Proteus* and *Morganella* spp. (Koronakis *et al.*, 1987). This lead to the suspicion that other Pasteurellaceae may possess the RTX toxin determinants. The *P. haemolytica* A1 *lkt* genes were used successfully as a probe in the identification and cloning of toxin determinants from *A. pleuropneumoniae* (Smits *et al.*, 1991), *A. actinomycetemcomitans* (Kolodrubetz *et al.*, 1989), *A. suis* (Burrows and Lo, 1992) and *A. equuli* (Burrows, 1993). To-date, three different (but related) RTX toxins have been identified in the twelve serotypes in *A. pleuropneumoniae* (Frey *et al.*, 1994). At least two RTX determinants have also been identified in *A. suis* (Burrows, 1993). The cloning and characterization of RTX determinants in these bacterium greatly facilitate the understanding of the role of these toxins in the pathogenesis of the *A. pleuropneumoniae* (Frey, 1994). In addition, several variations of the *lkt* determinant have been observed in the sixteen serotypes of *P. haemolytica* (Burrows *et al.*, 1993). These examples illustrate the type of molecular studies and information that could be gathered and produced from the characterization of an important virulence gene.

STUDIES WITH THE SIALOGLYCOPROTEASE GENE

The sialoglycoprotease of *P. haemolytica* A1 has been shown to have a very limited target cell specificity (Abdullah *et al.*, 1992). It only hydrolyses proteins which have O-linked glycosides, ie. suger moieties attached to the serine or threonine residues of the proteins. This enzyme have been shown to be a very useful tool for the characterization of O-linked glycoproteins. In addition, the sialoglycoprotease is being used in the development of an improved technique for the immunomagnetic selection of human bone-marrow stem-cells (Marsh, *et al.*, 1992). The purification of the primitive haematopoietic stem-cells is an important step in the development of autologous and allogeneic stem-cell bone-marrow transplantation. Up to now, most of the studies have been performed with the authentic sialoglycoprotease purified from *P. haemolytica* A1. At present, the recombinant sialoglycoprotease expressed in *E. coli* is lacking biological activities characteristic of the authentic enzyme. It is possible that fully active sialoglycoprotease requires the activation or modification by an additional *P. haemolytica* A1 factor. Continuing studies are being carried out to examine the involvement of modification or accessory proteins which are needed for the production of recombinant sialoglycoprotease that have all the important enzymatic properties of the authentic enzyme.

With respect to vaccine studies, high level expression experiments have been carried out with the cloned *gcp* gene similar to that for the *lkt* genes. A fusion gene between *gcp* and the 3'-end of *hlyA* has been constructed and placed behind the inducible *tac* promoter (Lo *et al.*, 1994). The hybrid protein (rGcpF) could be produced after induction of the promoter and secreted into the culture supernatant of *E. coli* in the presence of the specific HlyB/HlyD secretion system. This rGcpF could be easily recovered from the culture supernatant relatively free from other cellular components (Lo *et al.*, 1994). The recovered rGcpF protein has been used in vaccine trial and challenge experiments similar to those for the leukotoxin. The results suggest that rGcpF (hence the sialoglycoprotease) should also be considered as a component of an efficacious vaccine.

STUDIES WITH THE TRANSFERRIN BINDING PROTEIN GENES

Iron is in very limiting supply in the host environment, however, it is an essential nutrient requirement of bacteria since iron functions as co-factors in a number of important enzymes (Neilands, 1981). To acquire iron during growth in a host, pathogenic bacterium have evolved elaborate systems for the uptake of iron. One such system is the transferrin binding/uptake system (Williams and Griffiths, 1992; Otto, *et al.*, 1992). Specific transferrin binding proteins (Tbps) are produced by the bacterium during growth in iron limiting conditions, ie. inside a host. These Tbps specifically bind to the host's transferrin, resulting in uptake of the molecules and release of iron inside the bacterial cell (Otto, *et al.*, 1992). Tbp1 is an outer membrane protein which has been found in pathogenic bacteria including *N. meningitidis* (Schryvers and Morris, 1988), *N. gonorrhoeae* (Mickelsen and Sparling, 1981), *A. pleuropneumoniae* (Gonzalez, *et al.*, 1990), *H. influenzae* (Herrington and Sparling, 1985) and *P. haemolytica* A1 (Ogunnariwo and Schryvers, 1990). Recently, the genes coding for Tbp1 have been cloned and characterized from *N. gonorrhoeae* (Cornelissen, *et al.*, 1992) and *N. meningitidis* (Legrain *et al.*, 1993). Examination of the deduced amino acid sequences of the cloned genes suggest that these proteins belong to the TonB-dependent family of outer membrane proteins (Cornelissen, *et al.,* 1992). TonB is a periplasmic protein which is responsible for transduction of energy between the inner membrane and the outer membrane (Postle, 1990). Our data on the *P. haemolytica* A1 *tbpA* gene showed that the *P. haemolytica* A1 Tbp1 also belongs to the TonB-dependent family. The alignment in Figure 3 showed the N-terminal region of the *P. haemolytica* A1 Tbp1 and the Tbp1 proteins from *N. gonorrhoeae* and *N. meningitidis* which contain the conserved TonB box (Kadner, 1990). Presently, we have cloned and characterized the *tbp* genes from *P. haemolytica* A1. Upon completion of these studies, we will be able to examine the involvement of the Tbp proteins in pathogenesis and the development of vaccine components which include the Tbp proteins. In addition, we are also investigating the presence of TonB homologues in a number of *Pasteurellaceae* spp. These TonB proteins will be examined for their involvement in transduction of energy across the membrane which affects the uptake of macromolecules and survival of the bacteria.

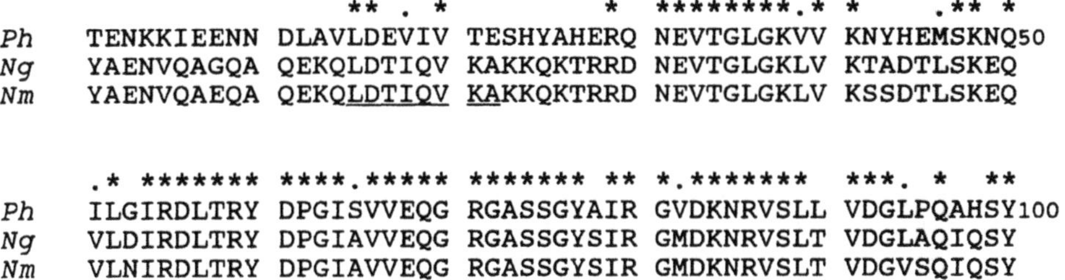

Figure 3. An alignment showing the deduced amino acid sequences from the N-terminal of Tbp1 proteins from *P. haemolytica* A1 (*Ph*), *N. gonorrhoeae* (*Ng*) and *N. meningitidis* (*Nm*). Only the first 100 amino acids are shown, with the residue positions indicated by the number to the right. Identical amino acids are marked by an asterisk and conserved amino acids are marked by a dot. The location of a putative TonB box is marked by the underline. The Tbp1 sequences of *Ng* and *Nm* are from Cornelissen *et al.*, 1992 and Legrain *et al.*, 1993, respectively.

In addition to the above described work, experiments are in progress to manipulate the other cloned genes of *P. haemolytica* A1 for the expression of recombinant antigens for vaccine trail and challenge experiments. These include the serotype-1 specific outer membrane protein and the lipoproteins. Each of the recombinant antigens will be examined either singularly, in combination with other antigens or as supplements to the culture supernatant vaccine Presponse. Eventually, it may be possible that a vaccine composed entirely of recombinant antigens could be developed.

CLONING OF LPS BIOSYNTHETIC GENES

The LPS of *P. haemolytica* A1 plays an important role in the generation of inflammation throughout the course of pneumonic pasteurellosis (Whitley *et al.*, 1990; Confer and Simons, 1986) and in damaging bovine pulmonary artery endothelial cells (Paulsen *et al.*, 1989; Paulsen *et al.*, 1990). The *P. haemolytica* A1 LPS is structurally similar to that of Gram-negative bacteria which is composed of a hydrophobic lipid A moiety, a core oligosaccharide, and an immunogenic O polysaccharide (Raetz, 1990; Schnaitman and Klena, 1993). The *P. haemolytica* A1 LPS also possesses biological properties typical of Gram negative endotoxin (Keiss *et al.*, 1964; Rimsay *et al.*, 1981). In addition, *P. haemolytica* A1 derived endotoxin, unlike that from *E. coli*, causes pathological changes in calves' lungs

(Slocombe *et al.*, 1990). The structure of the *P. haemolytica* A1 LPS oligosaccharide core has not yet been determined, but preliminary carbohydrate analysis indicates the presence of heptose (both L and D isomers), galactose and glucose (LaCroix *et al.*, 1993). The O polysaccharide structure has been elucidated and it is composed of the repeating trisaccharide -> 3)- -D-Gal*p*-(1 -> 3)- -D-Gal*p*NAc-(1 -> 4)- -D-Gal*p*-(1 -> (Severn and Richards, 1993). Although the clinical effects of *P. haemolytica* A1 LPS have been well characterized, the genetics of its biosynthesis have not yet been elucidated. From the extensive studies on the genetics and chemistry of LPS biosynthesis in *E. coli* and *S. typhimurium*, it is clear that this is a very complex process which involve numerous genes and functions. We have initiated a project to investigate into some of the genes involved in LPS biosynthesis in *P. haemolytica* A1.

From the screening experiments using the anti-LPS serum described above, we have identified a DNA region of *P. haemolytica* A1 which alters the LPS phenotypes of the *E. coli* host. Subcloning of this region narrowed the function to a 2 kbp DNA fragment. *E. coli* clones which carry this fragment on a recombinant plasmid show an altered LPS profile by SDS-PAGE electrophoresis. Further, phage sensitivity test on the *E. coli* clones using phage U3 (which is specific for the core region of *E. coli* LPS) showed that the receptor for U3 has been masked, most likely by the alteration of the LPS core of the *E. coli* host. This is supported by results of Western immunoblot analysis in which the altered LPS was recognized by a serum specific for the LPS core and not by a serum specific for the O-antigen of *P. haemolytica* A1. Nucleotide sequence analysis identified a gene on the DNA fragment which was designated *lpsA*. Homology comparison with sequence data banks showed significant similarity with a gene (*lex-1* or *lic2A*) from *H. influenzae* type b which is involved in LOS biosynthsis (Cope *et al.*, 1991; High *et al.*, 1993). Interestingly, the *P. haemolytica* A1 *lpsA* gene did not have the characteristic CAAT repetitive motif seen in the *H. influenzae* gene. This may be one reason which accounts for the lack of 'phase variation' in the LPS of *P. haemolytica* A1. Further experiments are in progress to examine the function of LpsA and its role in LPS biosynthesis in *P. haemolytica* A1.

The enzyme UDP-galactose-4-epimerase (GalE) is involved in the conversion of UDP-glucose into UDP-galactose. UDP-galactose is one of the main precursor molecules in the biosynthesis of LPS and capsule materials. It has been demonstrated in *E. coli*, *S. typhimurium*, *N. meningitidis* and *H. influenzae* that *galE* mutants exhibit a deep rough LPS phenotype and have reduced virulence compared to the wild type (Hone *et al.*, 1987). It is anticipated that a similar role of GalE in LPS biosynthesis should be found in *P. haemolytica* A1. Even though GalE is not directly involved in LPS biosynthesis, it has an important house keeping function and could lead to further understanding of part of the LPS biosynthetic pathway. Characterization of the *galE* gene may also lead to further studies into the synthesis of surface stuctures of the bacterium which may influence its survival in the host.

By the complementation of a defined *galE* mutant of *S. typhimurium* with the *P. haemolytica* A1 clone bank, we have isolated recombinant plasmids which encode the *galE* gene of *P. haemolytica* A1. *S. typhimurium galE* clones which carry this recombinant plasmid showed a complete restoration of the O-antigen chains. In addition, it is of interest to note that in *E. coli* and *S. typhimurium*, *galE* is part of the galactose operon, which includes the *galK-galT* genes (Adhya, 1987). However in *N. meningitidis* and *H. influenzae*, the *galE* gene is not part of the *gal* operon, but located within the genes involved in capsule biosynthesis (Jennings *et al.*, 1993; Robertson *et al.*, 1993). Our data also suggest that in *P. haemolytica* A1, *galE* is also not linked to the *galK-galT* genes. Continuing experiments are being carried out to determine the nature of the genes flanking *galE* and determine whether it is also located within the genes involved in capsule biosythesis in *P. haemolytica* A1.

MAPPING OF GENES ON THE *P. HAEMOLYTICA* A1 GENOME

Very little is known about the genomic organization of *P. haemolytica* A1. Reports in the literature estimated the genome to be between 2 to 3.6 Mbp (Mannheim, 1984). We applied the technique of Pulsed-Field Gel electrophoresis (PFGE) to examine the genomes of *P. haemolytica* A1 and *A. suis*.

In PFGE, the electric field undergoes periodic alterations (pulse) during agarose gel electrophoresis (Carle *et al.*, 1986). Resolution of a wide ranging size of DNA fragments (from 50 to 1,000 kbp) could be achieved by varying the pulse times to expand portions of the molecular size ranges (Chu *et al.*, 1986). For genomic mapping studies, the DNA is digested with a restriction endonuclease which cuts rarely on the DNA. This could be done by using restriction endonucleases whose recognition sites are 8 or more base-pairs, thus the sites appear less frequently on the DNA. For the HAP organisms which mostly have a low %G+C content, one could be further selective by using restriction endonucleases which have primarily G and C residues in the recognition sites. We have examined the genomic DNA of *P. haemolytica* A1 and *A. suis* using a panel of twelve different restriction endonucleases. From the data with some of the enzymes which produce fragments that could be resolved and counted accurately, we have estimated that the genome size of *P. haemolytica* and *A. suis* are approximately 2.4 Mbp. In particular, the enzymes *Ksp*I and *Bss*HII both produce approximately 25 fragments which could be separated and identified.

Some recently discovered intron-encoded endonucleases provide even more specific cleavage of the DNA and thus produce very few fragments. For example, I-*Ceu* I is an endonuclease encoded by a group I intron in the large subunit rRNA gene of *Chlamydomonas eugametos* (Gauthier *et al.*, 1991). This endonuclease has a recognition sequence of 26 bp, which is found within the highly conserved 23s rRNA genes (*rrn*) (Marshall and Lemieux, 1991). Therefore, digestion of genomic DNA with I-*Ceu* I produces very few fragments, the number of which should correspond to the number of *rrn* genes in the organisms. Using this enzyme, we have shown that there are at least six *rrn* genes in *P. haemolytica* and *A. suis*. The use of I-*Ceu* I, in conjunction with other rare cutting endonuclease will be valuable in the analysis of the genomes of *P. haemolytica* and other HAP organisms and in the construction of a physical map.

Since our observation that there are two RTX determinants in *A. suis*, a series of mapping experiments was done to determine the relative locations of the two determinants. The mapping data after PFGE and Southern hybridization on *A. suis* DNA showed that the two RTX determinants, *asxI* and *asxII* are located on separate large DNA fragments (Burrows, 1993). Continuing work is focused on positioning these fragments with respect to each other to get an estimation on the distance which separates the two RTX determinants on the *A. suis* DNA.

Finally, we have mapped or located six different cloned genes of *P. haemolytica* A1 on the large DNA fragments by Southern hybridization of PFGE separated fragments with the various cloned DNA fragments as probes. The preliminary mapping data showed that all of the cloned genes hybridized to the same DNA fragment (Figure 4). This suggests that a region of the *P. haemolytica* A1 genome (approx. 100 kbp) appears to code for all of these genes, which are all potential virulence factors. This may have implications with regard to the dissemination and evolution of these virulence associated genes. This data is supported by mapping of recombinant plasmids which showed that *gcp* and *lpsA* are closely positioned with respect to each other. The two genes were isolated independently in separate screening experiments.

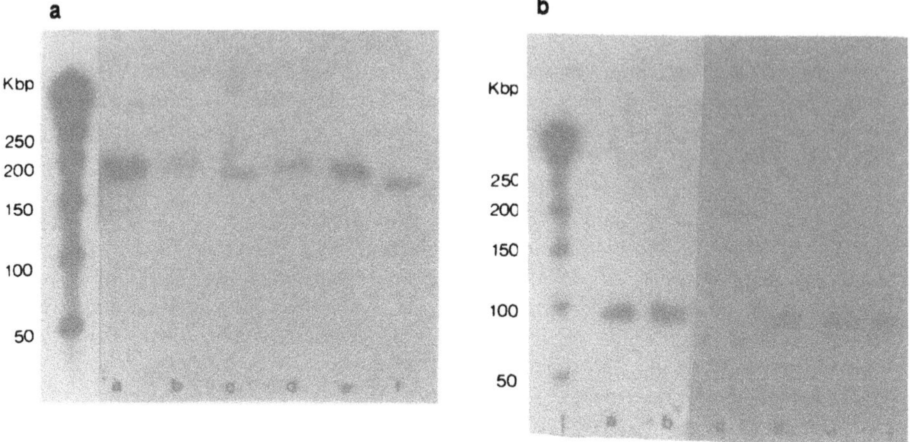

Figure 4. Southern hybridization using cloned DNA fragments coding for various genes of *P. haemolytica* A1 to genomic DNA digested with *Ksp*I (a) or *Bss*HII (b) and separated by PFGE. The probes used were from plasmids coding for the various antigens: lane a, an uncharacterized 17 kDa antigen in plasmid pPH20; lane b, the lipoproteins; lane c, the leukotoxin; lane d, the serotype-1 specific outer membrane protein, lane e, the LPS biosynthetic protein (LpsA); lane f, the sialoglycoprotease. Molecular size markers are indicated on the left.

EVOLUTION STUDIES ON THE HAP ORGANISMS

In addition to characterization of the antigens and their development as components in a vaccine, the molecular approach also provided some insight into the genetics and relatedness of the HAP organisms. The cloned genes could be used as specific probes in hybridization studies into the organization and distribution of the virulence antigens in the bacteria. For example, we have examined the distribution of six different antigen genes in the sixteen serotypes of *P. haemolytica*. The data showed that the A and T biotypes carry a very different genetic organization for these genes (Burrows, 1993). In most cases, different restriction fragments corresponding to the genes examined was observed. In other cases, the genes are completely absent, eg. the *lpsA* gene is not found in the T biotypes. These data are consistent with the recent suggestion that the T biotypes of *P. haemolytica* be renamed *P. trehalosi* (Sneath and Stevens, 1990).

SUMMARY

The data collected from these molecular studies prove invaluable in the building up of our knowledge on these microorganisms. Techniques and applications that we were not aware of since the last HAP International Conference in 1989 continue to be developed. With the continuing adoption of molecular techniques to study the HAP organisms, we hope to further our understanding of the genetic organization, gene expression and evolution of the HAP organisms.

ACKNOWLEDGEMENTS

I thank all the hard working people in my laboratory who contributed to the various aspects of this work. This work is funded by research grants from the Natural Sciences and Engineering Research Council of Canada and the Canadian Bacterial Diseases Network.

REFERENCES

Abdullah, K.M., 1991, Biochemical and genetic characterization of a *Pasteurella haemolytica* A1 glycoprotease. Ph.D. Thesis, University of Guelph.

Abdullah, K.M., Lo, R.Y.C., and Mellors, A., 1991, Cloning, nucleotide sequence, and expression of the *Pasteurella haemolytica* A1 glycoprotease gene. J. Bacteriol. 173: 5597-5603.

Abdullah, K.M, Udoh, E.A., Shewen, P.E., and Mellors, A., 1992, A neutral glycoprotease of *Pasteurella haemolytica* A1 specifically cleaves O-sialoglycoproteins. Infect. Immun. 60: 56-62.

Adhya, S., 1987, The galactose operon. In *Escherichia coli* and *Salmonella typhimurium*: Cellular and Molecular Biology. F.C. Neidhart *et al*.,. eds., American Society for Microbiology, Wachington, D.C.

Braun, V., and Focareta, T., 1991, Pore-forming bacterial protein hemolysins (cytolysins). Crit. Rev. Microbiol. 18: 115-158.

Burrows, L.L., 1993, Molecular characterization of the RTX cytolysin determinants from Gram-negative pathogens of veterinary significance. Ph.D. Thesis, University of Guelph, Guelph, Ontario, Canada.

Burrows, L.L., and Lo, R.Y.C., 1992, Molecular characterization of an RTX toxin determinant from *Actinobacillus suis*. Infect. Immun. 60: 2166-2173.

Burrows, L.L., Olah-Winfield, E., and Lo, R.Y.C., 1993, Molecular analysis of the leukotoxin determinants from *Pasteurella haemolytica* serotypes 1 to 16. Infect. Immun. 61: 5001-5007.

Carle, G.F., Frank, M, and Olson, M.V., 1986, Electrophoretic separation of large molecules by periodic inversion of the electric field. Science 232: 65-68.

Cavalieri, S.J., Bohach, G.A., and Snyder, I.S., 1984, *Escherichia coli* à-haemolysin: characteristics and probable role in pathogenicity. Microbiol. Rev. 48: 326-343.

Chang, Y.-F., Young, R., Moulds, T.L., and Struck, D.K., 1989, Secretion of the *Pasteurella* leukotoxin by *Escherichia coli*. FEMS Microbiol. Lett. 60: 169-174.

Chen, L., and Coleman, Jr., W.G., 1993, Cloning and characterization of the *Escherichia coli* K-12 (*rfaC*) gene, a gene required for lipopolysaccharide inner core synthesis. J. Bacteriol. 175: 2534-2540.

Cho, H.J. and Jericho, K.W.F., 1986, Induction of immunity against pneumonic pasteurellosis following experimental infection in calves. Can. J. Vet. Res. 50: 27-31.

Chu, G., Vollrath, D., and Davis, R.W., 1986, Separation of large DNA molecules by contour-clamped homogeneous electric fields. Science 234: 1582-1585.

Confer, A.W., and Simons, K.R., 1986, Effects of *Pasteurella haemolytica* lipopolysaccharide on selected functions of bovine leukocytes. Am. J. Vet. Res. 47: 154-157.

Conlon, J.A., Shewen, P.E., and Lo, R.Y.C., 1991, Efficacy of a recombinant leukotoxin in protection against pneumoniae challenge with *Pasteurella haemolytica* A1. Infect. Immun. 59: 587-591.

Cooney, B.J., and Lo, R.Y.C., 1993, Three contiguous lipoprotein genes in *Pasteurella haemolytica* which are homologous to a lipoprotein gene in *Haemophilus influenzae* type B. Infect. Immun. 61: 4682-4688.

Coote, J.G., 1992, Structural and functional relationships among the RTX toxin determinants of Gram-negative bacteria. FEMS Microbiol. Rev. 88: 137-162.

Cope, L.D., Yogev, R., Mertsola, J., Latimer, J.L., Hanson, M.S., McCracken, Jr., G.H., and Hansen, E.J., 1991, Molecular cloning of a gene involved in lipooligosaccharide biosynthesis and virulence expression by *Haemophilus influenzae* type B. Molec. Microbiol. 5: 1113-1124.

Cornelissen, C.N., Biswas, G.D., Tsai, J., Paruchuri, D.K., Thompson, S.A., and Sparling, P.F., 1992, Gonococcal transferrin-binding protein 1 is required for transferrin utilization and is homologous to TonB-dependent outer membrane receptors. J. Bacteriol. 174: 5788-5797.

Frey, J., 1994, RTX-toxins in *Actinobacillus pleuropneumoniae* and their potential role in virulence. In "Molecular Mechanisms of Bacterial Virulence", C.I. Kado and J.H. Crosa, eds., Kluwer Academic Publishers.

Frey, J., Beck, M., and Nicolet, J., 1994, RTX-toxins of *Actinobacillus pleuropneumoniae*. In "Bacterial Protein Toxins, Freer *et al*.,. eds., Gustav Fisher.

Gauthier, A., Turmel, M., and Lemieux, C., 1991, A group I intron in the chloroplast large subunit rRNA gene of *Chlamydomonas eugametos* encodes a double-strand endonuclease that cleaves the homing site of this intron. Curr. Genet. 19: 43-47.

Gonzalez, G.C., Caamano, D.L., and Schryvers, A.B., 1990, Identification and characterization of a porcine-specific transferrin receptor in *Actinobacillus pleuropneumoniae*. Molec. Microbiol. 4: 1173-1179.

Gonzaleze-Rayos, C., Lo, R.Y.C., Shewen, P.S., and Beveridge, T.J., 1986, Cloning of a serotype-specific antigen from *Pasteurella haemolytica* A1. Infect. Immun. 53: 505-510.

Herrington, D.A., and Sparling, P.F., 1985, *Haemophilus influenzae* can use human transferrin as a sole source for required iron. Infect. Immun. 48: 248-251.

High, N.J., Deadman, M.E., and E.R. Moxon, 1993, The role of a repetitive DNA motif (5'-CAAT-3') in the variable expression of the *Haemophilus influenzae* lipopolysaccharide epitope àGal(1-4) Gal. Molec. Microbiol. 9: 1275-1282.

Hone,D., Morona, R., Attridge, S., and Hackett, J., 1987, Construction of defined *galE* mutants of *Salmonella* for use as vaccines. J. Infect. Dis. 156: 167-174.

Houng, H.S.H., Kopecko, D.J., and Baron, L.S., 1990, Molecular cloning and physical and functional characterization of the *Salmonella typhimurium* and *Salmonella typhi* galactose utilization operons. J. Bacteriol. 172: 4392-4398.

Jennings, M.P., van der Ley, P., Wilks, K.E., Maskell, D.J., Poolman, J.T., and Moxon, E.R., 1993, Cloning and molecular analysis of the *galE* gene of *Neisseria meningitidis* and its role in lipopolysaccharide biosynthesis. Molec. Microbiol. 10: 361-369.

Kadner, R.J., 1990, Vitamin B_{12} transport in *Escherichia coli*: energy coupling between membranes. Molec. Microbiol. 4: 2027-2033.

Keiss, R.E., Will, D.H., and Collier, J.R., 1964, Skin toxicity and hemodynamic properties of endotoxin derived from *Pasteurella haemolytica*. Am. J. Vet. Res. 25:935-942.

Kolodrubetz, D., Dailey, T., Ebersole, J., and Kraig, E., 1989, Cloning and expression of the leukotoxin gene from *Actinobacillus actinomycetemcomitans*. Infect. Immun. 57: 1465-1469.

Koronakis, V., Cross, M., Senior, B., Koronakis, E., and Hughes, C., 1987, The secreted hemolysins of *Proteus mirabilis, Proteus vulgaris,* and *Morganella morganii* are genetically related to each other and to the alpha-hemolysin of *Escherichia coli*. J. Bacteriol. 169: 1509-1515.

Koronakis, V., Stanley, P., Koronakis, E., and Hughes, C., 1992, The HlyB/HlyD-dependent secretion of toxins by Gram-negative bacteria. FEMS Microbiol. Immunol. 105: 45-54.

LaCroix, R.P., Duncan, J.R., Jenkins, R.P., Leitch, R.A., Perry, J.A., and Richards, J.C., 1993, Structural and serological specificities of *Pasteurella haemolytica* lipolysaccharides. Infect. Immun. 61: 170-181.

Legrain, M., Mazarin, V., Irwin, S.W., Bouchon, B., Quentin-Millet, M., Jacobs, E., and Schryvers, A.B., 1993, Cloning and characterization of *Neisseria meningitidis* genes encoding the transferrin-binding proteins Tbp1 and Tbp2. Gene, 130: 73-80.

Lemaire, H.-G., and Muller-Hill, B., 1986, Nucleotide sequences of the *galE* gene and the *galT* gene of *E. coli*. Nucl. Acids Res. 14: 7705-7711.

Lo, R.Y.C., 1990, Molecular characterization of cytotoxins produced by *Haemophilus, Actinobacillus, Pasteurella*. Can. J. Vet. Res. 54: S33-S35.

Lo, R.Y.C., 1992, An analysis of the codon usage of *Pasteurella haemolytica* A1. FEMS Microbiol. Lett. 100: 125-132.

Lo, R.Y.C., and Cameron, L.A., 1986, A simple immunological detection method for the direct screening of genes form clone Banks. Can. J. Biochem. Cell Biol. 64: 73-76.

Lo, R.Y.C., Shewen, P.E., Strathdee, C.A., and Greer, C.N., 1985, Cloning and expression of the leukotoxin gene of *Pasteurella haemolytica* A1 in *Escherichia coli* K-12. Infect. Immun. 50: 667-671.

Lo, R.Y.C., Strathdee, C.A., and Shewen, P.E., 1987, Nucleotide sequence of the leukotoxin genes of *Pasteurella haemolytica* A1. Infect. Immun. 55: 1987-1996.

Lo, R.Y.C., Strathdee, C.A., Shewen, P.E., and Cooney, B.J., 1991, Molecular studies of Ssa1, a serotype-specific antigen of *Pasteurella haemolytica* A1. Infect. Immun. 59: 3398-3406.

Lo, R.Y.C., Watt, M.-A., Gyorffy, S., and Mellors, A., 1994, Preparation of recombinant glycoprotease of *Pasteurella haemolytica* A1 utilizing the *Escherichia coli* α-hemolysin secretion system. FEMS Microbiol. Lett. 116:225-230.

Mannheim, W. 1984, In Bergey's Manual of Systematic Bacteriology, Vol.1. (Kreig, N.R., Ed.), pp. 550-552, Williams and Wilkins Press, Baltimore/London.

Marshall, P., and Lemieux, C., 1991, Cleavage pattern of the homing endonuclease encoded by the fifth intron in the chloroplast large subunit rRNA-encoding gene of *Chlamydomonas eugametos*. Gene 104: 241-245.

Marsh, J.C.W., Sutherland, D.R., Davidson, J., Mellors, A., and Keating, A., 1992, Retention of progenitor cell fraction in $CD34^+$ cells purified using a noval O-sialoglycoprotease. Leukemia 6: 926-934.

Maskell, D.J., Szabo, M.J., Deadman, M.E., and Moxon, E.R., 1992, The *gal* locus from *Haemophilus influenzae*: cloning, sequencing and the use of *gal* mutants to study lipopolysaccharide. Molec. Microbiol. 6: 3051-3063.

Mickelsen, P.A., and Sparling, P.F., 1981, Ability of *Neisseria gonorrhoeae*, *Neisseria meningitidis* and commensal *Neisseria* species to obtain iron form transferrin and iron compounds. Infect. Immun. 33: 555-564.

Neilands, J.B., 1981, Microbial iron compounds. Annu. Rev. Microbiol. 50: 715-731.

Ogunnariwo, J.A., and Schryvers, A.B., 1990, Iron acquisition in *Pasteurella haemolytica*: expression and identification of bovine-specific transferrin receptor. Infect. Immun. 58: 2091-2097.

Otto, B.R., Verweij-van Vught, A.M.J.J., and MacLaren, D.M., 1992, Transferrins and heme-compounds as iron sources for pathogenic bacteria. Crit. Rev. Microbiol. 13: 217-233.

Paulsen, D.B., Mosier, D.A., Clinkenbeard, K.D., and Confer, A.W., 1989, Direct effects of *Pasteurella haemolytica* lipopolysaccharide on bovine pulmonary endothelial cells in vitro. Am. J. Vet. Res. 50:1633-1637.

Paulsen, D.B., Confer, A.W., Clinkenbeard, K.D., and Mosier, D.A., 1990, *Pasteurella haemolytica* lipopolysaccharide-induced arachidonic acid release from and neutrophil adherence to bovine pulmonary artery endothelial cells. Am J. Vet. Res. 51:1635-1639.

Postle, K., 1990, TonB and the Gram-negative dilemma. Molec. Microbiol. 4: 2019-2025.

Potter, M.D. and Lo, R.Y.C., 1993, Genetic characterization of *Pasteurella haemolytica* A1 LPS biosynthetic genes. 93rd Ann. Meet. Am. Soc. Microbiol., 1993, Atlanta, Georgia, Abstr. B354.

Raetz, C.R.H., 1990, Biochemistry of endotoxins. Annu. Rev. Biochem. 59: 129-170.

Rimsay, R.L., Coyle-Dennis, J.E., Lauerman, L.H., and Squire, P.G., 1981, Purification and biological characterization of endotoxin fractions from *Pasteurella haemolytica*. Am. J. Vet. Res. 42:2134-2138.

Robertson, B.D., Frosch, M., and van Putten, J.P.M., 1993, The role of *galE* in the biosynthesis and function of gonococcal lipopolysaccharide. Molec. Microbiol. 8: 891-901.

Schnaitman, C.A., and Klena, J.D., 1993, Genetics of lipopolysaccharide biosynthesis in enteric bacteria. Microbiol. Rev. 57: 655-682.

Schryvers, A.B., and Morris, L.J., 1988, Identification and characterization of the transferrin receptor from *Neisseria meningitidis*. Molec. Microbiol. 2: 281-288.

Severn, W.B., and Richards, J.C., 1993, Characterization of the O-polysaccharide of *Pasteurella haemolytica* serotype A1. Carbohydr. Res. 240:277-285.

Shewen, P.E., and Wilkie, B.N., 1988, Vaccination of calves with leukotoxic culture supernatant from *Pasteurella haemolytica*. Can. J. Vet. Res. 52: 30-36.

Shewen, P.E., Sharpe, A., and Wilkie, B.N., 1988, Efficacy testing of a *Pasteurella haemolytica* extract vaccine. Vet. Med. 10: 1078-1083.

Sirisena, D.M., Brozek, K.A., MacLachlan, P.R., Sanderson, K.E., and Raetz, C.R.H., 1992, The *rfaC* gene of *Salmonella typhimurium*. J. Biol. Chem. 267: 18874-18884.

Slocombe, R.F., Mulks, M., Killingsworth, C.R., Derksen, F.J., and Robinson, N.E., 1990, Effect of *Pasteurella haemolytica*-derived endotoxin on pulmonary structure and function in calves. Am. J. Vet. Res. 51:433-438.

Smits, M.A., Briaire, J., Jansen, R., Smith, H.E., Kamp, E.M., and Gielkens, A.L.J., 1991, Cytolysins of *Actinobacillus pleuropneumoniae* serotype 9. Infect. Immun. 59: 4497-4504.

Sneath, P.H.A. and Stevens, M., 1990, *Actinobacillus rossii* sp. nov., *Actinobacillus seminis* sp. nov., nom. rev., *Pasteurella bettii* sp. nov., *Pasteurella lymphangitidis* sp. nov., *Pasteurella mairi* sp. nov., and *Pasteurella trehalosi* sp. *Nov. Int. J. Syst. Bact.* 40: 148-153.

Strathdee, C.A., 1989, Molecular characterization of the *Pasteurella haemolytica* leukotoxin determinant. Ph.D. Thesis, University of Guelph, Guelph, Ontario, Canada.

Strathdee, C.A., and Lo, R.Y.C., 1987, Extensive homology between the leukotoxin of *Pasteurella haemolytica* A1 and the alpha-hemolysin of *Escherichia coli*. Infect. Immun. 55: 3233-3236.

Strathdee, C.A., and Lo, R.Y.C., 1989. Cloning, nucleotide sequence, and characteriztion of genes encoding the secretion function of the *Pasteurella haemolytica* leukotoxin determinant. J. Bacteriol. 171: 916-928.

Sutherland, A.D., and Donachie, W., 1986, Cytotoxic effect of serotypes of *Pasteurella haemolytica* on sheep bronchoalveolar macrophages. Vet. Microbiol. 11: 331-336.

Welch, R.A., 1991, Pore-forming cytolysins of Gram-negative bacteria. Molec. Microbiol. 5: 521-528.

Welch, R.A., Forestier, C., Lobo, A., Pellett, S., Thomas, Jr., W., and Rowe, G., 1992, The synthesis and function of the *Escherichia coli* hemolysin and related RTX exotoxins. FEMS Microbiol. Immunol. 105: 29-36.

Williams, P., and Griffiths, E., 1992, Bacterial transferrin receptors - structure, function and contribution to virulence. Med. Microbiol. Immunol. 181: 301-322.

Whiteley, L.O., Maheswaran, S.K., Weiss, D.J., and Ames, T.R., 1990, Immunohistochemical localization of *Pasteurella haemolytica* A1-derived endotoxin, leukotoxin, and capsular polysaccharide in experimental bovine pasteurella pneumoniae. Vet Pathol. 27: 150-161.

Woo, T.K.W., and Lo, R.Y.C., 1993, The transferrin binding protein of *Pasteurella haemolytica* is homologous to the gonococcal transferrin binding protein. 93rd Ann. Meet. American Society for Microbiol., Atlanta Georgia, Abstr. B294.

MODULATION OF LEUKOCYTES BY EXOTOXINS PRODUCED BY HAP ORGANISMS

C.J. Czuprynski

Dept. Pathobiological Sciences
School of Veterinary Medicine
University of Wisconsin
Madison, WI 53706

INTRODUCTION

The HAP group of organisms is comprised of small Gram-negative rods that can colonize the mucosal surface of the respiratory and genitourinary tracts. In some instances there is a clear relationship between a particular HAP species and its host of choice. In other cases the organism is more promiscuous and will colonize a number of host species. Depending on the HAP species involved, this colonization may be benign, resulting in little or no recognizable infection or host response in a healthy individual (Nicolet, 1990). Subsequent exposure to various types of stressors, or underlying infection with a virus or other pathogenic agent, can favor multiplication of HAP organisms that leads to clinical disease. Leukocytes provide a first line of cellular defense against invading microorganisms. As such, it is critically important for HAP organism to evade the antimicrobial activity of leukocytes if they are to multiply and cause clinical disease. HAP organisms have developed several strategies for doing this, including the production of antiphagocytic capsules and the release of exotoxins (leukotoxins) that inhibit leukocyte antimicrobial activity or kill the leukocytes (Czuprynski and Sample, 1990). More recently, it has become clear that smaller amounts of some of these leukotoxins can activate leukocytes to release oxygen radicals and other inflammatory mediators that likely contribute to the intense inflammation that characterizes many HAP infections (Czuprynski *et al.*, 1991; Breider *et al.*, 1991; Maheswaran *et al.*, 1993; Udeze and Kadis, 1992). The purpose of this chapter is to provide a brief overview of these exotoxins, their relatedness among various HAP species, their differential effects on leukocyte populations, and the possible contributions of other bacterial components (i.e., lipopolysaccharide) to their biological effects.

PRODUCTION OF RTX TOXINS BY HAP ORGANISMS

Investigators have been aware of the leukotoxic activity of HAP organisms for some time. Several laboratories independently identified the toxic effects of *Pasteurella haemolytica* on bovine leukocytes (Berggen *et al.*, 1981; Benson *et al.*, 1978; Kaehler *et al.*, 1980a, Maheswaran *et al.*, 1980). This was then shown to result from release of a toxic factor into the culture medium during the logarithmic growth phase of *P. haemolytica* (Baluyut *et al.*, 1981;). Other studies established that the toxin was protein in nature, heat-labile, and non-dialyzable (Chang *et al.*, 1986; Shewen and Wilkie, 1982). Various estimates of molecular size were made, although most indicated that the active moiety was approximately 100 kD (Chang *et al.*, 1987). Particularly noteworthy was the observation that the leukotoxin was lethal for ruminant leukocytes, but not those from other domestic or laboratory animal species, nor humans leukocytes (Kaehler *et al.*, 1980b; Shewen and Wilkie, 1982). It was shown that sera from healthy cattle varied in their ability to inhibit leukotoxic activity, and that the resistance of cattle to *P. haemolytica* infection often, but not always, correlated with the presence of anti-leukotoxin-neutralizing antibodies (Mosier *et al.*, 1989). These observations led investigators to conclude that the leukotoxin was an important, but not the sole, virulence mechanism of *P. haemolytica* (Confer *et al.*, 1990).

Similar observations were made for the porcine pathogen *Actinobacillus pleuropneumoniae*, which as its name implies, causes an intense fibrinous pleuropneumonia that is similar to fulminant *P. haemolytica* infection in cattle (Inzana, 1991). This observation was not altogether surprising, since *P. haemolytica* and *A. pleuropneumoniae* are known to be closely related organisms (Mutters *et al.*, 1989). The toxin status of *A. pleuropneumoniae* became somewhat more complex and confusing, however, as several laboratories independently reported at least three distinct hemolysins and cytotoxins that were produced by different serotypes of the organism (Inzana, 1991). These had molecular weights of approximately 100 to 110 kD and varied somewhat in the conditions required to express their production and release (Devenish *et al.*, 1992; Van Leengoed and Dickerson, 1992). Serologic cross-reactivity between the *P. haemolytica* leukotoxin and certain of the *A. pleuropneumoniae* toxins was noted under some conditions (Devenish *et al.*, 1989). However, unlike the *P. haemolytica* leukotoxin, the activity of *A. pleuropneumoniae* toxins did not demonstrate species-restricted activity (Inzana, 1991).

The human periodontal pathogen *Actinobacillus actinomycetemcomitans* was also shown to produce a leukotoxin that was lethal to neutrophils from humans and certain nonhuman primates (Tsai *et al.*, 1979; Tsai *et al.*, 1984; Iwase *et al.*, 1990; Iwase *et al.*, 1992; Taichman, *et al.*, 1991). Cells from other mammalian species were unaffected *in vitro*. Unlike the toxins produced by *P. haemolytica* and *A. pleuropneumoniae*, the *A. actinomycetemcomitans* leukotoxin remained associated with the bacterial cell, rather than being released into the culture medium (Tsai *et al.*, 1979). The toxin could be extracted from the bacteria by sonication or treatment with polymyxin B (Tsai *et al.*, 1984). More recent evidence suggests that the leukotoxin is secreted outside the periplasmic space, but remains adherent to nucleic acids that coat the outer surface of *A. actinomycetemcomitans* (Ohta *et al.*, 1991; Ohta *et al.*, 1993). This observation is consistent with knowledge that the *A. actinomycetemcomitans* leukotoxin is a basic protein (pI = 8.2 to 9.7) (Ohta *et al.*, 1991). The toxin otherwise exhibits a similar molecular size, and susceptibility to heat and protease treatment, as the *P. haemolytica* and *A. pleuropneumoniae* leukotoxins (Tsai *et al.*, 1984).

A breakthrough in our understanding of the biological role of these leukotoxins occurred when Lo and coworkers cloned and sequenced the leukotoxin operon and demonstrated that it was very homologous to the hemolysin operon for *Escherichia coli* (Strathdee and Lo, 1987). This was of significant because the *E. coli* hemolysin had been described previously as being lethal for leukocytes from humans and other mammalian species (Cavalieri and Snyder, 1982; Gadeberg and Orskov, 1984). Since that time a number of other exotoxins produced by Gram-negative bacteria have been described that utilize this same operon structure. These toxins share the property of having a varying number of glycine-rich Ca^{2+} -binding tandem repeats in the N-terminal end of the structural toxin molecule (Boehm et al, 1990). As a result, these toxins are referred to as RTX (repeat in toxin) toxins (Welch *et al.*, 1992). Continued investigation has shown that there is a large family of Gram-negative bacteria, including human, animal, and plant pathogens, that produce a full or partial RTX toxin (Burrows and Lo, 1992; Kolodrubetz *et al.*, 1989; Kraig *et al.*, 1990; Sebo and Ladant, 1993). For a detailed examination of the structure-function relationships of these toxins, the reader is referred elsewhere (Welch *et al.*, 1992). For the purposes of the present discussion, however, it is important to note that these toxins are thought to have a membrane-spanning domain in the C-terminus end of the structural toxin molecule that can insert itself into the cytoplasmic membrane. This results in formation of a pore that allows influx of Ca^{2+} and efflux of K^+ and macromolecules like ATP (Menestrina *et al.*, 1994).

EFFECTS OF HAP TOXINS ON NEUTROPHILS

An important early observation was made by Cavalieri and Snyder (1982), who reported that sublethal concentrations of the *E. coli* hemolysin stimulated human neutrophil chemiluminescence, whereas higher concentrations impaired neutrophil phagocytic and chemotactic activities. Later, Bhakdi and coworkers (1989) reported that sublethal concentrations of hemolysin caused rapid membrane defects in human neutrophils that resulted in loss of ATP, degranulation, and impaired phagocytic activity. Human neutrophils appeared to be even more sensitive than erythrocytes to the effects of hemolysin. Subsequent studies indicated that sublethal hemolysin concentrations stimulated human neutrophils to release superoxide anion (Bhakdi and Martin, 1991) and lipoxygenase-dependent eicosanoids (Grimminger *et al.*, 1991). These data suggested that RTX toxins are leukocyte-modulating agents that can activate, inhibit, or kill a leukocyte depending on the toxin concentration and the conditions present.

These findings stimulated several laboratories to investigate the effects of RTX toxins produced by HAP organisms on leukocytes from their normal host species. Several studies showed that *P. haemolytica* leukotoxin caused pore formation in bovine neutrophils (Clinkenbeard, *et al.*, 1989a), platelets (Clinkenbeard and Upton, 1991), and lymphoma cells (Clinkenbeard *et al.*, 1989b). The lethal effects of leukotoxin were temperature and calcium dependent, and could be blocked by the addition of osmotic protectants, such as sucrose (Clinkenbeard *et al.*, 1989b; Czuprynski and Noel, 1990; Styrt *et al.*, 1990). These results were interpreted as indicating the resulting membrane lesion (i.e., pore) was smaller than the diameter of the molecules with osmotic protective activity. These and similar studies have resulted in estimates of 0.9 to 2.0 nm for the pores formed by RTX toxins in the cytoplasmic membrane (Clinkenbeard *et al.*, 1989b; Lalonde *et al.*, 1989; Menestrina *et al.*, 1994). Low concentrations of *P. haemolytica* leukotoxin were found to stimulate bovine neutrophil chemiluminescence, degranulation, and shape change (Czuprynski *et al.*, 1991). This process was associated with a rapid increase in intracellular calcium; blocking calcium uptake with verapamil or propranolol dramatically reduced neutrophil activation (Ortiz-Carranza and Czuprynski, 1992). Likewise, addition of a MAb that neutralized the activating and lethal effects of *P. haemolytica* leukotoxin also prevented the increase in intracellular calcium. Flow cytometry analysis indicates that bovine neutrophils are killed rapidly (within minutes) by leukotoxin as assessed by uptake of propidium iodide and the inability of a neutralizing MAb to prevent neutrophil death if added within 5 minutes after the addition of leukotoxin (Stevens and Czuprynski, unpublished observations). Other investigators reported that *P. haemolytica* leukotoxin can stimulate bovine neutrophils to produce oxygen radicals that damage bovine endothelial cells *in vitro* (Breider *et al.*, 1991; Maheswaran *et al.*, 1993). This effect is interesting in that it is contradictory to the ability of neutrophils to prevent the direct toxic effects of *P. haemolytica* LPS on bovine endothelial cells (Breider *et al.*, 1991). The sum of these studies, however, indicate that bovine neutrophils are very sensitive to the activating and lethal effects of *P. haemolytica* leukotoxin, a finding that is consistent with knowledge that neutrophil depletion of cattle can reduce tissue damage in experimental pulmonary pasteurellosis (Slocombe *et al.*, 1990; Weiss, *et al.*, 1991).

Several studies have reported that the leukotoxins of *A. pleuropneumoniae* have similar effects on porcine neutrophils. Intact organisms, and high concentrations of culture filtrates, were lethal for porcine neutrophils or inhibited their phagocytic activity (Devenish *et al.*, 1992). Sublethal toxin concentrations stimulated a neutrophil oxidative burst and primed the neutrophils for a stronger oxidative response to phorbol myristate acetate (Dom *et al.*, 1992; Udeze and Kadis, 1992). The effects on neutrophils was calcium dependent and could be neutralized with convalescent sera from previously infected pigs (Leengoed and Dickerson, 1992; Byrd and Kadis, 1992).

The lethal effects of *A. actinomycetemcomitans* leukotoxin for human neutrophils has been described previously (Tsai *et al.*, 1979; Tsai et al. 1984). It was shown that this toxin caused a very rapid influx of calcium (within 5 sec) followed by a loss of membrane potential that resulted in the death of the cells within 10 to 15 min as assessed by flow cytometric analysis of propidium iodide staining (Taichman *et al.*, 1991). Neutrophil damage and calcium uptake could be blocked with a neutralizing MAb. A similar response was observed with human neutrophil cytoplasts (organelle depleted), suggesting that killing by the *A. actinomycetemcomitans* leukotoxin involved a direct effect on the neutrophil cytoplasmic membrane (Iwase *et al.*, 1992). This same laboratory observed that certain cell types (i.e., K562 cells) were resistant to killing, but could absorb leukotoxic activity (Iwase *et al.*, 1990). This suggests that membrane damage may involve more than just toxin binding. What mechanism might be involved remains to be elucidated. An earlier report that *A. actinomycetemcomitans* leukotoxin can be inactivated by the neutrophil myeloperoxidase system (Clark *et al.*, 1986), suggests that neutrophil activation by RTX toxins provides at least one mechanism for dampening the effects of the toxin, albeit after the damage may already have been done.

EFFECTS OF HAP TOXINS ON MONONUCLEAR PHAGOCYTES

There is much less information regarding the effects of HAP toxins on mononuclear phagocytes than on neutrophils. Many of the studies have used whole organisms rather than examined in detail the effects of the toxins themselves. Bovine alveolar macrophages have a limited ability to kill *P. haemolytica in vitro* (Maheswaran *et al.*, 1980). Killing occurred at low bacteria to macrophage ratios, but higher ratios resulted in cytotoxic damage to the macrophages, presumably as a result of leukotoxin production (Benson *et al.*, 1978). Other workers did not observe phagocytosis of *P. haemolytica* by alveolar macrophages *in vitro* (Walker *et al.*, 1980). *P. haemolytica* was also found to be cytocidal for bovine peripheral blood mononuclear cells (lymphocytes and monocytes) again presumably reflecting leukotoxin activity (Kaehler *et al.*, 1980a). Crude *P. haemolytica* leukotoxin has been reported to be toxic for bovine (Shewen and Wilkie, 1982) and ovine (Sutherland, 1985) peripheral blood mononuclear cells.

Bovine alveolar macrophages have been shown to produce various inflammatory mediators in response to *P. haemolytica*. These include superoxide anion (Dyer *et al.*, 1985), and procoagulant activity (Car *et al.*, 1991). Conversely, high concentrations of crude leukotoxin impaired release of neutrophil chemotactic factors from alveolar macrophages *in vitro* (Markham *et al.*, 1982). Alveolar macrophages were found to exacerbate damage to bovine endothelial cells *in vitro* (Sharma *et al.*, 1992) presumably as the result of production of inflammatory mediators. Serum levels of the inflammatory cytokine TNFα were elevated at 2 to 8 hours after intratracheal inoculation of *P. haemolytica* (Pace *et al.*, 1993). However, one cannot conclude from these data that alveolar macrophages were the sole source of the TNFα detected, since in other species neutrophils and other cells can also produce TNFα. There is evidence that bovine alveolar macrophages and blood monocytes are more susceptible than neutrophils to the lethal effects of *P. haemolytica* leukotoxin, although monocyte death proceeds much more slowly, suggesting it may involve apoptosis (Stevens and Czuprynski, 1995). In this same study, both monocytes and alveolar macrophages released TNF, and monocytes released IL-1, when incubated with leukotoxin. Addition of a neutralizing MAb prevented killing of monocytes, but paradoxically enhanced TNF release. Likewise, heating (60 C for 1 hr) eliminated the lethal but not the monokine stimulating activity of the leukotoxin. We therefore cannot exclude the possible contributions of LPS to the stimulatory effects of the leukotoxin. Similarly, Sharma *et al* (1991) found that alveolar macrophages mediated damage to endothelial cells *in vitro* was not eliminated by heat-treatment of crude leukotoxin, nor by addition of an anti-leukotoxin MAb.

These results are somewhat similar to those reported by Bhakdi *et al.*, (1990) who detected release of IL-1 but not TNF activity after human monocytes were incubated with hemolysin-producing *E. coli*. Because nonhemolytic *E. coli* did not have a comparable effect, these investigators concluded that the response was stimulated by the hemolysin and not LPS in the preparation. They conceded, however, that they could not exclude a possible interactive effect between hemolysin and LPS. We favor the possibility that a similar interactive stimulation occurs when bovine monocytes are incubated with *P. haemolytica* leukotoxin.

Porcine alveolar macrophages also appear to be very susceptible to the inhibitory and lethal effects of *A. pleuropneumoniae* leukotoxin. Cruijsen *et al.*, (1992) found that porcine alveolar macrophages could phagocytose but not kill *A. pleuropneumoniae in vitro*. In addition, the alveolar macrophages suffered damage from the intracellular actinobacilli, presumably reflecting leukotoxin production. Chung *et al.*, (1993) made similar observations, noting an impaired ability to ingest yeast, undergo an oxidative burst, and kill *Pasteurella multocida* by porcine alveolar macrophages exposed to *A. pleuropneumoniae in vitro*.

EFFECTS OF HAP TOXINS ON LYMPHOCYTES

A number of studies indicate that HAP leukotoxins functionally inhibit lymphocytes under conditions where outright cytolysis is not a factor. Majury and Shewen (1991) observed inhibition of the proliferative response of mitogen-stimulated bovine lymphocytes exposed to *P. haemolytica* leukotoxin. These workers noted that the response of human and dog lymphocytes were also inhibited; a surprising observation in light of the lack of cytolytic activity of *P. haemolytica* leukotoxin for these cells. Other investigators confirmed an inhibitory effect of leukotoxin on mitogen stimulated peripheral blood cells, without causing a substantial loss of cell viability (Ortiz-Carranza and Czuprynski, 1992). The lethal effects of leukotoxin for the BL-3 bovine lymphoblastoid cells is well recognized (Clinkenbeard *et al.*, 1989); whether sublethal concentrations of leukotoxin affects BL-3 cell proliferation has not been reported.

A series of papers documented the suppressive effects of *A. actinomycetemcomitans* leukotoxin on human lymphocyte proliferation. This may reflect in part killing of monocytes that are required for release of cofactors (Rabie *et al.*, 1988). However, a subsequent report found that approximately 70% of human T cells were killed during extended culture (2 days) with *A. actinomycetemcomitans* leukotoxin (Mangan *et al.*, 1991). These authors found evidence of nuclear fragmentation, suggesting that death may have occurred by apoptosis. A similar effect was noted for human T cells incubated with *E. coli* hemolysin (Jonas *et al.*, 1993).

EFFECTS OF HAP TOXINS ON TISSUE

Most studies of HAP toxins have focussed on the use of single-cell suspensions or adherent cell monolayers. *In vivo*, multiple cell types will be present, each of which might release mediators that could influence tissue damage or the responsiveness of cells to the leukotoxin and other stimuli. There is some evidence, in very simple co-culture systems, that this is an important issue. *P. haemolytica* leukotoxin stimulates bovine neutrophils to produce oxygen radicals that damage endothelial cells *in vitro* (Maheswaran *et al.*, 1993; Breider *et al.*, 1991). Conversely, neutrophils protect bovine endothelial cells against damage when alveolar macrophages are incubated with *P. haemolytica* LPS (Breider *et al.*, 1991). Our laboratory has obtained preliminary evidence that bovine lung parenchyma releases increased amounts of histamine and eicosanoids when preincubated with *P. haemolytica* leukotoxin and then exposed to a second stimulus like calcium ionophore A23187 (Skoyen *et al.*, 1993). It is hoped that continued investigation in this system will elucidate some of the complex interactions that occur in the lung during pulmonary pasteurellosis. Other investigators have also reported bovine pulmonary mast cell release of histamine in response to *P. haemolytica* leukotoxin *in vitro* (Adusu *et al.*, 1994).

A similar approach, using perfused rabbit lungs, has been used in several studies to show that *E. coli* hemolysin stimulates production of thromboxane and other eicosanoids that provoke pulmonary hypertension and edema (Grimminger *et al.*, 1990; Seeger *et al.*, 1991; Ermert *et al.*, 1992). The role of the *E. coli* hemolysin in lung injury model may include a direct effect on endothelial cells, perhaps via the release of nitric oxide (Suttorp *et al.*, 1993). These workers did not see a similar effect using nonhemolytic *E. coli*, suggesting that it was the hemolysin and not LPS that mediated the effect. However, these workers could not exclude possible interactive effects between the hemolysin toxin and LPS. A similar caveat exists for our preliminary studies with bovine lung parenchyma incubated with *P. haemolytica* leukotoxin.

This last problem remains one of some concern. Because of the lability of the RTX toxins, it has proven difficult to obtain toxin preparations that are free of LPS and other microbial components (Ostolaza *et al.*, 1991). The issue is probably not so important for neutrophil studies, since LPS in general does not directly activate neutrophil effector functions, and it is possible to neutralize the effects of *P. haemolytica* leukotoxin using certain anti-leukotoxin MAbs (Czuprynski *et al.*, 1991). The concern is greater when working with monocytes, for which it takes little LPS to stimulate vigorous release of inflammatory mediators. Therefore, when studying the effects of HAP toxins on mononuclear phagocytes the possible contributions of contaminating LPS are not trivial. It is our own experience that the cytolytic and stimulatory effects of *P. haemolytica* leukotoxin for bovine monocytes can be completely dissociated. Addition of a neutralizing MAb completely eliminated killing of monocytes (Stevens and Czuprynski, 1995), similar to what has been reported for bovine neutrophils (Czuprynski *et al.*, 1991). In contrast, addition of this same MAb enhanced rather than impaired monocyte release of IL-1 and TNF activities. Likewise, heating (60°C for 1 hr) eliminated the lethal, but not the stimulatory, effects of the leukotoxin for monocytes. Thus, one might conclude that the stimulating effects of leukotoxin were due solely to contaminating LPS. Establishing how much LPS is present in leukotoxin preparations is also problematic. We estimated a 1000-fold greater amount of LPS contamination with the chromogenic *Limulus* assay, than the standard *Limulus* coagulation assay that is so frequently used. At present, we are considering the possibility that this reflects the ability of the chromogenic *Limulus* assay to detect (by ELISA methodology) LPS complexed with leukotoxin. In contrast, if a LPS-leukotoxin complex exists, perhaps it is unable to activate the coagulation on which the standard *Limulus* assay is dependent.

There is additional evidence that the production, and perhaps biological activity, of LPS and RTX toxins are related. Bauer and Welch (1994) have presented preliminary evidence that rough LPS mutants of *E. coli* produce less biologically active hemolysin. Other investigators have reported additional evidence for the molecular relationship between LPS and RTX toxin synthesis (Bailey *et al.*, 1992; Stanley *et al.*, 1993; Wandersman *et al.*, 1993). Additional research is needed to clarify this point.

CONCLUSIONS

It is hoped that this chapter provides a useful overview of the biological effects of RTX toxins produced by HAP bacteria. Certain points seem clear and consistent. 1) Many HAP organisms produce RTX toxins that are molecularly and biologically similar to the prototype RTX toxin, the *E. coli* hemolysin. 2) These toxins are potent neutrophil modulating agents that can activate, inhibit, or kill the neutrophils depending on the toxin concentration and other culture conditions. 3) These toxins may stimulate release of inflammatory mediators from mononuclear phagocytes, although evaluation of this response is confounded by the presence of LPS in many toxin preparations. 4) The effects of these toxins on mononuclear cells occurs more slowly than for neutrophils, and may involve apoptotic events. 5) These toxins may stimulate release of eicosanoids and other inflammatory mediators in tissue exposed to them. 6) The possible role of LPS in the molecular regulation, release, and biological activities of these toxins is an important area that needs additional investigation.

Other questions come to mind when considering the biological effects of these toxins. What do they bind on the cell membrane? Are there specific receptors? Does binding necessarily lead to membrane perturbation? Recent evidence suggests that is not always the case. These and other questions will command our attention during the coming years as we strive to understand better the complex role these toxins play in modulating the cellular response to infection with HAP organisms.

REFERENCES

Adusu, T.E., Conlon, P.E., Shewen, P.E., and Black, W.D., 1994, *Pasteurella haemolytica* leukotoxin induces histamine release from bovine pulmonary mast cells, *Can. J. Vet. Res.* 58:1-5.

Bailey, M.J.A., Koronakis, V., Schmoll, T., and Hughes, C., 1992, *Escherichia coli* Hly T protein, a transcriptional activator of haemolysin synthesis and secretion, is encoded by the *rfaH (sfrB)* locus required for expression of sex factor and lipopolysaccharide genes, *Molec. Microbiol.* 6:1003-1012.

Baluyut, C.S., Simonson, R.R., Bemrick, W.J., and Maheswaran, S.K., 1981, Interaction of *Pasteurella haemolytica* with bovine neutrophils: Identification and partial characterization of a cytotoxin, *Am J. Vet. Res.* 42:1920-1926.

Bauer, M., and Welch, R.A., 1994, Is lipopolysaccharide directly involved in the expression or activity of *Escherichia coli* hemolysin? 94th Annu. Meet. Amer. Soc. Microbiol., Abst. B110.

Benson, M.L., Thomson, R.G., and Valli, V.E.O., 1978, The bovine alveolar macrophage. II. In vitro studies with *Pasteurella haemolytica*, *Can. J. Comp. Med.* 42:368-369.

Berggren, K.A., Baluyut, C.S., Simonson, R.R., Bemrick, W.J., and Maheswaran, S.K., 1981, Cytotoxic effects of *Pasteurella haemolytica* on bovine neutrophils, *Am. J. Vet. Res.* 42:1383-1388.

Bhakdi, S., Greulich, S., Muhly, M., Eberspacher, B., Becker, H., Thiele, A., and Hugo, F., 1989, Potent leukocidal action of *Escherichia coli* hemolysin mediated by permeabilization of target cell membranes, *J. Exp. Med.* 169:737-754.

Bhakdi, S., Muhly, M. Korom, S., and Schmidt, G., 1990, Effects of *Escherichia coli* hemolysin on human monocytes, *J. Clin. Invest.* 85:1746-1753.

Boehm, D.F., Welch, R.A., and Snyder, I.S., 1990, Domains of *Escherichia coli* hemolysin (HlyA) involved in binding of calcium and erythrocyte membranes, *Infect. Immun.* 58:1959-1964.

Breider, M.A., Kumar, S., and Corstvet, R.E., 1991, Interaction of bovine neutrophils in *Pasteurella haemolytica* mediated damage to pulmonary endothelial cells, *Vet. Immunol. Immunopathol.* 27:337-350.

Burrows, L.L., and Lo, R.Y.C., 1992, Molecular characterization of an RTX toxin determinant from *Actinobacillus suis*, *Infect. Immun.* 60:2166-2173.

Byrd, W., and Kadis, S., 1992, Preparation, Characterization, and immunogenicity of conjugate vaccines directed against *Actinobacillus pleuropneumoniae* virulence determinants, *Infect. Immun.* 60:3042-3051.

Car, B.D., Slauson, D.O., Suyemoto, M.M., Dore, M., and Neilsen, N.R., 1991, Expression and kinetics of induced procoagulant activity in bovine pulmonary alveolar macrophages, *Exp. Lung Res.* 17:939-957.

Cavalieri, S.J., and Snyder, I.S., 1982, Effect of *Escherichia coli* alpha-hemolysin on human peripheral leukocyte function in vitro, *Infect. Immun.* 37:966-974.

Chang, Y., Renshaw, H.F., and Richards, A.B., 1986, *Pasteurella haemolytica* leukotoxin: Physicochemical characteristics and susceptibility of leukotoxin to enzymatic treatment, *Am. J. Vet. Res.* 47:716-723.

Chang, Y., Young, R., Post, D., and Struck, K.K., 1987, Identification and characteristics of the *Pasteurella haemolytica* leukotoxin, *Infect. Immun.* 55:2348-2354.

Chung, W.-B., Backstrom, L., McDonald, J., and Collins, M.T., 1993, *Actinobacillus pleuropneumoniae* culture supernatants interfere with killing of *Pasteurella multocida* by swine pulmonary alveolar macrophages, *Can. J. Vet. Res.* 57:190-197.

Clark, R.A., Leidal, K.G., and Taichman, N.S., 1986, Oxidative inactivation of *Actinobacillus actinomycetemcomitans* leukotoxin by the neutrophil myeloperoxidase system, *Infect. Immun.* 53:252-256.

Clinkenbeard, K.D., Mosier, D.A., and Confer, A.W., 1989, Effects of *Pasteurella haemolytica* leukotoxin on isolated bovine neutrophils, *Toxicon* 27:797-804.

Clinkenbeard, K.D., Mosier, D.A., and Confer, A.W., 1989, Transmembrane pore size and role of cell swelling in cytotoxicity caused by *Pasteurella haemolytica* leukotoxin, *Infect. Immun.* 57:420-425.

Clinkenbeard, K.D., and Upton, M.L., 1991, Lysis of bovine platelets by *Pasteurella haemolytica* leukotoxin, *Am. J. Vet. Res.* 52:453-457.

Confer, A.W., Panciera, R.J., Clinkenbeard, K.D., and Mosier, D.A., 1990, Molecular aspects of virulence of *Pasteurella haemolytica*, *Can. J. Vet. Res.* 54:S48-S52.

Cruijsen, T.L.M., Van Leengoed, L.A.M.G., Dekker-Nooren, T.C.E.M., Schoevers, E.J., and Verheijden, J.H.M., 1992, Phagocytosis and killing of *Actinobacillus pleuropneumoniae* by alveolar macrophages and polymorphonuclear leukocytes isolated from pigs, *Infect. Immun.* 60:4867-4871.

Czuprynski, C.J., and Sample, A.K., 1990, Interactions of *Haemophilus-Actinobacillus-Pasteurella* bacteria with phagocytic cells, *Can. J. Vet. Res.* 54:S36-S40.

Czuprynski, C.J., Noel, E.J., Ortiz-Carranza, O., and Srikumaran, S., 1991, Activation of bovine neutrophils by partially purified *Pasteurella haemolytica* leukotoxin, *Infect. Immun.* 59:3126-3133.

Czuprynski, C.J., and Ortiz-Carranza, O., 1992, *Pasteurella haemolytica* leukotoxin inhibits mitogen-induced bovine peripheral blood mononuclear cell proliferation in vitro, *Microb. Pathog.* 12:459-463.

Devenish, J., Brown, J.E., and Rosendal, S., 1992, Association of the RTX proteins of *Actinobacillus pleuropneumoniae* with hemolytic, CAMP, and neutrophil-cytotoxic activities, *Infect. Immun.* 60:2139-2142.

Devenish, J., Rosendal, S., Johnson, R., and Hubler, S., 1989, Immunoserological comparison of 104-kilodalton proteins associated with hemolysis and cytolysis in *Actinobacillus pleuropneumoniae, actinobacillus suis, Pasteurella haemolytica, and Escherichia coli*, *Infect. Immun.* 57:3210-3213.

Dom, P., Haesebrouck, F., Kamp, E.M., and Smits, M.A., 1992, Influence of *Actinobacillus pleuropneumoniae* serotype 2 and its cytolysins on porcine neutrophil chemiluminescence, *Infect. Immun.* 60:4328-4334.

Dyer, R.M., Benson, C.E., and Boy, M.G., 1984, Production of superoxide anion by bovine pulmonary macrophages challenged with soluble and particulate stimuli, *Am. J. Vet. Res.* 46:336-341.

Ermert, L., Rousseau, S., Schutte, H., Birkemeyer, R.G., Grimminger, Ff., Bhakdi, S., Duncker, H.R., and Seeger, W., 1992, Induction of severe vascular leakage by low doses of *Escherichia coli* hemolysin in perfused rabbit lungs, *Lab. Invest.* 66:362.

Gadeberg, O.V., and Orskov, I., 1984, In vitro cytotoxic effect of à-hemolytic *Escherichia coli* on human blood granulocytes, *Infect. Immun.* 45:255-260.

Grimminger, F., Walmrath, D., Birkemeyer, R.G., Bhakdi, S., and Seeger, W., 1990, Leukotriene and hydroxyeicosatetraenoic acid generation elicited by low doses of *Escherichia coli* hemolysin in rabbit lungs, *Infect. Immun.* 58:2659-2663.

Henricks, P.A., Binkhorst, G.J., Drijver, A.A., and Nijkamp, F.P., 1992, *Pasteurella haemolytica* leukotoxin enhances production of leukotriene B_4 and 5-hydroxyeicosatetraenoic acid by bovine polymorphonuclear leukocytes, *Infect. Immun.* 60:3238-3243.

Inzana, T.J., 1991, Virulence properties of *Actinobacillus pleuropneumoniae*, *Microb. Pathog.* 11:305-316.

Iwase, M., Korchak, H.M., Lally, E.T., Berthold, P., and Taichman, N.S., 1992, Lytic effects of *Actinobacillus actinomycetemcomitans* leukotoxin on human neutrophil cytoplasts, *J. Leukoc. Biol.* 52:224-227.

Iwase, M., Lally, E.T., Berthold, P., Korchak, H.M., and Taichman, N.S., 1990, Effects of cations and osmotic protectants on cytolytic activity of *Actinobacillus actinomycetemcomitans* leukotoxin, *Infect. Immun.* 58:1782-1788.

Jonas, D., Schultheis, B., Klas, C., Krammer, P.H., and Bhakdi, S., 1993, Cytocidal effects of *Escherichia coli* hemolysin on human T lymphocytes, *Infect. Immun.* 61:1715-1721.

Kaehler, K.L., Markham, R.J.F., Muscoplat, C.C., and Johnson, D.W., 1980a, Evidence of cytocidal effects of *Pasteurella haemolytica* on bovine peripheral blood mononuclear leukocytes, *Am. J. Vet. Res.* 41:1690-1693.

Kaehler, K.L., Markham, R.J.F., Muscoplat, C.C., and Johnson, D.W., 1980b, Evidence of species specificity in the cytocidal effects of *Pasteurella haemolytica*, *Infect. Immun.* 30:615-616.

Kolodrubetz, D., Dailey, T., Ebersole, J., and Kraig, E., 1989, Cloning and expression of the leukotoxin gene from *Actinobacillus actinomycetemcomitans*, *Infect. Immun.* 57:1465-1469.

Kraig, E., Dailey, T., and Kolodrubetz, D., 1990, Nucleotide sequence of the leukotoxin gene from *Actinobacillus actinomycetemcomitans*: homology to the alpha-hemolysin/leukotoxin gene family, *Infect. Immun.* 58:920-929.

Lalonde, G., McDonald, T.V., Gardner, P., and O'Hanley, P.D., 1989, Identification of a hemolysin from *Actinobacillus pleuropneumoniae* and characterization of its channel properties in planar phospholipid bilayers, *J. Biol. Chem.* 264:13559-13564.

Maheswaran, S.K., Berggren, K.A., Simonson, R.R., Ward, G.E., and Muscoplat, C.C., 1980, Kinetics of interaction and fate of *Pasteurella hemolytica*: bovine alveolar macrophages, *Infect. Immun.* 30:254-262.

Maheswaran, S.K., Kannan, M.S., Weiss, D.J., Reddy, K.R., Townsend, E.L., Yoo, H.S., Lee, B.W., and Whiteley, L.O., 1993, Enhancement of neutrophil-mediated injury to bovine pulmonary endothelial cells by *Pasteurella haemolytica* leukotoxin, *Infect. Immun.* 61:2618-2625.

Majury, A.L., and Shewen, P.E., 1991, The effect of *Pasteurella haemolytica* A1 leukotoxic culture supernate on the in vitro proliferative response of bovine lymphocytes, *Vet. Immunol. Immunopathol.* 29:41-56.

Mangan, D.F., Taichman, N.S., Lally, E.T., and Wahl, S.M., 1991, Lethal effects of *Actinobacillus actinomycetemcomitans* leukotoxin on human T lymphocytes, *Infect. Immun.* 59:3267-3272.

Markham, R.J.F., Ramnarine, M.C.R., and Muscoplat, C.C., 1982, Cytotoxic effect of *Pasteurella haemolytica* on bovine polymorphonuclear leukocytes and impaired production of chemotactic factors by *Pasteurella haemolytica*-infected alveolar macrophages, *Am. J. Vet. Res.* 43:285-288.

Menestrina, G., Moser, C., Pellet, S., and Welch, R., 1994, Pore-formation by *Escherichia coli* hemolysin (Hly A) and other members of the RTX toxins family, *Toxicology* 87:249-267.

Mosier, D.A., Confer, A.W., and Panciera, R.J., 1989, The evaluation of vaccines for bovine pneumonic pasteurellosis, *Res. Vet. Sci.* 47:1-10.

Mutters, R., Mannheim, W., and Bisgaard, M., 1989, Taxonomy of the group, *In*: Pasteurella and Pasteurellosis, C. Adlam and J.M. Rutter, ed., Academic Press, NY.

Nicolet, J., 1990, Overview of the virulence attributes of the HAP-group of bacteria, *Can. J. Vet. Res.* 54:S12-S15.

Ohta, H., Hara, H., Fukui, K., Kurihara, H., Murayama, Y., and Kato, K., 1993, Association of *Actinobacillus actinomycetemcomitans* leukotoxin with nucleic acids on the bacterial cell surface, *Infect. Immun.* 61:4878-4884.

Ohta, H., Kato, K., Kokeguchi, S., Hara, H., Fukui, K., and Murayama, Y., 1991, Nuclease-sensitive binding of an *Actinobacillus actinomycetemcomitans* leukotoxin to the bacterial cell surface, *Infect. Immun.* 59:4599-4605.

Ortiz-Carranza, O., and Czuprynski, C.J., 1992, Activation of bovine neutrophils by *Pasteurella haemolytica* leukotoxin is calcium dependent, *J. Leukoc. Biol.* 52:558-564.

Ostolaza, H., Bartolome, B., Serra, J.L., de la Cruz, F., and Goni, F.M., 1991, α-Hemolysin from *E. coli*. Purification and self-aggregation properties, *FEBS Lett.* 280:195-198.

Pace, L.W., Kreeger, J.M., Bailey, K.L., Turnquist, S.E., and Fales, W.H., 1993, Serum levels of tumor necrosis factor-α in calves experimentally infected with *Pasteurella haemolytica* A1, *Vet. Immunol. Immunopathol.* 35:353-364.

Rabie, G., Lally, E.T., and Shenker, B.J., 1988, Immunosuppressive properties of *Actinobacillus actinomycetemcomitans* leukotoxin, *Infect. Immun.* 56:122-127.

Sebo, P., and Ladant, D., 1993, Repeat sequences in the *Bordetella pertussis* adenylate cyclase toxin can be recognized as alternative carboxy-proximal secretion signals by the *Escherichia coli* α-haemolysin translocator, *Mol. Microbiol.* 9:999-1009.

Seeger, W., Obernitz, R., Thomas, M., Walmrath, D., Suttorn, N., Holland I.B., Grimminger, F., Eberspacher, B., Hugo, F., and Bhakdi, S., 1991, Lung vascular injury after administration of viable hemolysin-forming *Escherichia coli* in isolated rabbit lungs, *Am. Rev. Respir. Dis.* 143:797-805.

Sharma, S.A., Olchowy, T.W.J., and Breider, M.A., 1991, Alveolar macrophage and neutrophil interactions in *Pasteurella haemolytica*-induced endothelial cell injury, *J. Infect. Dis.* 165:651-657.

Shewen, P.E., and Wilkie, B.N., 1982, Cytotoxin of *Pasteurella haemolytica* acting on bovine leukocytes, *Infect. Immun.* 35:91-94.

Skoyen, S., Czuprynski, C.J., Saban, R., 1993, Effects of the *Pasteurella haemolytica* leukotoxin and LPS on bovine lung parenchymal tissue in vitro, 74th Annu. Conf. Res. Work. Anim. Dis., Abst. P123.

Slocombe, R.F., Malark, F., Ingersoll, R., Derksen, F.J., and Robinson, N.E., 1985, The importance of neutrophils in the pathogenesis of acute pneumonic pasteurellosis, *Am. J. Vet. Res.* 46:2253-2258.

Stanley, P.L.D., Diaz, P., Bailey, M.J.A., Gygi, D., Juarez, A., and Hughes, C., 1993, Loss of activity in the secreted form of *Escherichia coli* haemolysin caused by an rfaP lesion in core lipopolysaccharide assembly, *Mol. Microbiol.* 10:781-787.

Stevens, P., and Czuprynski, C., 1995, Dissociation of cytolysis and monokine release by bovine mononuclear phagocytes incubated with *Pasteurella haemolytica* partially-purified leukotoxin and lipopolysaccharide. *Can. J. Vet. Res.* (in press, April).

Strathdee, C.A., and Lo, R.Y.C., 1987, Extensive homology between the leukotoxin of *Pasteurella haemolytica* A1 and the alpha-hemolysin of *Escherichia coli*, *Infect. Immun.* 55:3233-3236.

Styrt, B., Walker, R.D., Dahl, L.D., and Potter A., 1990, Time and temperature dependence of granulocyte damage by leukotoxic supernatants from *Pasteurella haemolytica* A1, *J. Gen. Microbiol.* 136:2173-2178.

Sutherland, A.D., Effects of *Pasteurella haemolytica* cytotoxin on ovine peripheral blood leucocytes and lymphocytes obtained from gastric lymph, *Vet. Microbiol.* 10:431-438.

Suttorp, N., Fuhrmann, M., Tannert-Otto, S., Grimminger, F., and Bhakdi, S., 1993, Pore-forming bacterial toxins potently induce release of nitric oxide in porcine endothelial cells, *J. Exp. Med.* 178:337-341.

Taichman, N.S., Iwase, M., Lally, E.T., Shattil, S.J., Cunningham, M.E., and Korchak, H.M., 1991, Early changes in cytosolic calcium and membrane potential induced by *Actinobacillus actinomycetemcomitans* leukotoxin in susceptible and resistant target cells, *J. Immunol.* 147:3587-3594.

Tsai, C.-C., McArthur, W.P., Baehni, P.C., Hammond, B.F., and Taichman, N.S., 1979, Extraction and partial characterization of a leukotoxin from a plaque-derived gram-negative microorganism, *Infect. Immun.* 25:427-439.

Tsai, C.-C., Shenker, B.J., DiRienzo, J.M., Malamud, D., and Taichman, N.S., 1984, Extraction and isolation of leukotoxin from *Actinobacillus actinomycetemcomitans* with polymyxin B, *Infect. Immun.* 43:700-705.

Udeze, F.A., and Kadis, S., 1992, Effects of *Actinobacillus pleuropneumoniae* hemolysin on porcine neutrophil function, *Infect. Immun.* 60:1558-1567.

van Leengoed, L.A.M.G., and Dickerson, H.W., 1992, Influence of calcium on secretion and activity of the cytolysins of *Actinobacillus pleuropneumoniae*, *Infect. Immun.* 60:353-359.

Walker, R.D., Corstvet, R.E., and Panciera, R.J., 1979, Study of bovine pulmonary response to *Pasteurella haemolytica*: Pulmonary macrophage response, *Am. J. Vet. Res.* 41:1008-1014.

Wandersman, C., and Letoffe, S., 1993, Involvent of lipopolysaccharide in the secretion of *Escherichia coli* α-haemolysin and *Erwinia chrysanthemi* proteases, *Molec. Microbiol.* 7:141-150.

Weiss, D.J., Bauer, M.C., Whiteley, L.O., Maheswaran, S.K., Ames, T.R., 1991, Changes in blood and bronchoalveolar lavage fluid components in calves with experimentally induced pneumonic pasteurellosis, *Am. J. Vet. Res.* 52:337-343.

Welch, R.A., Forestier, C., Lobo, A., Pellett, S., Thomas Jr, W., and Rowe, G. 1992, The synthesis and function of the *Escherichia coli* hemolysin and related RTX exotoxins, *Microbiol. Immunol.* 105:29-36.

ANTI-IDIOTYPE MONOCLONAL ANTIBODIES AS LPS INTERNAL IMAGES AND IMMUNOLOGIC PROTECTION AGAINST ENDOTOXIN LETHALITY IN MICE

S.K. Field[1], M. Pollack[2] and D.C. Morrison[3]

[1]Department of Molecular Biology, Bristol-Meyers Squibb Pharmaceutical Research Institute, Princeton, NJ 08534 USA
[2]Department of Medicine, Uniformed Services University of the Health Sciences, Bethesda, MD 20814 USA
[3]Cancer Center, University of Kansas Medical Center, Kansas City, KS 66160 USA

INTRODUCTION

Gram negative bacterial infections continue to be a major cause of morbidity and mortality both in the United States and throughout the world (Anon). The existing evidence would strongly support the concept that, in many cases of Gram negative bacteremia and sepsis, endotoxic lipopolysaccharides (LPS) released from the microbial outer cell membrane are a major contributing factor to the pathophysiology of disease. In this respect, while LPS is generally regarded not to be directly toxic to host cells and tissues, it is potentially capable of stimulating cells to release a variety of inflammatory mediators, including tumor necrosis factor (TNF), many of the interleukins (Il-1, Il-6, Il-8, Il-10), products of phospholipid metabolism such as prostaglandins, leukotrienes and platelet activating factor (PAF) and procoagulant activity in the form of tissue factor (Morrison and Ryan, 1979). These biologically active mediators serve to promote a more generalized and often uncontrolled systemic inflammatory response which results in the hypotension, disseminated intra-vascular coagulation and multisystem organ failure characteristic of endotoxin shock.

Because of the important role which has been ascribed to endotoxic LPS in Gram negative infectious disease, investigators have long argued that therapeutic strategies which would result in significant reductions in circulating levels of endotoxin should produce lower mortality. Although immunologic approaches to address therapeutically the neutralization and/or elimination of bacterial toxins have been established since the time of Pasteur, von Behring and Kitasoto, the chemistry of the LPS macromolecule has presented unusual problems with respect to immunologic uniqueness of the chemically and immunologically dominant O-antigen polysaccharide which is structurally different in each strain of Gram negative microorganism. Further, the fact that the LPS macromolecule is generally considered to be a T-cell independent antigen (Jacobs and Morrison, 1975), eliciting primarily low affinity IgM antibody and little, if any, memory response, did not encourage efforts toward active immunization protocols for immunologic protection against the lethal effects of LPS.

Haemophilus, Actinobacillus, and Pasteurella
Edited by W. Donachie *et al.*, Plenum Press, New York, 1995

Experimental efforts to overcome these problems focused, therefore, upon two major directions, first the use of so called rough chemotype LPS preparations as immunogens and second, the development of passive immunity as the method of choice to counteract endotoxemia. The first approach was predicated upon the fact that all LPS macromolecules retain conserved chemical structures termed inner core polysaccharide and covalently attached lipid A, wherein reside the endotoxic principals of LPS, and it was argued that antibodies directed against these conserved structures should manifest broad cross protective immunity against LPS from Gram negative bacteria (eg McCabe and Greely, 1972). Resolution of the second problem was achieved by passive administration of inner core/lipid A specific antibody to experimental animals subsequently challenged with otherwise lethal doses of endotoxic LPS (eg Braude and Douglas, 1972).

Early successes in a number of experimental animal models provided strong support for the concept that antibody to conserved inner core 2-keto-3-deoxyoclulosonate (Kdo) and/or lipid A residues on LPS would, in fact, reduce mortality both with viable microorganisms and with purified bacterial endotoxin (reviewed in Cross and Opal, 1994). These well-documented results provided the foundation for several large placebo-controlled clinical trials in patients with presumed Gram-negative sepsis. Unfortunately while the initial studies with both polyclonal (Ziegler *et al.*, 1982) and monoclonal (Ziegler *et al.*, 1991; Greenman *et al.*, 1991) antibodies indicated promise for such treatments, more recent studies have not fully confirmed these earlier findings (reviewed in Cross and Opal, 1994), thus triggering speculation as to the real effectiveness of anti-endotoxin strategies.

It is clear that conclusions with respect to the above-cited studies are confounded by a number of potential problems, including the isotype and/or affinity of the antibody passively administered to patients, the time of administration of antibody relative to time of onset of endotoxemia, the nature of the underlying disease, the requirement of assessment of outcome at 28 days post therapy and the levels of circulating endotoxin at the time of initiation of anti-endotoxin therapy. As a consequence, the role of anti-endotoxin antibody in treating Gram negative sepsis remains unresolved.

While not actively pursued by many investigators, some of the early studies to generate cross-reactive anti-endotoxin immunity involved active rather than passive immunity. Thus, McCabe and Greely, (1972) were able to demonstrate that immunization of mice with Re-chemotype LPS provided significant levels of protection against subsequent challenge with either *Klebsiella* or *E. coli* bacteria. In these studies, however, the duration of the protective efficacy was not assessed.

During the past decade, interest has developed in the potential use of anti-idiotype monoclonal antibodies as candidate vaccines to induce protective immunity against a variety of microbial pathogens. Such studies have been predicated upon the concept that certain subclasses of anti-idiotype antibodies (termed Ab2) reflect an internal image of the binding site of antibodies directed against microbial antigens and, thus, can serve as molecular mimics of the original antigen. In this respect, several recent reports (Westerink *et al.*, 1990; Monafo *et al.*, 1987; Shreiber *et al.*, 1991; McNamara *et al.*, 1984; Stein and Soderstrom 1984; McNamara *et al.*, 1992; Kato *et al.*, 1990) have demonstrated the induction of antibodies which can react with microbial antigens following anti-idiotype hyperimmunization, including anti-lipid A antibodies.

Our own efforts have focused upon the generation of a candidate Ab2 anti-idiotype antibody which reflects the internal image of an LPS inner core specific mouse monoclonal antibody. For these studies, Armenian hamsters were immunized with an IgG3 mouse monoclonal antibody with documented specificity for the Kdo-lipid A region of LPS. One monoclonal termed MAb4G2 was isolated and characterized and shown to manifest all of the properties required to characterize it as a true Ab2 (Field *et al.*, 1993; Field and Morrison, 1994). Immunization of mice with MAb4G2 results in the induction of anti-LPS specific antibody which reacts with R-chemotype LPS but not lipid A. MAb4G2 immunized mice are shown to be significantly protected against an otherwise lethal challenge with LPS (Field and Morrison, 1994).

MATERIALS AND METHODS

Antiidiotype monoclonal antibody MAb4G2

Armenian hamsters were immunized subcutaneously with 50 µg of a mouse IgG3 anti-Kdo/lipid A monoclonal antibody MAbY1-4A6 in Freunds Complete Adjuvant and boosted at two-week intervals with the same antigen in Freunds Incomplete Adjuvant. Splenocytes from immunized hamsters were fused with the mouse myeloma cell line Sp2/0 and resultant hybridomas initially screened for binding to MAbY1-4A6 but not control IgG3. Candidate clones were subcloned and further analyzed for binding properties as described in the following text. More detailed information can be obtained in the original publication describing this antiidiotype MAb4G2 monoclonal antibody (Field *et al.*,).

Immunoassays

Enzyme-linked immunosorbant assays (ELISA) were used for all antibody binding determinations. To detect LPS binding, polystyrene microtiter plates (96 well, Immunlon 1, Dynatech Labs, Alexandria, VA) were pretreated with 100 µg/ml of poly-l-lysine in phosphate buffered saline (PBS, pH 7.4) followed by addition of LPS (12.5-50 æg/ml, PBS) for 1 hour at 37°C. Plates were then rinsed three times with PBS, and non-specific binding sites blocked with 10% fetal bovine serum overnight at 4°C. Plates were again washed with PBS followed by addition of various dilutions of a source of anti-LPS antibody or control antibody for 1 hour at 37°C. After again washing with PBS (containing 0.05% tween 20), plates were developed using peroxidase-conjugated anti-immunoglobulin antibodies and substrate (1:1 ratio of hydrogen peroxide:3, 3', 5, 5' tetramethylbenzidine)(Kirkegaard and Perry Labs, Gaithersburg, MD USA). Similar protocols were employed to detect immunoglobulin antigens bound to microtiter plates except that the initial poly-l-lysine step was omitted.

Affinity Chromatography

To assess the specific binding properties of antibodies generated following immunization with MAb4G2, an affinity chromatography assay was developed in which MAb4G2 was conjugated to Sepharose 4B. For these studies, cyanogen bromide activated Sepharose 4B was added to a solution of either MAb4G2 or a control hamster monoclonal IgG (Jackson Immunoresearch, West Grove, PA USA) in coupling buffer (0.1M $NaHCo_3$, pH 8.3 containing 0.5 M NaCl). at room temperature and rocked end-over-end for two hours. Buffer was then decanted and excess binding sites blocked with 1.0 M ethanolamine using similar procedures as described above. The Sepharose 4B was then washed an additional several times with coupling buffer, acetone, and finally coupling buffer with azide (0.01%). Approximately 10 mg of monoclonal antibody was bound per 100 ml of packed Sepharose 4B.

Monoclonal MAb4G2 Immunization Protocols

To assess immune responses to the antiidiotype monoclonal MAb4G2 antibody, Syrian hamsters were administered 100 µg of MAb4G2 subcutaneously in CFA followed by two additional immunizations 7 days apart with antigen in IFA. After each seven day interval, those hamsters were exsanguinated by cardiac puncture and serum samples analyzed for specific antibody as described above.

Mouse Lethality Studies

To assess the capacity of mice immunized with MAb4G2 or LPS to be protected against the lethal effects of endotoxin, groups of 8-10 mice were given MAb4G2 (150 μg) or Re-LPS (50 μg intraperitoneally on day 0 and day 14. Seven days later, mice were sensitized to the lethal effects of LPS by treatment with D-galactosamine and given an approximate lethal dose 50 (LD_{50}) of RE-LPS. Survival was monitored over 72 hours. Control mice were treated with either saline or normal hamster monoclonal IgG antibody.

RESULTS

Anti-idiotype Monoclonal Antibody 4G2

Immunization of Armenian hamsters with the mouse IgG_3, MAb, Y1-4Ab, a Kdo/lipid A specific antibody, and subsequent fusion of splenocytes from immunized animals with a mouse myeloma cell line resulted in the generation of more than 600 hybridomas. Analysis of the specific binding of antibody in culture supernatants to the immunizing MABY1-4Ab, but not to control mouse IgG_3, allowed the identification of 23 candidate anti-idiotype hybridomas. Of these, eight with the highest apparent binding affinity were subcloned and retested for both specific binding to MAbY1-4Ab and for their ability to inhibit the binding of the latter to Re-LPS. Partial results for the latter assay are shown in Figure 1 and indicate that one monoclonal antibody, secreted by hybridoma clone 4G2.1, would completely inhibit binding of MAbY1-4Ab to Re-LPS. The other eight clones, for which 9F5.1 is representative, did not inhibit binding of MA6Y1-4Ab to Re-LPS, nor did control goat antimouse IgG.

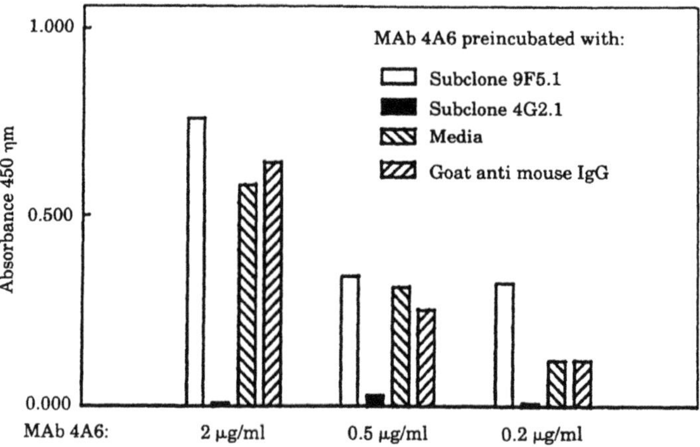

Figure 1. Inhibition of MAb4G2 binding of a Kdo-specific monoclonal antibody to *Salmonella* Re 595 chemotype LPS. Three concentrations of MAbY1-4A6 were incubated with 50 μg of RPMI 1640, goat anti-mouse IgG monoclonal or culture supernatants from nine hybridomas secreting antibodies specific for the variable region of MAbY1-4A6. Residual binding of MAbY1-4A6 to Re-LPS coated on ELISA plates was then determined as described in Materials and Methods (one of the nine culture supernatants from hybridomas secreting antibodies specific for the variable region of MAbY1-4A6, subclone 9F5.1, is represented).

Binding properties of MAb4G2.1 for Monoclonal and Polyclonal Anti Re-LPS Antibodies

The data shown in Figure 1 suggest that MAb4G2.1 may function as an anti-idiotype antibody for MABY1-4A6 and serve as an internal image of the binding site for this antibody, that is, as a structural analog of the Kdo region of Re-LPS. If so, the MAb4G2.1 monoclonal antibody should bind to other anti Re-LPS specific monoclonals as well as to some of the antibodies present in a polyclonal antiserum raised against Re-LPS. To address the former question, the capacity of MAb4G2.1 monoclonal to inhibit the binding to Re-LPS of three additional anti Re-LPS monoclonals (MAbY1-F11 derived from a Balb/c mouse, and two monoclonals, MAbR22C3 and MAbF4, from outbred CF1 mice). The results of this study, summarized in Figure 1, demonstrate that, similar to data shown above for MAbY1-4Ab, all three additional monoclonals are also inhibited in Re-LPS binding by MAb-4G2.1.

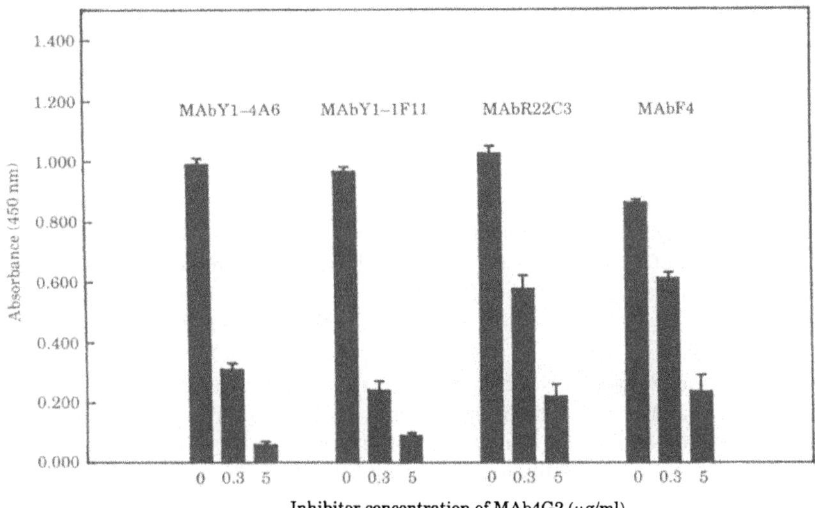

Figure 2. Inhibition of binding of mouse anti-Re-LPS monoclonal antibodies to *Salmonella* Re 595 LPS by MAb4G2. Serial dilutions of MAb4G2 were incubated overnight at 4°C with the BALB/c monoclonal antibodies MAbY2-4A6 and MAbY1-1F11, or two Re-LPS-specific CF-1 monoclonal antibodies, MAbF4 and MAbR22C3. After microcentrifugation, aliquots of supernatants were added to Re-LPS-coated microtiter plates and ELISAs were developed according to standard procedures.

Experiments were also carried out to assess the ability of MAb4G2.1 to bind to anti Re-LPS antibodies present in serum of mice previously immunized with Re-LPS. For these studies, C3HeB/FeJ mice were immunized with 50 μg Re-LPS in Freunds complete adjuvant three times 7 days apart and serum obtained by retroorbital bleeding 7 days after the second immunization. Soluble antigen-antibody competitive inhibition assays were then carried out in which various dilutions of either MAb4G2 monoclonal or Re-LPS were preincubated with the mouse polyclonal anti Re-LPS antiserum and residual binding to Re-LPS in microtiter plates then determined. The results of this experiment, shown in Figure 3, establish that both Re-LPS and MAb4G2.1 monoclonal antibody will inhibit binding of anti Re-LPS polyclonal antibodies to Re-LPS. In control experiments (data not shown), the Re-LPS, but not MAb4G2.1 will inhibit binding of a lipid A-specific monoclonal antibody to Re-LPS.

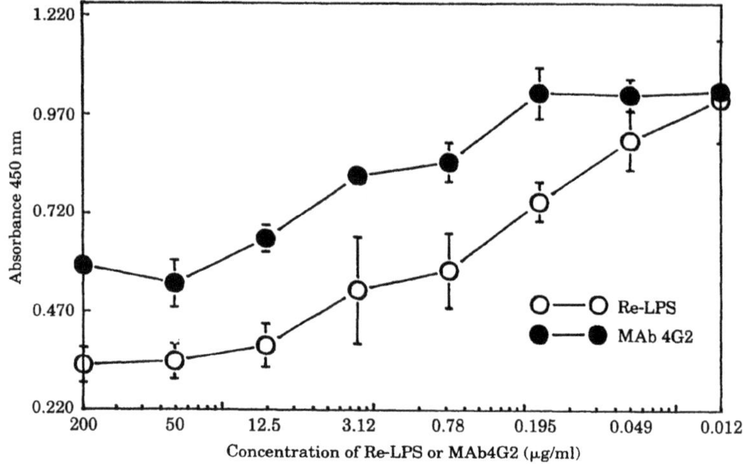

Figure 3. MAb4G2 recognition of anti-Re-LPS polyclonal sera from mice immunized with Re-LPS. Re-LPS or MAb4G2 was serially diluted and then added to mouse anti-Re-LPS polyclonal antiserum. Mixtures were incubated overnight at 4§C, microcentrifuged and aliquots added to microtiter plates coated with Re-LPS. ELISAs were developed according to standard procedures.

While these data would support the concept that MAb4G2.1 is, in fact, a true anti-idiotype antibody, it was also necessary to insure that there was nothing unique about the interaction of the MAb4G2.1 monoclonal antibody with mouse anti Re-LPS antibodies (i.e., the presence of a highly conserved heavy chain variable region). It was, therefore, necessary to demonstrate that the MAb4G2.1 would also bind to Re-LPS specific antibodies present in serum from animals of a species other than mouse. A rabbit polyclonal anti-Re-LPS antiserum (generously provided by Dr. Ben Appelmelk, Vrije Universiteit, Amsterdam, The Netherlands) was, therefore, evaluated for its ability to bind to either Re-LPS, MAb4G2.1 monoclonal antibody, or a control preparation of hamster IgG antibody, all of which were immobilized on microtiter plates. The result of this study, in which binding of various dilutions of rabbit anti Re-LPS polyclonal antibody were determined, are shown in Figure 4 and clearly demonstrate specific binding to the MAb4G2.1. Collectively, therefore, these results showing binding of MAb4G2.1 to four monoclonal antibodies with specificity for Re-LPS, as well as polyclonal anti Re-LPS antibody from both mice and rabbits.

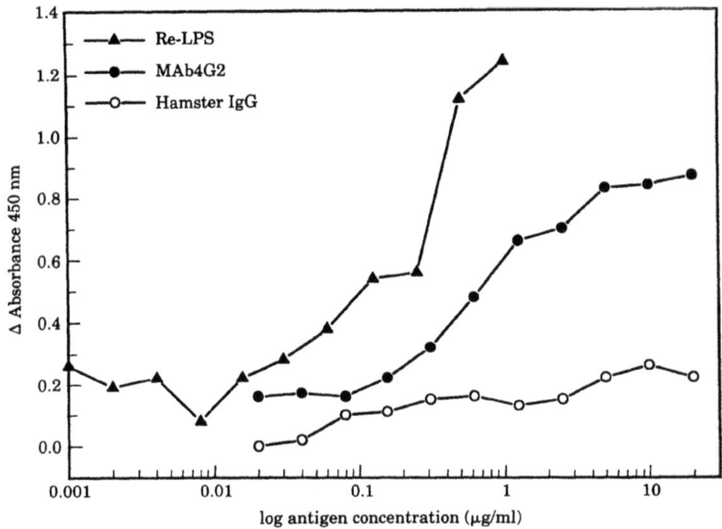

Figure 4. Re-LPS, MAb4G2 or an isotype-matched hamster IgG monoclonal antibody were serially diluted onto microtiter plates. After blocking, rabbit anti-*E. coli* J5 antisera was added to each well and ELISAs were developed using a goat anti-rabbit IgG peroxidase-conjugated antibody.

Generation of hamster anti-MAb4G2 antiserum (Ab3)

If the MAb4G2 monoclonal antibody truly represents a molecular mimic of the Kdo determinant in the inner core region of LPS, then immunization of experimental animals with MAb4G2 should result in the production of anti-LPS antibody. Syrian hamsters were, therefore, immunized three times with MAb4G2 in adjuvant, as described in Methods, and serum obtained seven days following each immunization. The presence of antibody specific for Re-LPS was then assessed by standard ELISA, and the data are shown in Figure 5. A single immunization results in low level anti-LPS antibody, and titers are increased following the second and third immunization.

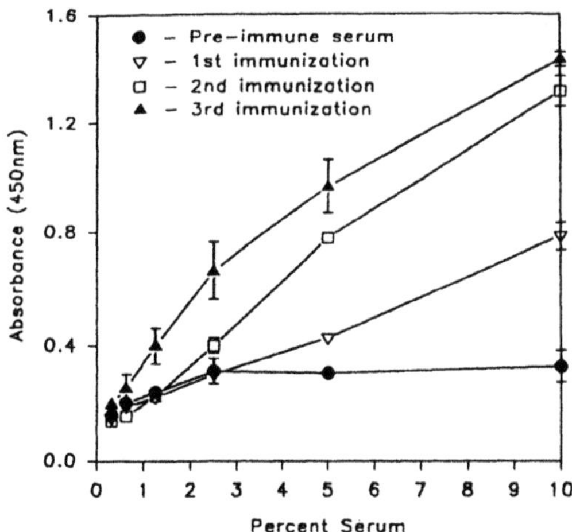

Figure 5. Syrian hamsters were immunized with 100 μg of MAb4G2 on days 0, 7 and 14 in adjuvant. Sera obtained from 3 immunized animals on days 0, 7, 14 and 21 were pooled and assayed in triplicate by ELISA for specific binding to Re-LPS. The data represent the mean ± SEM of triplicate determinations.

To assess the binding specificity of the antibodies generated in hamsters following immunization with MAb4G2 (Ab3) to that of the MAb4A6 monoclonal antibody (Ab1) used to generate Mab4G2 (Ab2), an affinity column using MAb4G2 immobilized to Sepharose 4B was prepared as described in Material and Methods. As a control, normal hamster IgG was also immobilized on Sepharose 4B. When either the hamster anti-MAb4G2 (Ab3) or mouse MAbY4A6 (Ab1) was passed through the MAb4G2 (Ab2) affinity column, antibody immunoreactive with Re-LPS by ELISA was selectively removed (data not shown). No significant reduction in LPS specific antibody was observed when similar studies were carried out using the control hamster IgG affinity column. Thus, both hamster anti-MAb4G2 sera and MAbY4A6 contain antibodies which react both with MAb4G2 and Re-LPS.

Experiments were then carried out to determine whether the antigenic specificity of the anti-MAb4G2 antibodies was the same as that of the MAbY4A6 antibody. For this study, the MAb4G2 affinity column was preabsorbed with an excess of MAbY4A6 antibody. Control studies confirmed that additional MabY4A6 antibody added to this column no longer bound to the affinity column (Figure 6, top). When hamster anti-MAb4G2 antiserum was then added to this column, the antibody specific for Re-LPS was no longer retained by the affinity column (Figure 6, bottom) indicating that MAbY4A6 (Ab1) and hamster anti-MAb4G2 (Ab3) compete for the same binding site on MAb4G2 (Ab2). These findings provide formal proof of the Ab2 specificity of MAb4G2.

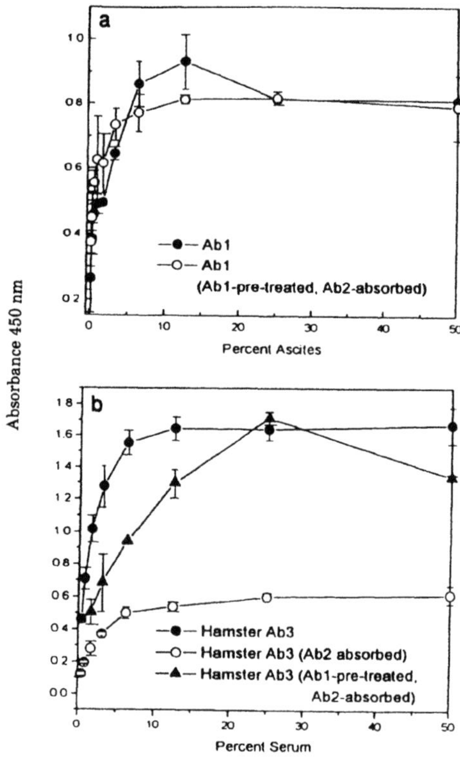

Figure 6. Ab1-like properties of hamster αMAb4G2 IgG serum antibodies. (a) the MAb4G2-Sepharose affinity column was saturated with the Ab1 as described in the text. Then additional Ab1 was added and the unbound material analyzed for residual binding to Re-LPS by ELISA. (b) Hamster αMAb4G2 immune serum was added to the Ab1-saturated, MAb4G2-Sepharose affinity column, and the unbound material tested for binding to Re-LPS coated onto microtiter plates as previously described. The data represent the mean determinations of duplicate samples ñ the range of error.

Active immunization of mice with MAb4G2 provides protection against LPS lethality

To assess the potential protective efficacy of MAb4G2 as an immunogen, mice were immunized with either Re-LPS (50 µg), MAb4G2 (150 µg), a control hamster IgG monoclonal IgG antibody (150 µg), or physiological saline on day 0 and 14. On day 21, all mice were challenged with an approximately 50% lethal dose of Re-LPS, and survival was monitored over the next three days. The results of three experiments are summarized in Table 1 and confirm that immunization with either LPS or MAb4G2 will provide highly significant protection against LPS lethality.

Table 1. Protection of D-galactosamine-treated mice from lethal challenge with Re-LPS by pretreatment with MAb4G2

Treatment	No of dead mice				
	In expt 1 (n = 10)	In expt 2 (n = 10)	In expt 3 (n = 8)	Total (n = 28)	% Survival
150µg of MAb4G2	0	2	0	2	93*
150µg of hamster IgG	2	5	3	10	64
50µg of Re-LPS	0	0	0	0	100
Saline	16	4	11	61	

*<0.005 versus hamster IgG or saline pretreatment groups.

We have also demonstrated (data not shown) that both LPS- and MAb4G2-immunized mice develop circulating IgM and IgG antibodies specific for Re-LPS (17, 19).

DISCUSSION

In this manuscript we have described the isolation and characterization of a hamster IgG monoclonal antibody (MAb4G2) which binds to the antigen combining site of a mouse monoclonal IgG antibody (MAbY4A6), the latter of which has binding specificity for the inner core Kdo/lipid A region of LPS. Evidence from a variety of experimental approaches indicates that MAb4G2 is, in fact, a molecular mimic of the inner core region of LPS. Immunization of both mice and hamsters with MAb4G2 results in the production of anti-LPS antibody, and immunized mice are protected against lethality mediated by LPS. Thus, MAb4G2 is a true anti-idiotype antibody of the Ab2 type.

The ability of antibodies induced against Mab4G2 to protect against lethality from "homologous" *S. minnesota* R595 LPS, while of profound theoretical significance, may be of limited practical importance. It might be much more important, in practical terms, whether such antibodies are cross-reactive with heterologous LPS of the types identified with clinical disease and cross-protective against heterologous live challenge. Earlier studies in which mice immunized with rough Re-LPS from *S. minnesota* were protected against infection with *E. coli* and *K. pneumoniae* (McCabe and Greely, 1972) suggest the possibility, at least, that animals immunized with the anti-idiotype MAb4G2, which contains the internal image of Re-LPS, would induce similar cross-protective immunity.

Since our studies have demonstrated that mice immunized with MAb4G2 generate primarily IgG specific antibody (Field *et al.*, 1993, Field and Morrison, 1994), it is likely that it is this antibody which mediates the observed protection. This latter finding is of potential interest in that both major clinical trials for passive immunotherapy with monoclonal anti-lipid A antibody have been of the IgM isotype (Ziegler *et al.*, 1991, Greenman *et al.*, 1991).

As pointed out in the Introduction, while studies with experimental animals using passive immunotherapy with antibody to LPS have usually (but not always) provided significant protection against the lethal effects of LPS, efforts to translate such therapy to patients with Gram negative sepsis have been less effective (Cross and Opal, 1994). While the reasons for this are undoubtedly complex, it is likely that one reason involves the timing of administration of antibody relative to initiation of endotoxin-related pathophysiological responses. It is possible that active immunization under conditions such that more long-term immunity might be achieved, as with the antiidiotype monoclonal, might provide a state of immunologic readiness for production of cross reactive anti-LPS antibody early on in infection. For example, temporal profiles of Re-LPS-specific IgG and IgM titers in mice immunized with Re-LPS *vs* MAb4G2 illustrates the feasibility of this approach. Animals immunized and later boosted with MAb4G2 maintained elevated serum titers of Re-LPS-specific IgG antibodies for at least three weeks after treatment and, at the same time, were resistant to the lethal properties of endotoxin (Field and Morrison, 1994). In groups of mice immunized with MAb4G2 and later with Re-LPS, the IgG and IgM profiles were much like those of animals immunized with MAb4G2 on successive occasions and, likewise, these animals experienced enhanced resistance to lethal dosages of endotoxin. Hence, the prophylactic use of MAb4G2 might shift adaptive immunity to an LPS-specific IgG immune response subsequent to exposure to Gram-negative bacteria.

The observation that most core/lipid A-specific monoclonal antibodies do not readily bind to smooth LPS has raised doubts as to the clinical efficacy of these reagents. However, the fact that MAb4G2-specific serum antibodies do recognize smooth LPS present in supernatants from antibiotic-treated Gram-negative bacteria indicates that these antibodies may prove useful to bacteremic patients who are routinely administered antibiotics. Experiments are currently ongoing in our laboratory to assess the duration of immunologic priming with MAb4G2 to subsequent LPS challenge.

ACKNOWLEDGMENTS

This research was supported by NIH Grants R37A23447 (DCM), PO1-CA54474 (DCM), RO1AI22706 (MP) and Navy Medical Research and Development Contract 63706N.M00095.00 (M.P.) and 63706N.M00095.001.9218 (M.P.). S.K. Field was a scholar of the Kansas Health Foundation Cancer Research and Training Program. The authors thank Ms. Kathy Rode for preparation of this manuscript.

REFERENCES

Braude, A.I. and Douglas, H. Passive immunization against the local Shwartzman reaction, *J. Immunol.* 108:505-512 (1972).

Cross, A.S. and Opal, S. Therapeutic intervention in sepsis with antibody to endotoxin: is there a future?, *J. Endotoxin Research*, 1:57-69 (1994).

Field, S., Pollack, M. and Morrison, D.C. Development of an anti-idiotype monoclonal antibody mimicking the structure of lipopolysaccharide (LPS) inner-core determinants, *Microb. Pathog.* 15:103-120 (1993).

Field, S.K. and Morrison, D.C. A hamster anti-idiotype monoclonal antibody, mimicking the inner-core region of Gram-negative bacterial lipopolysaccharide (LPS), stimulates LPS inner-core-specific serum antibodies in hamsters, *J. Endotoxin Research* 1:120-130 (1994).

Field, S.K. and Morrison, D.C. An anti-idiotype antibody which mimics the inner-core region of lipopolysaccharide protects mice against a lethal challenge with endotoxin, *Infect. Immun.* 62:3994-3999 (1994).

Greenman, R., Schein, R.M.H., Martin, M.A., Wenzel, R.P., MacNtyre, N.R., Emmanuel, G., Herman, C., Kohler, R.B., McCarthy, M., Plouffe, J., Russel, J.A. and the XOMA Study Group. A controlled clinical trial of E5 murine monoclonal anti-lipid A antibody in bacteremic and nonbacteremic patients. *JAMA* 266:1097-1102 (1991).

Increase in national hospital discharge survey rates for septicemia-United States, 1979-1987. *MMWR* 39:31-34 (1990).

Jacobs, D.M. and Morrison, D.C. Stimulation of a T-independent primary anti-hapten response *in vitro* by TNP-lipopolysaccharide, *J. Immunol.* 114:360-366 (1975).

Kato, T., Takazoe, I. and Okuda, K. Protection of mice against the lethal toxicity of a lipopolysaccharide (LPS) by immunization with anti-idiotype antibody to a monoclonal antibody to lipid A from *Eikenella corrodens* LPS, *Infect. Immun.* 58:416-420 (1990).

McCabe, W.R. and Greely, A. Immunization with R mutants of *S. minnesota*. I. Protection against challenge with heterologous Gram-negative bacilli, *J. Immunol.* 108:601-610 (1972).

McNamara, M.K., Ward, R.E. and Kohler, H. Monoclonal idiotype vaccine against *Streptococcus pneumoniae* infection, *Science* 226:1325-1326 (1984).

Monafo, W.J., Greenspan, N.S., Cebra-Thomas, J. and Davie, J.M. Modulation of the murine response to streptococcal group A carbohydrate by immunization with monoclonal anti-idiotype, *J. Immunol.* 139:2702-2707 (1987).

Morrison, D.C. and Ryan, J.L. A review: bacterial endotoxins and host immune function, *Adv. Immunol.* 28:293-430 (1979).

Shreiber, J.R., Nixon, K.L., Tosi, M.F., Pier, G.B. and Patawaran, M.B. Anti-idiotype-induced, lipopolysaccharide-specific antibody response to *pseudomonas aeruginosa*. II. Isotype and functional activity of the anti-idiotype-induced antibodies. *J. Immunol.* 146:188-193 (1991).

Stein, K.E. and Soderstrom, T. Neonatal administration of idiotype or anti-idiotype primes for protection against *E. coli* K13 infection in mice, *J. Exp. Med.* 166:1001-1011 (1984).

Su, S., McNamara, W., Apicella, M.A. and Ward, R.E. A nontoxic, idiotype vaccine against gram-negative bacterial infections, *J. Immunol.* 148:234-238 (1992).

Westerink, M.A., Giardina, P., Campagnari, J.A. and Apicella, M.A. The thymus-dependent nature of the murine antibody response to a monoclonal anti-idiotypic antibody to the *Neisseria meningitidis* serogroup C capsular polysaccharide, *Microb. Pathog.* 8:411-419 (1990).

Ziegler, E.J., Fisher, C.J., Sprung, C.L., Straube, R.C., Sadoff, J.C., Foulke, J.C., Wortel, C.H., Fink, M.P., Dellinger, R.P., Teng, N.N.H., Allen, I.E., Berger, H.J., Knatterud, G., Lobuglio, A.F., Smith, C.R., and the HA-1A Study Group. Treatment of Gram-negative bacteremia and septic shock with HA-1A human monoclonal antibody against endotoxin, *N. Engl. J. Med.* 324:429-436 (1991).

Ziegler, E.J., McCutchan, J.A., Fierer, A., Glauser, M.P., Sadoff, J.C., Douglas, H., and Braude, A.I. Treatment of gram-negative bacteremia and shock with human antiserum to a mutant *Escherichia coli*, *N. Engl. J. Med.* 307:1225-1230 (1982).

HOST RESPONSE TO INFECTION WITH HAP: IMPLICATIONS FOR VACCINE DEVELOPMENT

P.E. Shewen

Department of Veterinary Microbiology and Immunology
Ontario Veterinary College
University of Guelph
Guelph, Canada N1G 2W1

Organisms of the HAP (Haemophilus, Actinobacillus, Pasteurella) Group exist as commensals of the mucosae of mammals, where for the most part they cause no adverse effect. This inoffensive colonization results from a balance between bacterial growth and host response such that deep colonization is prevented or controlled by non-specific defense mechanisms in concert with an essentially local immune response (Brandtzaeg, 1992). Under conditions of stress, when such defenses are compromised or impaired, the opportunity for infection to penetrate more deeply may arise. The disease which often follows results from the combined pathological effects of bacterial virulence factors and the ensuing host response. In many, if not most cases, observed lesions can be attributed to the inflammation which results from the host's attempts to control infection. The scenario for induction of bovine pneumonic pasteurellosis by *Pasteurella haemolytica* A1 typifies this model for HAP induced disease (Figure 1, column 1).

Similar models for disease induction could be drawn for *P. multocida* pneumonia in several species, *Haemophilus somnus* in cattle, *Actinobacillus pleuropneumoniae* in pigs, or *Haemophilus influenzae* in children. Likewise, septicemic HAP infections result when the mucosal surface is breached or rendered penetrable, most likely in a deep mucosal site such as the lung or small intestine where the physical barrier between outside and inside is slight. While this pattern of disease has broad applicability, it is important to recognise that not all animals at risk actually develop disease. In any given outbreak some will succumb completely, some become ill but recover spontaneously, while others remain apparently unaffected. The existence of the latter "immune" animals provides for the hypothesis that disease may be preventable by immune intervention through vaccination. Comparison of the immune response in resistant and susceptible animals has the potential to yield valuable insight about protective immunity. Furthermore, the commensal colonization which provided for protective immunity in some animals, may be successfully exploited in their more susceptible counterparts as a priming response for immunization.

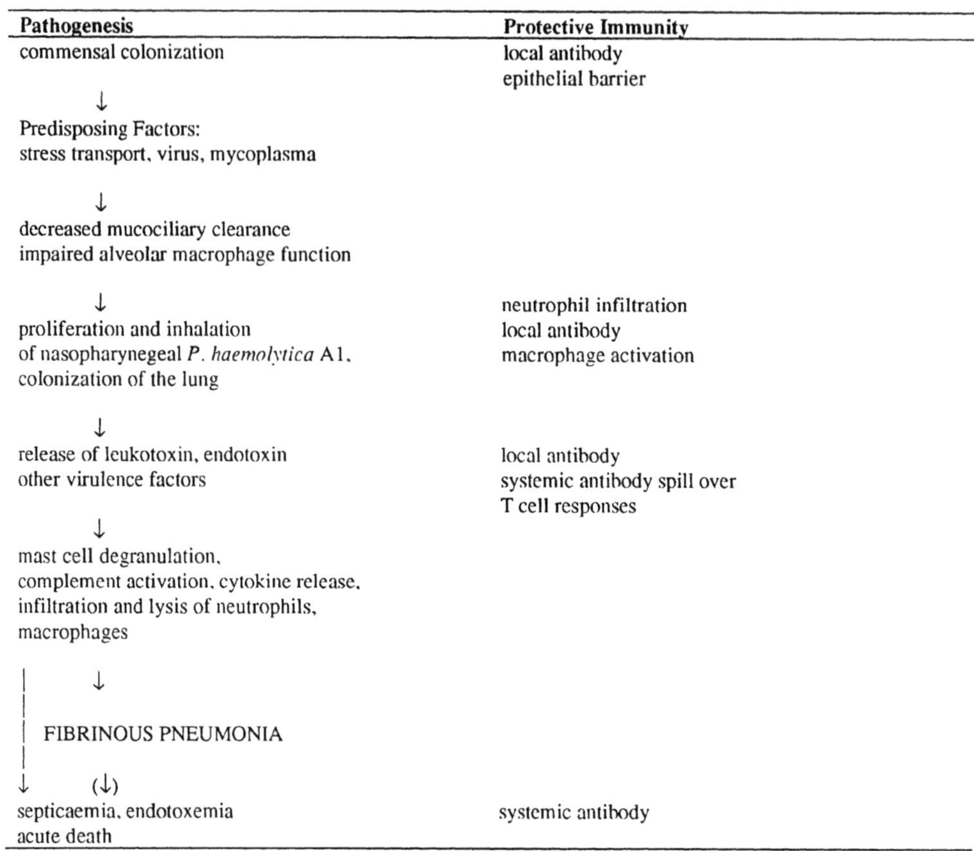

Figure 1. A model for the pathogenesis of bovine pneumonic pasteurellosis, following commensal colonization with *Pasteurella haemolytica*, is outlined in the first column. Immune mechanisms which may operate to control infection and prevent disease at various levels in the pathogenesis are listed at the corresponding level in the second column.

In attempting to prevent disease in commensally infected animals one must be cognizant of the complex interplay between the host and the bacteria which commences with initial nasopharyngeal colonization (Figure 1). Absorption of some bacterial antigens, particularly soluble molecules, across the nasopharyngeal mucosa may induce a local immune response with potential for systemic spill-over. For *Pasteurella haemolytica* this is evidenced by the presence of organism-specific secretory immunoglobulin, predominately IgA, in the nasal mucus of healthy cattle commencing at a very young age (Wilkie and Markham, 1979). This is often, accompanied by circulating antibodies to bacterial surface antigens. More importantly, the presence of organisms in the tonsillar region permits the induction of immune response to bacterial antigens without the need for bacterial penetration or insult to the mucosal surface (Brandtzaeg, 1984). Studies in several species have shown that memory cells induced by tonsillar exposure "prime" deeper mucosal sites including the lung, although this may not be apparent or detectable prior to pulmonary challenge (Berman *et al.*, 1990). Similarly there is evidence for migration of antigen-bearing tracheal dendritic cells to bronchial lymph nodes, again providing for pulmonary priming. Occasional low grade inhalation of bacteria deeper into the respiratory system may result in both local lung (BALT) and systemic immune stimulation whether or not there is clinical pneumonia. No doubt it is intermittent exposure of this type that leads to the eventual "maturation" of bronchus associated lymphoid tissue (BALT) as animals age (Anderson *et al.*, 1986). Support for its occurrence comes from studies of clinically normal calves, which reveal that as many as 60 percent of apparently healthy animals actually have sub-clinical pneumonia when examined rigorously using radiography and lung biopsy (Pringle *et al.*, 1988). The resulting spontaneous or "natural" priming gives the animal considerable advantage when it is subsequently faced with a more severe pneumonic challenge.

This advantage may be exploited successfully in vaccination, providing the appropriate antigens are included as immunogens. Experience with the *P. haemolytica* culture supernate vaccine illustrates this well, since many calves respond anamnestically to a single dose of vaccine, and are protected against subsequent intrabronchial challenge (Conlon *et al.*, 1995). Likewise, several field studies have demonstrated the immunogenicity and protective efficacy of a single dose of this and other *P. haemolytica* vaccines given on entry to the feedlot (Bateman, 1988; Jim *et al.*, 1988; Van Donkersgoed *et al.*, 1993). In these cases vaccine induced protection may relate as much to refreshment of an animal's natural immune response as to stimulation of response de novo.

Although the time of initial colonization of calves with *Pasteurella haemolytica* is unknown, circumstantial evidence suggests that this occurs quite early, possibly from the frequent nose to nose contact between calf and dam and among calves (Hodgins and Shewen, 1992). Colonization permits exposure to bacterial antigens associated with adherence to the nasopharyngeal mucosa. Thus capsular and somatic surface antigens are those most likely to be available for stimulation of an immune response in the upper respiratory tract (Figure 2). Because of the thick mucosal barrier and continual drainage of mucus from this region, soluble factors such as leukotoxin, iron scavenging molecules, various enzymes or endotoxin are less likely to cross the epithelium and stimulate a response. In most animals deeper colonization would be needed for these antigens to be recognised. In young unstressed calves inhalation of small numbers of bacteria would be expected but this may be insufficient to establish pulmonary colonization, even transiently. Under stressful conditions, bacterial replication in the nasopharynx results in increased numbers of organisms in inspired tracheal air (Grey and Thomson, 1971), enhancing the likelihood of deeper penetration. While this pulmonary infection may be transient and readily cleared in otherwise healthy calves it may be sufficient to induce protective memory.

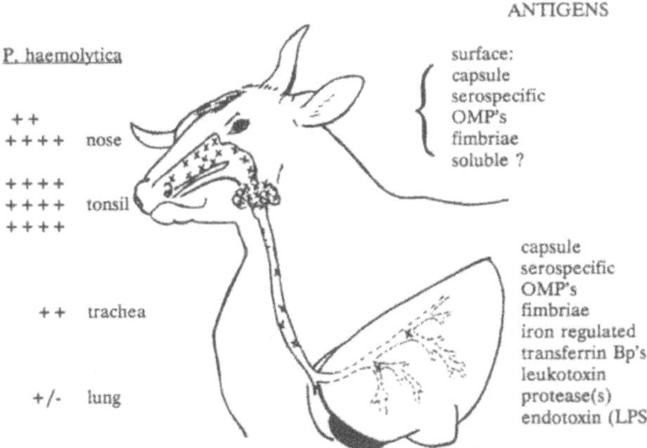

Figure 2. Natural exposure to *Pasteurella haemolytica* A1 in the commensally colonized conventional calf, indicating the usual level of colonization at various levels of the respiratory tract on the left (x), and the antigens which may potentially be recognised at the corresponding level on the right.

A recent serological survey focused on young dairy calves in southern Ontario has revealed that virtually all appear to develop an active immune response to *P. haemolytica* surface antigens (detected by bacterial agglutination or capsular enzyme immunoassay), co-incident with the decline in passive maternal immunoglobulins (Hodgins and Shewen, 1992). The onset of response in individuals varied according to the amount of antigen specific IgG_1 acquired passively. This is consistent with findings in other North American jurisdictions, and supports the notion of response to commensal colonization. In contrast, there was a remarkable seasonal influence on the induction of an active immune response to leukotoxin. Calves born in Spring and early Summer did not seroconvert until Fall (Figure 3). Perhaps the stress of changing environmental or possibly management factors, coincident with summer's end created the appropriate conditions for deeper colonization and immune exposure. Seroconversion in the production of toxin neutralizing antibodies in the Fall and early Winter was predictably dependent on the decline of passive maternal antibodies. This seasonal dichotomy in response might hold true for other soluble antigens as well. A similar situation in beef calves would render those animals particularly susceptible to *P. haemolytica* pneumonia and compound the stress induced by Fall shipment to feedlots. Early vaccination might be essential to alter response in these calves.

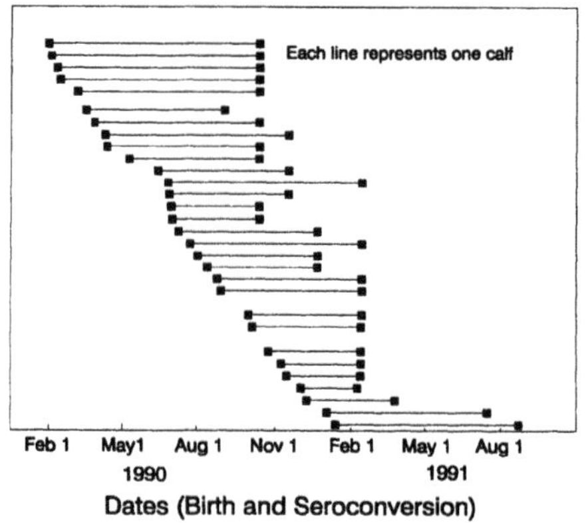

Figure 3. Seroconversion for leukotoxin neutralizing antibodies in neonatal dairy calves. The left dot (■) indicates the date of birth, the right dot indicates the date at which leukotoxin neutralization was first detected in serum. Note that this interval is delayed until Fall for calves born in Spring or early summer, while seroconversion in Fall and Winter calves occurs co-incident with the decline of passive maternal antibodies. (Hodgins and Shewen, 1992).

Not only the induction but also the character of the immune response appears to be significantly affected by the calf's prior immunological exposure. Studies in calves and humans have demonstrated that once mucosal priming has been induced in GALT (gut associated lymphoid tissue), subsequent systemic vaccination induced anamnesis at mucosal sites (Bowersock *et al.*, 1989; Svennerholm *et al.*, 1980). Similarly, while conventional dogma suggests that parenteral vaccination induces a predominately systemic response in naive individuals, evidence from experimental trials in typical naturally "conditioned" calves has demonstrated that parenteral vaccination is as effective as intrabronchial instillation of antigen in inducing a pulmonary immune response to *P. haemolytica* (Wilkie and Markham, 1979; Wilkie *et al.*, 1980) . This is only partially accounted for by pulmonary spill-over from the vasculature and is thought to reflect local synthesis as well. While, it is recognised that lymphocytes derived from tonsillar, tracheal and BALT priming recirculate predominately to pulmonary lymphoid tissues, a small proportion (perhaps 10 percent) remain in the systemic immune compartment. These memory cells are available in peripheral lymph nodes and spleen for stimulation by parenteral antigen exposure, once triggered to proliferate they have the potential to provide numerous daughter cells which will "home" to pulmonary tissues. Furthermore, it is important to recognise that in most species both IgA and IgG producing cells are found throughout the respiratory tract with a shift towards the latter in deeper tissues (Anderson *et al.*, 1986b). Numerous studies have demonstrated that IgG predominates in lung and other deep mucosal sites such as uterus (Wilkie and Markham, 1981; Pringle *et al.*, 1988). This is particularly true in cattle where IgG_1 has been shown to bind secretory component for selective mucosal secretion. The shift towards IgG in the lung is further exacerbated by pulmonary inflammation which leads to increased vascular permeability in a highly vascular organ. Both IgG_1 and IgG_2 are found in the lungs of healthy cattle and both are recognised to play a role in protection against pneumonic pathogens including *P. haemolytica.* Mechanisms involved in IgG mediated immunity include opsonization to enhance bacterial uptake by alveolar macrophages, toxin and enzyme neutralization and, together with IgA, inhibition of bacterial adherence to the respiratory epithelium. A relationship between the antibody response to *P. haemolytica* A1 in serum and lung washing fluids has been demonstrated in several vaccination and challenge trials in calves (Wilkie *et al.*, 1980; Shewen and Wilkie, 1988). In these experiments, vaccination resulted in a similar increase in indirect bacterial agglutination titres and leukotoxin neutralizing antibodies in both compartments following immunization, whether vaccine was administered by the parenteral or intrabronchial route. Thus, if the aim of immune intervention is prevention of pneumonia, then the ability to induce pulmonary immunoglobulin through parenteral vaccination of naturally primed calves provides considerable advantage for vaccine delivery.

There is also evidence for an effect of commensal colonization on the magnitude and bias of the antigen responsiveness of calves. Again, conventional dogma suggests that animals respond poorly to immunization with polysaccharide antigens. Many studies in naive animals have demonstrated that bacterial capsular polysaccharide is a poor immunogen, which when immunogenic at all induces a "T-independent" response, resulting in production of only IgM antibodies (Mosier and Subbaro, 1982). However, the response of calves to commensal colonization with *P. haemolytica* results in circulating anti-capsular antibodies of the IgM, IgG_1, and IgG_2 isotypes (Hodgins and Shewen, 1994). Similarly, vaccination of calves with purified capsular polysaccharide caused an increase in serum antibodies of all isotypes as well, even though circulating antibodies were not detected by capsule specific enzyme immunoassay (EIA) prior to immunization (Conlon and Shewen, 1993). The rapid switch to production of IgG isotypes following a single inoculation of capsule strongly suggests that priming by prior natural exposure is responsible for both the level of immunogenicity and the T dependent nature of the response. A similar phenomenon may operate in the well recognised age-dependent ability of children to respond to the capsular polysaccharide of another HAP organism, *Haemophilus influenzae.* Such natural priming is not however, entirely without negative consequences. In the trial described above, 36% (9/25) of calves which were injected with purified capsular polysaccharide experienced systemic anaphylaxis at vaccination which could be shown to be related to the presence of circulating capsule specific IgE. Interestingly, calves which received capsule in its native form, as a component of culture supernate, showed no such reaction. That natural priming is an important prerequisite for response has also be suggested by studies showing that isolation-reared colostrum deprived calves respond poorly to surface antigens of *P. haemolytica* and not at all to the leukotoxin, while immunized conventional calves of similar age respond readily providing there is no passive interference (Hodgins and Shewen, 1993). The poor responsiveness in naive animals may account for the exquisite sensitivity of colostrum deprived calves and sheep to experimental challenge.

In addition to consideration of the fundamentals of immune responsiveness described above, efficacious vaccination requires that the response be focused on factors which appear to relate to subsequent protection against disease. For *P. haemolytica* the spectrum of potential antigens is large (Figure 2). To determine which antigens are crucial for protection and in order ascertain when immune intervention is likely to be beneficial, it is useful to examine the immune status of calves in the field. An examination of the serological profiles of healthy calves can provide clues to each animal's immunological experience vis a vis the organism of interest, and may also furnish predictors of its disease resistance. Early serological studies of feedlot aged animals often yielded conflicting or at least confusing data, which were due in part to a failure to recognise that the feedlot gate or the sales barn did not mark the first encounter with the organism (Frank and Smith, 1983). The absence of an understanding of the kinetics of the response at the time samples were drawn frequently resulted in confusion and misleading interpretations of data. Yet much potentially valuable information can be derived from a careful examination of field material. Comparison of reactivity in naturally resistant calves with that in susceptible ones, coupled with examination of the kinetics of response following a pulmonary challenge in individual calves, has the potential to reveal not only each individual's immunological reactivity, but also which antigens or virulence factors should be targeted in vaccine development. For example, early studies on the role of *P. haemolytica* leukotoxin in bovine pneumonic pasteurellosis included an investigation of the antibody response in sera of feedlot calves which revealed a strong correlation between the presence of anti-leukotoxin and resistance to fibrinous pneumonia (Shewen and Wilkie, 1983). In addition seroconversion in the majority of calves following shipping verified the production of leukotoxin *in vivo*. We have since used sera collected from feedlot-aged calves which resist pneumonia to indicate potential protective antigens. The spectrum of antigens recognised on Western immunoblot by sera from most resistant calves is broader both before challenge and in response to pulmonary infection. Recognition of specific antigens, once identified, may be correlated retrospectively with response to challenge. Furthermore, characterization of the antigens producing unique bands has the potential to reveal new crucial vaccine immunogens. The same approach can be used with colony blot EIA to screen clone banks for detection of recombinant genes which express these antigens. The recombinant antigens can then be tested alone or in various combinations in experimental vaccination and challenge trials to determine their protective efficacy. This strategy was used successfully to suggest a role for *P. haemolytica* sialoglycoprotease as a protective antigen (Lee *et al.*, 1994). Subsequent production of a recombinant sialoglycoprotease fusion protein (rGcp-F) and its use in a vaccination and challenge experiment verified the potential of this antigen as a protective immunogen (Shewen *et al.*, 1994). While it is probable that no single antigen will confer protection, the strategy of examination of natural response in concert with experimental validation of hypothetically protective antigens provides for an efficient mechanism for future vaccine development.

Efficacious vaccination thus arises from exploitation of the host response to commensal colonization. Characterization of this response can lead to identification of antigens and immune mechanisms relevant in protection. The priming which occurs naturally can be used to advantage in induction of protective immunity and may bias the response to vaccination with regard to the antigens recognised, the immunoglobulin isotypes produced and the sites where responses occur. Rational approaches to the design and administration of vaccines will benefit from recognition of the ongoing interactions between the host and the HAP organism.

REFERENCES

Anderson, M.L., Moore, P.F., Hyde, D.M., and Dungworth, D.L., 1986a, Bronchus associated lymphoid tissue in the lungs of cattle: relationship to age. *Res. Vet. Sci.* 41:211.

Anderson, M.L., Moore, P.F., Hyde, D.M., and Dungworth, D.L., 1986b, Immunoglobulin containing cells in the tracheobronchial tree of cattle: relationship to age. *Res. Vet. Sci.* 41:221.

Bateman, K.G., 1988, Efficacy of a *Pasteurella haemolytica* vaccine/bacterial extract in the prevention of bovine respiratory disease in recently shipped feedlot calves. *Can. Vet. J.* 29:838.

Berman, J.S., Dennis, J.B., Theodore, A.C., Kornfeld, H., Bernardo, J., and Center, D.M., 1990, Lymphocyte recruitment to the lung. *Am. Rev. Resp. Dis.* 142:238.

Bowersock, T.L., Walker, R.D., McCracken, M.D., Hopkins, F.M., and Moore, R.N., 1989, Induction of pulmonary antibodies to *Pasteurella haemolytica* following intraduodenal stimulation of the gut associated lymphoid tissue in cattle. *Can. J. Vet. Res.* 53:371.

Brandtzaeg, P., 1984, Immune functions of human nasal mucosa and tonsils in health and disease, *in* "Immunology of the lung and upper respiratory tract.," J. Bienenstock, ed., McGraw Hill,New York.

Brandtzaeg, P., 1992, Humoral immune response patterns of human mucosae: Induction and relation to bacterial respiratory infections. *J. Infect. Dis.* 165:S167.

Conlon, J.A., Gallo, G.F., Shewen, P.E., and Adlam, C., 1995, A comparison of the protection conferred by a *Pasteurella haemolytica* bacterial extract vaccine when used in a single or in a double injection schedule in a challenge model in cattle. *Can. J. Vet. Res.* (accepted for publication).

Conlon, J.A., and Shewen P.E., 1993, Clinical and serological evaluation of a *Pasteurella haemolytica* A1 capsular polysaccharide vaccine. *Vaccine* 11:767.

Frank, G.H. and Smith, P.C., 1983, Prevalence of *Pasteurella haemolytica* in transported calves. *Am. J. Vet. Res.* 44:981.

Grey, C.L. and Thomson, R.G., 1971, *Pasteurella haemolytica* in the tracheal air of calves. *Can. J. Comp. Med.* 35:121.

Hodgins, D.C. and Shewen, P.E., 1992, Immune response of colostrum fed dairy calves to antigens of *Pasteurella haemolytica*. Abst. *73rd. Conf. Res. Work Animal. Dis.*, Chicago. Ill.

Hodgins, D.C. and Shewen, P.E., 1993, Immune response of neonatal colostrum deprived dairy calves to antigens of *Pasteurella haemolytica*. Abst. *74th. Conf. Res. Work Animal. Dis.*, Chicago. Ill.

Hodgins, D.C. and Shewen, P.E., 1994, Immune response of colostrum fed dairy calves to capsular polysaccharide of *Pasteurella haemolytica* A1. Abst. *Haemophilus, Actinobacillus and Pasteurella, Intl. Conf.*, Edinburgh, U.K.

Jim, K., Guichon, T. and Shaw, G., 1988, Protecting calves from pneumonic pasteurellosis. *Vet. Med.* 83:1084.

Lee, C.W., Shewen, P.E., Cladman, W.M., Conlon, J.A., Mellors, A. and Lo, R.Y.C., 1994, Sialoglycoprotease of *pasteurella haemolytica* A1: detection of anti-sialoglycoprotease antibodies in sera of calves. *Can. J. Vet. Res.* 58:93.

Mosier, D.E. and Subbarao, T., 1982, Thymus independent antigens: complexity of B-lymphocyte activation revealed. *Immunol. Today* 3:217.

Pringle, J.K., Viel, L., Shewen, P.E., Willoughby, R.A., Martin, S.W. and Valli, V.E.O., 1988, Bronchoalveolar lavage of cranial and caudal lung regions in selected normal calves: Cellular, microbiological, immunoglobulin, serological and histological variables. *Can. J. Vet. Res.* 52:239.

Shewen, P.E., Perets, A., Lee, C.W. and Lo, R.Y.C., 1994, Evaluation of a *Pasteurella haemolytica* recombinant sialoglycoprotease fusion protein (Rgcp-F) vaccine. Abst. *Haemophilus, Actinobacillus and Pasteurella, Intl. Conf.*, Edinburgh, U.K.

Shewen, P.E. and Wilkie, B.N., 1983, *Pasteurella haemolytica* cytotoxin neutralizing activity in sera from Ontario beef cattle. *Can. J. Comp. Med.* 47:497.

Shewen, P.E. and Wilkie, B.N., 1988, Vaccination of calves with leukotoxic culture supernatant from *Pasteurella haemolytica*. *Can. J. Vet. Res.* 52:30.

Svennerholm, A.M., Hanson, L.A., Holmgren, J., Lindblad, B.S., Nilsson, B. and Overeshi, F., 1980, Different secretory immunoglobulin A antibody responses to cholera vaccination in Swedish and Pakistani women. *Infect. Immun.* 30: 1032-1040.

Van Donkersgoed, J., Schumann, F.J., Harland, R.J., Potter, A.A. and Janzen, E.D., 1993, The effect of route and dosage of immunization on the serological response to a *Pasteurella haemolytica* and *Haemophilus somnus* vaccine in feedlot calves. *Can. Vet. J.* 34:731.

Wilkie, B.N. and Markham, R.J.F., 1979, Sequential titration of bovine lung and serum antibodies after parenteral or pulmonary inoculation with *Pasteurella haemolytica*. *Am. J. Vet. Res.* 40:1690.

Wilkie, B.N. and Markham, R.J.F., 1981, Bronchoalveolar washing cells and immunoglobulins of clinically normal calves. *Am. J. Vet. Res.* 42:241.

Wilkie, B.N., Markham, R.J.F., and Shewen, P.E., 1980, Response of caves to lung challenge exposure with *Pasteurella haemolytica* after parenteral or pulmonary immunization. *Am. J. Vet. Res.* 41:1773.

COMMERCIAL DEVELOPMENT OF *HAEMOPHILUS, ACTINOBACILLUS* AND *PASTEURELLA* VACCINES

S.B. Houghton

Vaccine Research & Development
Hoechst Animal Health
Walton Manor
Walton
Milton Keynes
MK7 7AJ

INTRODUCTION

It is now 115 years since the work of Louis Pasteur on fowl cholera heralded the dawning of a new era in the control of bacterial diseases by vaccination. The causative organism was later identified as *Pasteurella multocida* which can be considered as the elder statesman of the group of bacteria we now refer to as HAP. In view of the prominent role of a member of this group in those pioneering days it would be reasonable to assume that todays vaccines reflect the sum total of 115 years of continued research and development. Pasteur's studies on fowl cholera involved the evaluation of the virulence of *P. multocida* isolates in infected birds (Pasteur, 1880). Serendipity probably played a key role in this discovery of the concept of attenuation whereby strains maintained in his laboratory not only lost the ability to produce disease but also induced protective immunity against subsequent infection with a virulent strain. Nevertheless, Pasteur realised the implications of these results and went on to apply this approach to other diseases, notably anthrax and rabies. Soon afterwards, it was shown that inactivated vaccines could also induce protective immunity. On the face of it, we have made very little progress in vaccine technology since Pasteurs day as evidenced by the type of vaccines currently available i.e. live vaccines attenuated by non specific procedures and killed bacterins. Moreover, *P. multocida* continues to be a significant problem for poultry producers and in addition is implicated in other diseases particularly of cattle and pigs. Similarly, the vaccines on the market for other members of the HAP group seemingly owe more to Pasteurs research than subsequent work. One of the contributing factors to this state of affairs is the emphasis that has been placed on antimicrobial therapy as a means of controlling disease rather than prophylactic immunisation, although increasing concerns over antibiotic resistance and residues may be factors in the renewed interest in vaccine development.

The notable exception to this observation is *Haemophilus influenzae* in humans which has been the subject of considerable interest in recent years leading to well defined sub unit vaccines of proven efficacy and safety (Jones 1993, Adams *et al.*, 1993), although there is no room for complacency since the possibility of the emergence of new invasive strains of Haemophilus can not be ignored (Kroll, 1995).

This paper will be limited to HAP species of veterinary importance and will consider the relative importance of vaccines, describe the factors involved in vaccine research and development, and look to the future.

THE ANIMAL HEALTH MARKET

The ex-manufacturers sales total for animal health and nutrition products was estimated at $11,660 million in 1992 with a 2.3% annual increase in recent years being forecast to continue to 1998 (Phillips, 1994). Of the total market, biologicals accounted for 13.9% or $1,615 million in 1992 (Figure 1).

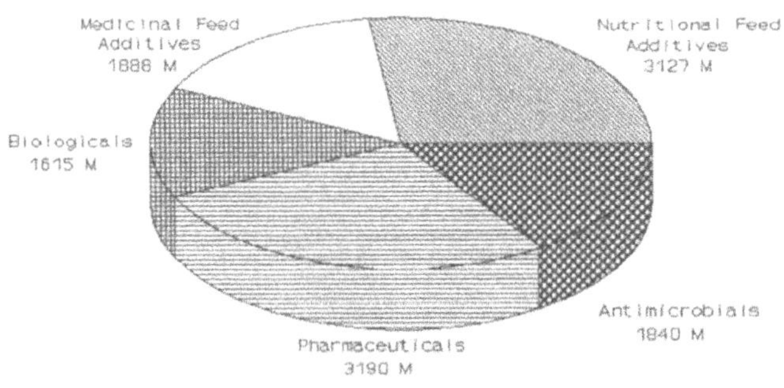

Figure 1. The world animal health market (1992)

The term biologicals is often used to describe a group of products, primarily Immunological Veterinary Medicinal Products (IVMP). IVMPs include vaccines, antisera and some diagnostics. The majority of the sales value referred to under the term biologicals is accounted for by vaccines. This is the fastest growing sector with real growth of 7% per annum forecast to continue until 1998 (Figure 2; Phillips, personal communication).

Two products account for 27% of the biologicals market, namely foot and mouth vaccines and rabies vaccines with sales of $250 million and $195 million respectively (Figure 3). Over half the overall market is in North America and Western Europe (Figure 4) with the developing nations representing a significant proportion of the market due to the fairly widespread use of FMD and rabies vaccines. The main sectors relevant to HAP vaccines are cattle, sheep, pigs and poultry which together comprise a market of $1022 million (Figure 5), excluding FMD. Viral vaccines account for almost 90% of the poultry market, the remainder being made up principally with sales of fowl cholera vaccines. In cattle, sales of respiratory viral vaccines were $145 million or 45% of the total. Sales of porcine viral vaccines represent 68% of the total for this segment of the market. The sheep market contrasts with other livestock sectors in that sales are largely attributable to bacterial vaccines.

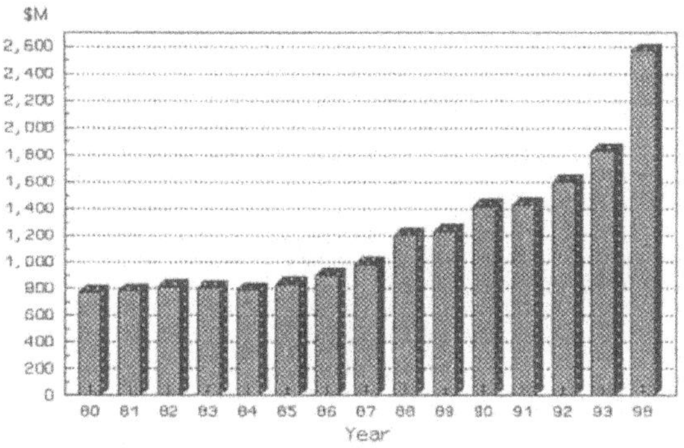

Figure 2. World sales of biological products. Actual figures to 1993 with projected sales for 1998 following a steady rate of increase. 1998 figure reflects real growth and is in 1993 dollar value with no allowance made for inflation.

The total market for bacterial vaccines for livestock and poultry is estimated at 25% of the total biologicals sales in these sectors, or about 3.5% of the total animal health markets (Figure 6). There are many bacterial species involved in various diseases of clinical and economic signficance in livestock and poultry for which vaccines are available and contribute to the sales figures shown. For example clostridial vaccines, particularly for use in sheep, have been available for over 50 years and will probably be major contributors to sales in these sectors for many years to come. Other bacterial pathogens such as Salmonella, *E. coli*, Leptospira, Mycobacterium, Chlamydia, Erysipelothrix etc, many of which contain several different species or serotypes, should also be considered when putting into perspective the role of HAP vaccines in global terms.

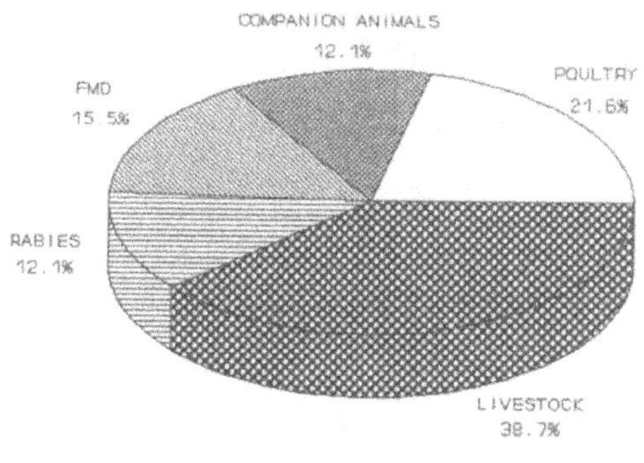

Figure 3. World sales of biologicals by market sectors

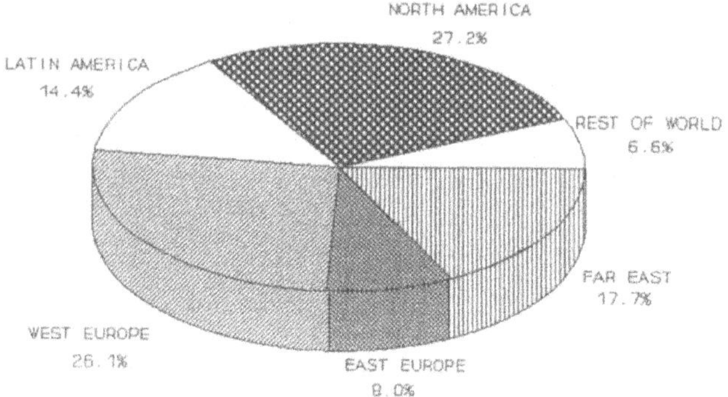

Figure 4. World animal health market by area.

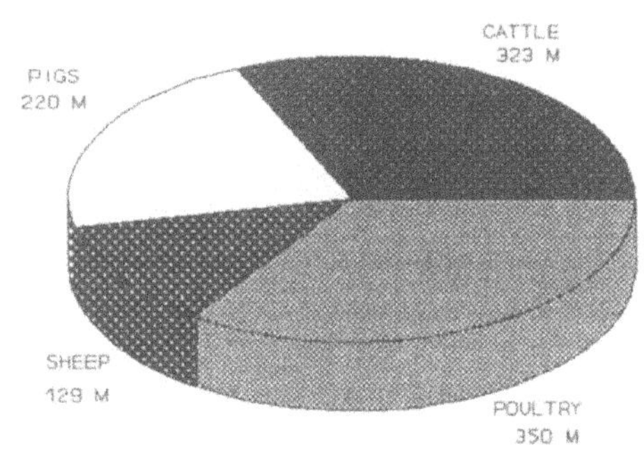

Figure 5. World animal health sales in the livestock sector (excluding FMD)

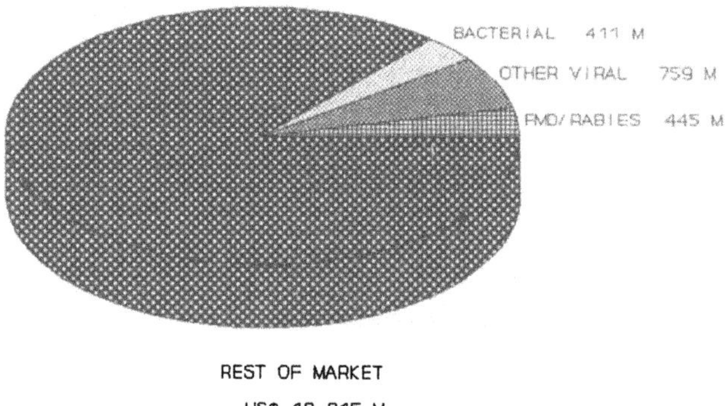

Figure 6. World biologicals market in relation to total market.

The nature of these markets and their price sensitivity have a significant effect on the research and development programmes in the animal health industry. Nevertheless, there is no doubt that there has been strong growth in the biologicals market and diseases caused by HAP organisms continue to be a source of significant economic loss. It is clear that there is a need to develop improved products and a significant proportion of the talks and posters at this conference involved studies directly relating to vaccine development.

VACCINE RESEARCH & DEVELOPMENT

Current Products

It has already been said that today's vaccines are not substantially different from those developed in Pasteur's era. It is true that most products are bacterins or crudely modified live strains but why should this be so given that research in HAP organisms is generally active and productive? There are one or two products beginning to come through which apply current scientific knowledge such as toxin based vaccines for *A. pleuropneumoniae* (Van den Bosch *et al.*, 1995) and *P. haemolytica* (Conlon *et al.*, 1991). These are, however, notable exceptions and there seems to be a long lag period between discovery and application. In order to appreciate the apparent lack of innovative new products it is necessary to consider the steps and constraints involved in product development from the perspective of the animal health industry.

Project Selection

The factors involved in the decision making process on whether or not to proceed with the development of a particular product can be considered under three headings.

Market Research. Successful animal health companies have research and development programmes which are market led, aiming to develop products which will meet the requirements of the relevant market sector. This requires a detailed understanding of the epidemiology of the relevant diseases, the farming practices involved, including husbandry and current disease control strategies, and the economic impact of all these factors. It is important to look for trends in disease prevalence because of the time taken to introduce new products to the market and to recoup the costs of such development through product sales.

There are a number of companies who offer market research services which the animal health industry uses but most will also conduct their own surveys. Major animal health companies are by necessity likely to be more interested in products with an international potential. Development of vaccines with markets in just one or two countries does happen but such products will be more vulnerable to local changes in disease epidemiology, control policies and competition. The source of such data needs to be as relevant and reliable as possible. It is common practice to use many sources and develop an overview but the most useful information often comes from the farmers themselves and veterinary surgeons specialised in the area of interest, together with disease surveillance reports such as those produced by the Central Veterinary Laboratory in the UK.

The aim of market research is to identify the countries where there is likely to be a market for a product, to estimate the sizes of the various markets, to establish the way in which it would be used, to evaluate the competitor products (if any), to identify how best to differentiate the new product from others, and determine the price that the end user would be prepared to pay.

Technical Feasibility. Having established the market potential for a vaccine it is obviously necessary to determine whether is it feasible to develop a product which will meet the required specification. Consideration is given to whether current products reflect the available scientific knowledge, particularly recent advances. Ideally new products should be of superior efficacy but this is not the only relevant factor. Significant advantages could be gained by improving product safety, changing production processes enabling price competition, or offering novel product combinations which reduce the number of times farmers have to handle the animals. Predicting the outcome of research programmes is anything but an exact science and it is often necessary to review progress in relation to initial concepts.

Economics. Estimates have to be made of cost considerations at a very early stage to determine the viability of a development programme. Market research will provide the background necessary to describe the product profile, the likely selling price in different markets and the distribution costs. The technical evaluation should give enough of an insight into likely protective antigens to estimate production costs and to identify any royalties which may be payable as part of collaborative research agreements or the existence of relevant patents. All of these factors should be considered not only at the start of a development programme but also continually reviewed to assess the impact of any deviations from the ideal product profile.

ANTIGEN CHARACTERISATION AND VACCINE FORMULATION

An ideal vaccine is one which presents known protective antigens to the hosts immune system in an appropriate way resulting in an effective response, together with the absence of any significant adverse reactions. Rapid onset of long lasting immunity, preferably from a single dose, and low production costs could be added to this description. The more we know about the whole process of infection and immunity the more likely it is that progress can be made towards well defined safe effective products.

Live vaccines offer the potential to induce a similar level of immunity that often results in animals recovered following natural infection without the necessity to characterise the protective antigens. Of more concern is safety and it is a requirement that live vaccine strains are altered to make them less virulent and preferably avirulent, even in animals whose immune system could be comprised, for example by poor nutritional status. Live vaccine strains therefore have to be sufficiently attenuated so as not to cause disease whilst retaining the ability to replicate in the animal to a level which will induce an effective immune response.

Alterations to wild type strains to produce vaccine strains should be specific and well defined preferably involving gene deletion. Reversion to virulence has to be a low frequency event to the extent that it is primarily a mathematical risk and unlikely to occur *in vivo*, even after sequential passage through susceptible animals. Historically strain manipulation has been crude, leading to vaccines which are poorly tolerated and prone to revert to virulence. Advances in the field of molecular biology have made the design of vaccine strains more of a reality. Gene manipulation also offers the opportunity to use marker antigens enabling vaccine strains, and the immune responses to them, to be differentiated from cases of natural infection. Such strains may also be used as vectors for immunisation against other pathogens, particularly where conventional products prove less then satisfactory, or are difficult to produce.

In characterising the antigenic profile of bacteria with a view to defining inactivated vaccines it may be misleading to examine *in vitro* cultures. Bacteria are adept at modifying their physiology and biochemistry in relation to their environment, with antigens produced *in vivo* not necessarily being produced *in vitro* and vice versa. The characterisation of antigens produced *in vivo* is most useful particularly where it is known that natural infection leads to a high level of immunity to re-infection. Examination of *in vivo* grown bacteria and the hosts immune response may offer insights into which antigens are involved and the type of immune response to aim for. An example of this is to be found in the development of vaccines containing iron regulated outer membrane proteins (IRP) of *Pasteurella haemolytica*. It was shown that convalescent sheep were immune to re-infection (Table 1) and passive protection studies demonstrated that humoral antibody is protective (Donachie *et al.*, 1986 a,b). Recovery of *P. haemolytica* from pleural fluid allowed analysis of a protective antibody response to bacterial cells as they are presented to the sheep during infection (Donachie and Gilmour, 1988). This identified IRP's as playing a significant role in the immunity of convalescent sheep.

Table 1. Ovine Pasteurellosis : Response to repeated challenge infection with serotype A2.

	Clinical Response	Died/ Killed	Lung Lesions	*P. haemolytica* isolated from Lungs
Uninfected Controls (n=7)	6	4	5	6
Previously Infected and Recovered (n=7)	0	0	0	0

The identification and characterisation of protective antigens will influence the possible vaccine formulations, examples of which are given in Figures 7 and 8. Through genetic manipulation it is now possible to consider the inclusion of antigens in inactivated vaccines which would otherwise be difficult or impossible. Whatever choices are made in the development of the product the implications of these in terms of production feasibility, quality and regulatory requirements have to be evaluated.

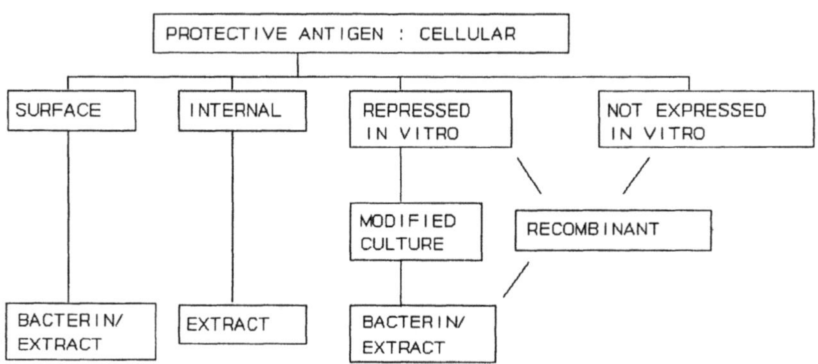

Figure 7. Vaccine formulations for cellular antigens

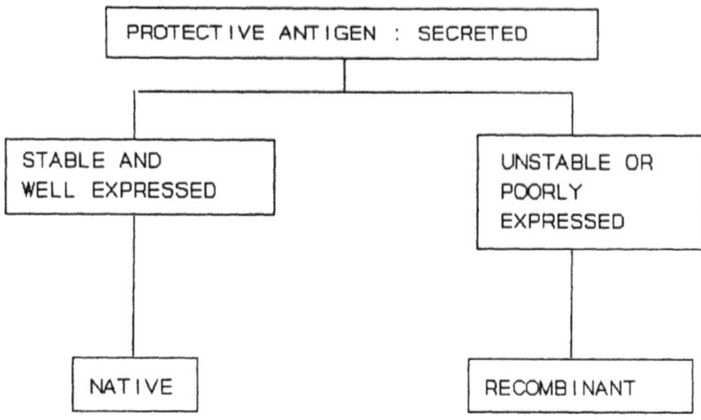

Figure 8. Vaccine formulations for soluble antigens

Experimental Studies and Field Trials

Challenge Infection Models

The availability of suitable disease models is necessary for the evaluation of the efficacy of candidate vaccine formulations, to determine the minimum antigen dose required for effective immunity, to assess the compatibility of multi component vaccines and to generate the data which will support the recommended vaccination schedules and product claims. The use of models involving challenge infection of laboratory animals may be of some benefit in the early stages of development for some bacterial vaccines but the relevance of such studies to the target species is always questionable.

Vaccine efficacy has to be demonstrated in the target species which can be achieved either by field trials or in experimental models. The generation of experimental efficacy data is complicated by a number of factors. With the majority of the diseases involving HAP bacteria the incidence of severe clinical symptoms in affected populations is generally low. Experimental studies performed in research facilities have to take account of practical considerations such as animal and accommodation availability which will dictate the size of the experiments. The lower the disease incidence in controls the greater the number of animals required in each group to produce statistically valid data. It is often not feasible to perform such large studies and therefore the severity of the disease has to be much higher, typically 80% or greater in susceptible control animals, to obtain significant results.

Under natural conditions it is common to find HAP organisms involved in disease complexes. For example respiratory disease in cattle may involve not only *P. haemolytica* but IBR, RSV, PI_3 or BVD and in pigs *A. pleuropneumoniae* infection may be complicated by concurrent infection with *Mycoplasma hyopneumoniae* or *P. multocida*.

Challenge models have been developed which involve infection with more than one agent, such as IBR + *P. haemolytica* (Jericho and Langford 1978), *M.bovis* and *P. haemolytica* (Houghton and Gourlay, 1983) in calves, and PI_3 and *P. haemolytica* in sheep (Gilmour *et al.*, 1983). If such models are used to evaluate the efficacy of vaccines against just one of the components then account must be taken of the degree of severity and consistency of the clinical response to the other challenge organism. It is preferable, where possible, to use challenge systems where just one pathogen is involved.

The route of administration and the method of preparation of the challenge inoculum should also be considered in relation to its relevance to the natural infection. Respiratory pathogens should therefore be administered by the respiratory route but there are still many variations which may affect the subsequent disease process. Aerosol administration is perhaps the most natural route involving the colonisation of the respiratory tract followed by multiplication. Other routes of administration have been used such as intratracheal, transthoracic and endobronchial deposition. Consideration should be given with such models as to whether this could affect the role of particular virulence determinants in the disease produced and if this could compromise the evaluation of vaccine efficacy. An example of this would be the direct inoculation of cultures of *P. haemolytica* into the lung. This could result in the introduction of significant amounts of leukotoxin into alveoli whereas in natural infection this would have been preceded by the process of colonisation and multiplication before leukotoxin could reach such levels. Vaccines which produce a response which inhibits colonisation or multiplication may not be seen to be effective in such systems but could still be highly effective against natural infection. It is therefore crucial to understand the pathogenesis of the natural disease and to be aware of how this may be modified in attempting to establish artificial challenge systems.

Development of challenge models that are more severe than the natural disease is a complicating factor in generating experimental data on duration of immunity. The immune response to a vaccine will peak following a primary vaccination course and than decline with time unelss the immune system is stimulated again by natural infection or re-vaccination. It is likely that there is a direct relationship between the level of immunity and the severity of infection that an individual animal can withstand. If the experimental disease is more severe than the natural infection than a higher level of immunity would be required to achieve protection, and the apparent duration of immunity will be shorter. The data generated using severe experimental systems may therefore be of great importance in establishing vaccine efficacy but may not always give a realistic indication of how the vaccine will perform on the farm. Furthermore it is common for vaccines to be described in terms of percentage efficacy with ideal products being those that give at least 80% efficacy. This needs to be considered in relation to field efficacy and not experimental efficacy. It would be unrealistic to expect a vaccine which is 90% effective against a natural infection affecting 20% of the population to offer the same percentage protection in a model which produces disease in more than 80% of susceptible animals.

Field Trials

Field trials are performed in order to generate efficacy and safety data in animals reared under commercial farming conditions. Trial sites should be used which have a relevant disease history. The vaccine is preferably compared in a blind study with a placebo although it is not always possible to persuade farmers to leave 50% of their stock unprotected and it may be necessary to compare a new vaccine with one already on the market. The data that this would generate would be less than ideal since it would be very difficult to interpret unless one or both products are clearly ineffective. Similarly, trials which involve 100% vaccination would not produce efficacy data which could be used to support product usage, but they may be of benefit if performed on a sufficient number of farms over a prolonged period of time in convincing farmers and their veterinary surgeons that the vaccine is of some benefit. In addition to any efficacy data that may or may not result from field trials it is important that any new vaccine is shown to be well tolerated under commercial conditions when used according to the recommended schedules, since experimental evaluation of safety may not always be representative of the farm situation where husbandry practices, including administration of other vaccines or treatments, may differ.

Quality and Production

Production facilities are inspected by licensing authorities and must meet the required standards for the manufacture and sale of vaccines. Manufacturers are required to establish control procedures which demonstrate that each batch of product satisfies the criteria described for consistency of specific antigen production, potency and safety. A significant proportion of the research and development effort is directed towards the provision of appropriate validated assays which will be used in batch control testing.

The production process itself is optimised, usually on a pilot scale, before scale up to production. This is documented in detail and includes information on media formulation and preparation, fermentation parameters, harvest criteria and downstream processing. Although procedures are specified we are of course dealing with biological systems with inherent variability, and quality assurance and quality control are essential activities.

In process controls should include assays which will quantify the amount of specified antigens produced. These antigens should be relevant to the principle mode of action of the product. Vaccines claimed to be effective through the induction of immunity to toxins for example are formulated to include consistent amounts of these antigens in the final product. The further a batch goes down the production process the greater its value, reflecting the use of production capacity, materials and operator time. The assays used to determine the quality and quantity of antigen production should preferably be quick to perform in order that any poor batches can be identified early in the process and appropriate action taken.

Each vaccine batch, although produced using defined procedures and with a known content of specified antigens, is subjected to a potency test to determine efficacy. Potency tests are required to be relevant, validated assays which can be correlated with efficacy in the target species. They must be able to detect substandard batches. It is not common practice to use the target species in batch potency testing for HAP vaccines, with the exception of *P. multocida* in turkeys (USA Code of Federal Regulations). Apart from any logistical problems in terms of the management of sufficient number of animals to perform routine testing, the majority of the HAP bacteria are commonly found in healthy animals. Prior, variable exposure and natural immunity may therefore compromise the evaluation of response to vaccination.

Potency tests are therefore usually performed in laboratory animal models. Tests involving challenge infection may prove difficult to establish since virulence and pathogenesis are likely to vary considerably from the target species models. Serological response in laboratory animals is commonly used for batch testing involving the quantitative measurement of response to specific protective antigens. If these antigens have a known biological function which can be measured *in vitro* then the potency test preferably involves assessment of inhibition of this function. For example toxin based vaccines can be evaluated on the basis of the ability to produce a neutralising antibody response. Similarly vaccines containing iron regulated outer membrane proteins can be controlled by measuring the ability of sera from vaccinated animals to inhibit iron uptake *in vitro*. Whatever tests are used it is necessary to include standards and establish pass criteria based on a comparison with vaccines shown to be effective in target species experiments.

REGISTRATION AND REGULATION

The animal health industry is well regulated in the EC and in North America with the requirements imposed on manufacturers having become much more stringent in recent years. Although the regulations cover similar aspects there have been, and still are, discrepancies in specific requirements of different national authorities. In the EC there has been considerable progress towards establishing common procedures (Lee, 1993). From 1995 the EC Medicines Control Agency will coordinate licensing activities on a centralised basis. It will not handle all applications for marketing authorisation, although all applications for biotechnology products must be made to the agency. Other procedures will still be available involving national registration with a system of mutual recognition between member states. The regulations in the USA are different to those of the EC (Espeseth & Greenberg, 1993) although discussions have taken place on how best to proceed towards international harmonisation. It is likely to be sometime until this is achieved which leaves manufacturers in the position of having to be conversant with the local requirements and interpretation of regulations in the markets where they hope to sell their products.

Registration of a product involves the provision of detailed documentation on those aspects covered under Quality and Production to satisfy the authorities in relation to batch consistency. In some countries the licensing authorities may perform their own tests on products to assess quality, safety and efficacy although in the UK the so-called QA/QC scheme has been introduced to replace such testing. This entails a detailed inspection of the quality systems in place at the manufacturing premises but does not reduce the requirements for batch release tests on the final product performed by the manufacturer.

Product safety has to be established with multiple dosing (10 times the recommended dose for live vaccines, double for killed) and repeat dosing in the most susceptible type of animal. This would be defined by the recommended vaccination schedule involving for example the vaccination of pregnant animals. Safety studies are now required in the EC to be conducted in compliance with Good Laboratory Practice which necessitates manufactures either achieving GLP compliance themselves or contracting work out to compliant laboratories. Safety testing each batch is also required on a small number of animals of the target species but the requirements for this are not as strict in terms of the type of animal used. Any claims made for the vaccine have to be supported by efficacy data including justification for the recommended vaccination schedule.

The conduct of field trials will become more involved and require more resources with the application of Good Clinical Practice to vaccine trials likely from 1995 onwards in the EC (Note for Guidance, 1994).

THE FUTURE

The time and costs taken to put a vaccine on the market have increased in line with stricter licensing controls and regulations. A relatively straightforward programme is likely to take four years at a cost of £1 million with more complex, technically difficult products taking several more years and millions to develop. It is inevitable that the animal health industry will be less likely to commit limited resources to the development of products where the market is relatively small or possibly perceived to be of limited duration.

It is clear, however, that there is much activity in relation to diseases caused by HAP bacteria both on the scientific front and in the development of improved vaccines. The key to this activity is the increased wealth of knowledge on virulence determinants, pathogenesis and immunity with this conference publication being a good reflection of the progress being made. An improved understanding of the structure and function of potentially protective antigens (Table 2) will enable the development of defined vaccines, together with appropriate control tests to enable registration. Conventional inactivated vaccines will continue to play a significant role in disease control programmes, particularly where it is known that several antigens are involved in protection. Surface antigens may well be presented to the immune system in the most appropriate way by the use of whole cell vaccines, and manipulation of strains and production processes can be applied to optimise these products. Biotechnology is likely to have a significant impact on future developments. Adjuvants are essential components of inactivated vaccines but their use is always a compromise between tolerability and effect. Highly effective adjuvants such as Freunds are not widely used in commercial vaccines containing HAP bacterial components. Such products would be likely to cause considerable reactions at the site of injection not to mention possible concerns over mineral oil residues in the tissue. Delivery systems which could influence vaccination strategies include immunisation of animals via the respiratory or gastrointestinal tracts to stimulate the local immune system, involving either live cells or inactivated antigens, and delayed or slow release systems which eliminate the need for repeat dosing. The use of immunomodulators to affect either response to vaccination or to natural infection is another area of research (Rossi-Campos *et al.*, 1992) which may lead to improved disease control in the future.

Table 2. Putative protective antigens of HAP bacteria

Haemophilus	Actinobacillus	Pasteurella
Capsule	APX I	Leukotoxin
LPS	APX II	Iron Regulated OMP's
OMP's	APX III	Capsule
	Iron Regulated OMP's	Proteases
	LPS	LPS
		Fimbriae
		Neuraminidase

The HAP group of bacteria are responsible for a number of diseases of livestock and poultry which continue to be of significant economic impact. Their importance is reflected by the number of products available throughout the world and perhaps more significantly in the number of commercial vaccine development projects identified in a recent survey (Table 3), using various sources of information.

Table 3. Number of vaccine development projects involving animal health companies.

	Actinobacillu	Haemophilus	Pasteurella
Modified Live	4		2
Sub Unit	4	3	9
Modified Bacterin	2		6
Antidiotype	1		
Unclassified	1		4

ACKNOWLEDGEMENTS

I am grateful to Matthew Phillips of Wood Mackenzie Consultants Ltd for permission to reproduce the market survey data.

REFERENCES

Adams, W.G., Deaver, K.A., Cochi, S.L., Plikaytis, B.D., Zell, E.R., Broome, C.V., and Wenger, J.D., 1993, Decline of childhood Haemophilus influenzae type b (Hib) disease in the Hib vaccine era, *JAMA*. 269:221-226.

Conlon, J.A., Shewen, P.E., and Lo RYC., 1991, Efficacy of recombinant leukotoxin in protection against pneumonic challenge with live *Pasteurella haemolytica*, *Inf Imm*. 59:587-591.

Donachie, W., and Gilmour, N.J.L., 1988, Sheep antibody response to cell wall antigens expressed *in vivo* by *Pasteurella haemolytica* serotype A2, *FEMS Microbiol. Letters*. 56:271-276.

Donachie, W., Burrells, C., Sutherland, A.D., Gilmour, J.S., and Gilmour, N.J.L, 1986, Immunity of specific pathogen free lambs to challenge with an aerosol of *Pasteurella haemolytica* (biotype A, serotype 2): Pulmonary antibody and cell responses to primary and secondary infections, *Vet. Imm. and Immunopath*. 11:265-279.

Donachie, W., Sutherland, A.D., and Jones, G.E., 1986b, Assessment of immunity to *Pasteurella haemolytica* in sheep by *in vitro* methods, *Developments in Biological Standards*. 64:63-69.

Espeseth, D.A., and Greenberg, J.B., 1993, Licensing and regulation in the USA, *in* "Vaccines for Veterinary Applications", A.R.Peters, ed., Butterworth-Heinemann, Oxford.

Houghton, S.B., and Gourlay, R.N., 1983, Synergism between *Mycoplasma bovis* and *Pasteurella haemolytica* in calf pneumonia, *Vet. Rec*. 113:41-42.

Gilmour, N.J., Martin, W.B., Sharp, J.M., Thompson, D.A., Wells, P.W., and Donachie, W., 1983, Experimental immunisation of lambs against pneumonic pasteurellosis, *Res. Vet. Sci*. 35:80-86.

Jericho, K.W.F., and Langford, E.V., 1978, Pneumonia in calves produced with aerosols of bovine herpes virus 1 and *Pasteurella haemolytica*, *Can. J Comp. Med*. 42:269:277.

Jericho, K.W.F., Darcel, C.Q., and Langford, E.V., 1982, Respiratory disease in calves produced with aerosols of parainfluenza 3 virus and *Pasteurella haemolytica*, *Can. J. Comp. Med*. 46:293-301.

Jones, D.M., 1993, Current and future trends in immunisation against meningitis, *J.Antimicrob. Chemotherapy*. 31: suppl.B 93-99.

Kroll, J.S., 1995, Genetics of pathogenicity of Haemophilus influenzae, *HAP Conference, Edinburgh 1994*.

Lee, A., 1993, Registration and regulation in the EC, *in*: "Vaccines for Veterinary Applications", A.R.Peters, ed., Butterworth-Heinemann, Oxford.

Note for Guidance, 1994, Good clinical practice for the conduct of clinical trials for veterinary medicinal products. Committee for veterinary medicinal products working party on efficacy, *European Commission Directorate General III Industry*.

Pasteur, L., 1880, De L'atténuation du virus du choléra des poules, *CR Acad. Sci*. 91:673.

Phillips, M., 1994, Animal Health and Nutrition Service, Wood Mackenzie Consultants Ltd, Edinburgh. USA Code of Federal Regulations 1989 pub Office of the Federal Register, National Archives and Records Administration, Washington.

Rossi-Campos, A., Campos, M., Potter, A.A., Harland, R., Bielefeldt-Ohmann, H., and Babiuk, L.A., 1992, *Actinobacillus pleuropneumoniae*: Modulation of pathogenesis by recombinant interferons, *Proc. Int. Pig Vet. Soc*. 1:206.

van den Bosch, J.F., Pubben, A.N.B., Jongenelen, I.M.C. and Segers, R.P.A.M., 1995, An Actinobacillus phenopneumoniae subunit vaccine inducing protection in pigs against all serotypes, *HAP Conference*, Edinburgh 1994.

GENETIC MANIPULATIONS OF MEMBERS OF THE FAMILY *PASTEURELLACEAE*

J. Frey[1] and J.I. MacInnes[2]

[1]Institute for Veterinary Bacteriology
University of Berne
Laenggasstr. 122
CH-3012 Berne
Switzerland

[2]Department of Veterinary Microbiology & Immunology
University of Guelph
Guelph, Ont. N1G 2W1
Canada

INTRODUCTION

The family *Pasteurellaceae* is comprised of three genera: *Haemophilus*, *Actinobacillus*, and *Pasteurella* (HAP). These bacteria cause important diseases in humans and animals and are of great medical and economical importance. Recently, the HAP bacteria have been the subject of intensive research in the fields of molecular mechanisms of pathogenicity, immunogenicity, and vaccine development. A considerable amount of work has been done in the molecular analysis of several virulence components including capsule, lipopolysaccharide, outer membrane proteins, and exotoxins. While cloning and expression of genes from the HAP bacteria in established hosts like *E. coli* K12 strains appears to be easy, very little work has been described on the genetic engineering of the HAP bacteria themselves.

For effective and direct analysis of virulence factors and their impact in pathogenicity, as well as for the construction of strains for the development of live attenuated vaccines, efficient gene transfer methods that can be applied directly to the HAP bacteria are necessary. Systems are needed that will allow manipulation of HAP bacteria by targeted gene mutagenesis using allelic replacement, insertional inactivation, or deletion formation. Strategies for stable gene cloning and expression are also needed. The following short overview is an attempt to review both published and unpublished work which was discussed during the "HAP Plasmids: Future Vectors?" workshop at the H.A.P 94 International Conference held July 31 to August 4, 1994 in Edinburgh, U.K. Due to the interests of the participants, much of the discussions were focused on *A. pleuropneumoniae* and *P. haemolytica*.

SELECTABLE MARKERS

In order to do insertional mutagenesis by transposition or by allelic gene replacement or to simply introduce plasmids, markers which permit the selection of transformed strains are required. Antibiotic resistance genes and genes conferring heavy metal resistance are the markers of choice. Their efficient use requires 1) that the host strain is not yet resistant (or inert) to the agent used, 2) that it does not easily revert to such resistance, 3) that the resistance genes are expressed in the host, and 4) that the host can provide the necessary cofactors to confer the resistant phenotype. Additional factors such as host range and restriction may also be involved depending on the system used.

The gene encoding aminoglycoside 3'-phosphotransferase I activity from transposon Tn*903* (Oka *et al.*, 1981) confers kanamycin resistance (Km^r) in *A. pleuropneumoniae* (Tascon *et al.*, 1993; J. Frey, unpublished results) and also seems to be a useful marker in other *Pasteurellaceae* (Workshop). In contrast, the aminoglycoside 3'-phosphotransferase II activity from Tn*5* (Beck *et al.*, 1982), which has been widely used in a variety of other gram-negative bacteria, does not confer Km^r to *A. pleuropneumoniae* and works poorly in other *Pasteurellaceae* (Willson, 1990; J. Frey, Workshop). Variants of Tn*5* can, however, confer high levels of resistance to kanamycin, bleomycin, and streptomycin to *A. pleuropneumoniae* (Willson, 1990).

The gene for gentamycin resistance (Gm^r) from the IncP plasmid, R1033, (Hirsch *et al.*, 1986) confers very high resistance to gentamycin (100 µg/ml) in *A. pleuropneumoniae* and has the potential to serve in this species as a selectable marker (J. Frey, Workshop). The Gm^r gene from Tn*1696* can also be expressed in *A. pleuropneumoniae* (West *et al.*, 1995). Similarly, the chloramphenicol resistance gene from an *A. pleuropneumoniae* plasmid, pYG10, which can also confer Cm^r to *E. coli* and *Salmonella typhimurium*, has been used as a selective marker in shuttle vectors described by Lalonde and O'Hanley, 1989. The type II Cm^r gene from the *Shigella flexneri* plasmid pS-a (Hedges and Datta, 1971) or Tn*9* (West *et al.*, 1995) also appear to be useful. These genes confer chloramphenicol resistance to *A. pleuropneumoniae*, *P. haemolytica*, and *P. multocida*, albeit at low levels ranging from 1 to 5 µg/ml (Frey, 1992; Azad *et al.*, 1994; C.L. Wright, Workshop). The Cm^r genes from pS-a and Tn*9* have been used as selective markers in cloning vectors for *Pasteurellaceae* (Frey, 1992; C. Prideaux, Workshop).

ROB-1 beta-lactamase (*bla*) genes are present on endogenous plasmids found in several species of *Pasteurellaceae*. Chang et al. (1992) reported that the deduced amino acid sequence of the ROB-1 *bla* from an *A. pleuropneumoniae* plasmid is identical to that found in *P. haemolytica* and differs from that found in *H. influenzae* by only a single amino acid. Not surprisingly, ROB-1 type *bla* genes originating from *P. haemolytica* (Azad *et al.*, 1992; Craig *et al.*, 1989; Livrelli *et al.*, 1988) express ampicillin resistance in *A. pleuropneumoniae* (J. Frey, unpublished), *P. haemolytica*, and *P. multocida* (Azad *et al.*, 1994). These genes confer moderate to high levels of resistance (15 to 100 µg/ml). The Tn*3* gene encoding Ap^r, such is found on *E. coli* plasmids like pBR322, appears only to be expressed if provided with an endogenous promoter (West *et al.*, 1995).

Genes encoding dihydrophteroate synthase, which confers sulfonamide resistance (Su^r) and streptomycin kinase, which confers streptomycin resistance (Sm^r), are present on a *P. haemolytica* plasmid, pYFC1 (Chang *et al.*, 1992). These genes are identical to the Su^r and Sm^r genes of RSF1010 (Scholz *et al.*, 1989) and might to be the same as the Su^r and Sm^r genes from the *P. multocida* plasmid T31 which were used in the construction of the vector pIG317 (C.Prideaux, Workshop). The Su^r and Sm^r genes of pYFC1 confer resistance to *P. haemolytica*, *P. multocida*, and *A. pleuropneumoniae*.

The tetracycline resistance (Tc^r) gene from Tn*10* is expressed very poorly in *A. pleuropneumoniae* and cannot be used as selectable marker (Tascon *et al.*, 1993; J.Frey, unpublished). However, a Tc^r gene originating from *Streptococcus agalactiae* can confer resistance to low concentrations of tetracycline (0.2 µg/ml) in *A. pleuropneumoniae* and has been used as a selectable marker (Jansen *et al.*, 1995).

Mercury resistance cannot be used as a selectable marker for *A. pleuropneumoniae* since all strains tested are naturally resistant to 100 µg/ml $HgCl_2$ at a level higher than that conferred by the *mer* operon of Tn*501* (J. Frey, unpublished).

PLASMID VECTORS

Because of their ease of use and the wealth of available mutants, it is likely that most cloning and manipulations of genes from HAP bacteria will continue to be done initially in *E. coli*. There is, however, a real need to study HAP genes in the homologous rather than heterologous host. To do this it is necessary to develop shuttle vectors for *Pasteurellaceae* that are able to replicate, and be selected for, in both *E. coli* and the relevant *Pasteurellaceae* species. Researchers are currently working on such shuttle vectors but there are still a large number of obstacles to overcome before they can be generally used.

A number of plasmids that are able to replicate in both *E. coli* and several species of the family of *Pasteurellaceae* are described below. Most of these plasmids have been derived from endogenous plasmids of *A. pleuropneumoniae* or *P. haemolytica*. Unfortunately, the incompatibility groups of these plasmids are not known and most have never been systematically tested in different *Pasteurellaceae* species. Since the original plasmids which served for the constructions of most of these vectors are only poorly described, it is difficult to compare the origins of replications. From comparisons of the available restriction maps, several of the replicons used seem to be very similar or even identical. It is possible that they able to replicate in all or most *Pasteurellaceae* but this remains to be proven. Further characterization of the replication features of these plasmids is needed.

The shuttle vector, pYG53, constructed from an *A. pleuropneumoniae* wild type plasmid is reported to replicate in *E. coli* and *A. pleuropneumoniae*. A mobilizable variant, pYG54, was constructed by cloning the *oriT* and *mob* functions of plasmid RK2 into pYG53 (Lalonde and O'Hanley, 1989). Plasmid pLS88, which is compatible with RSF1010 (Scholz *et al.*, 1989), contains genes encoding Sm^r, Su^r, and Km^r (similar to Tn*903*). It can be stably maintained and confers resistance to all three antibiotics in *E. coli*, *A. pleuropneumoniae*, *H. influenzae*, and *H. ducreyi* (Willson *et al.*, 1989). Plasmid pYFC1, a 4.2 kb plasmid similar to RSF1010 which contains Sm^r and Su^r genes, has been isolated from *P. haemolytica* (Chang *et al.*, 1992). C. Prideaux (Workshop) reported the construction of a small 2.3 kb plasmid pIG317 containing a Sm^r gene and a multiple cloning site which is able to replicate in *E. coli* and *A. pleuropneumoniae* serovars 1 and 7. This plasmid was derived from a 4.2 kb *A. pleuropneumoniae* plasmid which was not further described.

A 4.3kb plasmid pAB2 isolated from *P. haemolytica* encoding a ROB-1 beta lactamase and mob functions, similar to those found on the *E. coli* plasmid ColE1 was shown to be able to replicate in *E. coli*. It could be transferred to *P. haemolytica* by conjugative mobilization or by electroporation and thus represents an ideal shuttle vector for gene cloning in *P. haemolytica* (Wood *et al.*, P16). According to the DNA segment cloned on pAB2, transfer to *P. haemolytica* was inhibited or reduced, this was assumed to be caused by a *P. haemolytica* restriction barrier.

Two expression vectors, pJFF224-NX and pJFF224-XN, which contain the minimal broad host range replicon of RSF1010, a Cm^r gene from plasmid pS-a, and a broad host range promoter cassette from *E. coli* bacteriophage T4 gene 32 have been described (Frey, 1992). These plasmids can express foreign genes and be stable maintained in *A. pleuropneumoniae* for over 100 generations in the absence of antibiotic selection (Frey, 1992). Plasmid pJFF224-NX carrying the *apxICABD* operon has been used to complement (*in trans*) an apathogenic Δ*apxI* deletion mutant of *A. pleuropneumoniae* serotype 1 and fully restore virulence (Remer *et al.*, 1993; J. Frey, Workshop). These vectors also replicate in *P. haemolytica* and *P. multocida* (Frey, 1992 and C.L. Wright, Workshop).

A native 4.3 kb *P. haemolytica* plasmid, pPH33, conferring Ap resistance was reported to replicate in *E. coli* and to be mobilizable from *E. coli* to *P. haemolytica* using transfer function of IncP plasmids. In the presence of Cm, it can be amplified in *E. coli* (Azad *et al.*, 1992; Craig *et al.*, 1989). Another 4.3 plasmid pPH843 from *P. haemolytica*, which is extremely similar to pPH33 but which shows somewhat higher mobilization frequencies, was used for the construction of a mobilizable shuttle vector pAKA16. This vector can replicate in *E. coli*, *P. multocida*, and *P. haemolytica*. A derivative of it, pAKA19, containing the origin of replication from pBR322 and another derivative pAKA16 containing the R6K origin (which is dependent on the bacteriophage λ *pir* gene product) has been constructed for use as a mobilizable suicide vector for gene replacements in *P. haemolytica* and *P. multocida* (Azad *et al.*, 1994). Examples of gene replacements or expression of foreign genes have not been reported using these vectors, however.

Recently, several cloning vectors designated pGZRS-18/19 containing Ap^r from Tn*3* as selective marker and pGZRS-38/39 containing Km^r from Tn*903* have been constructed from an endogenous 4.3 kb *A. pleuropneumoniae* Sm^r plasmid (West *et al.*, 1995). These vectors replicate in *E. coli*, various serotypes and strains of *A. pleuropneumoniae*, *P. haemolytica*, and *Actinobacillus actinomycetemcomitans*. They contain a multiple cloning site in the *lacZ'* gene from pUC18 allowing insertional inactivation for screening of cloned DNA segments in *E. coli*. The p_{lac} promoter in these vectors, however, does not allow gene expression in *P. haemolytica* and *A. pleuropneumoniae*. These vectors are derived from a plasmid which seems to be very similar to plasmid T31 which was used as replicon in pIG317.

A cosmid vector has also been developed for use in *A. pleuropneumoniae* (S.E.H. West, personal communication). This vector, pSW206, is a derivative of the broad host range plasmid RK2 and contains the Km^r gene from Tn*903* for selection. Mobilization functions are provided *in trans* by the helper plasmid pRK2013. Plasmid pSW206 can be introduced into serotypes 1, 5a, and 7 at a frequency of $1 - 5 \times 10^{-1}$ transconjugates per recipient and into serotype 2 and 3 less efficiently. In addition to the low copy number, another attractive feature of pSW206 is the fact that it can be stable maintained in the presence of the pGZRS vectors. The properties and levels of antibiotic resistance of some of the better characterized plasmids are summarized in Table 1.

INSERTIONAL INACTIVATION BY TRANSPOSON MUTAGENESIS

Transposon mutagenesis is a very powerful tool for the genetic analysis of bacteria and has been used to study numerous bacterial features including virulence. Many transposon mutagenesis systems, however, do not appear to work in *A. pleuropneumoniae* (J. Frey, unpublished results) or other HAP bacteria (Workshop). Transposon mutagenesis with the conjugative, self-transmissilbe element Tn*916* (Gawron-Burke and Clewell 1982) has been reported recently in *A. actinomycetemcomitans* and *H. influenzae* (Sato *et al.*, 1992; Holland *et al.*, 1992). A transposition mutagenesis procedure, involving the use of a mobilizable, suicide plasmid pLOF/Km carrying a mini Tn*10* transposon carrying the Km^r gene from Tn*903* and an IPTG inducible transposase located outside the mobile element (Herrero *et al.*, 1990), gives reasonable transposition frequencies (approx. 10^{-5}) in *A. pleuropneumoniae* (Tascon *et al.*, 1993). Single, stable, and random insertions into the chromosome of *A. pleuropneumoniae* serotypes 1 to 11 can be made with this system. This system has been used to isolate insertions in *apxIA* or *apxIB* of the *A. pleuropneumoniae* serotype 1 strain 4074 as well as for the creation of tryptophan auxotrophs (Tascon *et al.*, 1993; Tascon *et al.*, 1994). Since the Km^r genes of pLOF/Km confer reasonable levels of Km resistance to other *Pasteurellaceae* and the genes for conjugative mobilization of the donor plasmid work well in a wide range of species, it is likely that pLOF/Km could be a valuable tool for the study of many of the HAP bacteria.

GENE REPLACEMENT SYSTEMS

Allelic gene exchange using non-replicating or conditionally-replicating plasmids has been used widely of the mutagenesis of both prokaryotes and eukaryotes. Allelic exchange can be used to create deletions of specific genes, parts of genes, or even complete operons. These knockout mutants are extremely useful for the study of mechanisms of pathogenicity as well as for the creation of safe and stable strains for use as live attenuated vaccines. Until very recently, allelic exchange was only possible in *H. influenzae* despite the efforts of numerous groups. Gene replacement systems are now available for *A. pleuropneumoniae* and *P. haemolytica*.

An efficient system for targeted mutagenesis of *A. pleuropneumoniae* has recently been described by Jansen et al. (1995). It is based on a conditionally-replicating, thermosensitive, broad-host-range plasmid, pVE6063, which was originally used in studies of gram-positive bacteria (Maguin *et al.*, 1992). This plasmid does not replicate in *A. pleuropneumoniae* even at permissive temperature and insertional mutagenesis occurs by homologous recombination. *apx* toxin genes, inserted in pVE6063, were deleted by replacement with either Cm^r genes from *S. aureus* or by Tc^r genes from *S. agalactiae*. These constructs were subsequently introduced in *A. pleuropneumoniae* serotype 1 by electroporation. Selection for Cm^r unexpectedly resulted in the isolation of double crossover events where the wild type gene was directly replaced by the genetically manipulated deletion-replacement gene (Jansen *et al.*, 1995). Although the direct formation of double crossovers cannot be explained, this system for targeted mutagenesis appears to have broad applicability since it was used to insert antibiotic markers in three different genes at different chromosomal locations. Using suitable selectable markers, this system might also be useful for targeted mutagenesis of other HAP species.

Table 1. Cloning vectors and plasmids for transposon and gene replacement delivery systems for HAP bacteria. Abbrieviations used: *A.pp, A.pleuropneumoniae*; *H.duc, H.ducrei*; *H.inf. H.influenzae; P.mul., P.multocida; P.hae., P.haemolytica*; MCS, multiple cloning site. The vectors and systems are explained in the text. 1) Plasmid antibiotic resistance gene not used for selection in recipients. 2) with A.pp. promoter.

Plasmid	Properties, origin	Antibiotic resistance (type/source)	Resistance selection (μg/ml)	Reference
pAKA16	Moblilizable shuttle vector derived from 4.3 kb *P.haemolytica* plasmid pHP843 with MCS and *lacZ'*	Ap[r] (pHP834)	*P.hae.* 100 *P.mul.* 100 *E.coli* 100	Azad *et al.*, 1994
pAKA19	Moblilzable suicide vector derived from pAKA16 with *oriV* of pBR322	Ap[r] (pHP834)	*E.coli* 100 1)	Azad *et al.*, 1994
pAKA22	Mobilizable suicide vector derived from pAKA16 with *oriV* of R6K (λ *pir* dependent)	Ap[r] (pHP834)	*E.coli* 100 1)	Azad *et al.*, 1994
pGZRS-1	Endogenous 4.3 kb plasmid from *A.pleuropneumoniae*	Sm[r]	*A.pp.* 100 *P.hae.* 100 *E.coli* 100	West *et al.*, 1995
pGZRS-18 /19	Shuttle vector derived from pGZRS-1 with MCS and *lacZ'*	Ap[r] (Tn*3*) 2)	*A.pp.* 25 *P.hae.* 50 *E.coli* 50	West *et al.*, 1995
pGZRS-38 /39	Shuttle vector derived from pGZRS-1 with MCS and *lacZ'*	Km[r] (Tn*903*)	*A.pp.* 25 *P.hae.* 50 *E.coli* 50	West *et al.*, 1995
pJFF224-NX/XN	Mobilizable broad host range expression shuttle vectors based on RSF1010 replicon with MCS and promoter cassette (phage T4 gene 32)	Cm[r] pS-a	*A.pp.* 1-2 *P.hae.* 1-2 *E.coli* 25	Frey , 1992
pLOF/Km	mobilizable transposon delivery plasmid carrying mini-Tn*10*/Km with inducible transposase (*oriV* of R6K, λ *pir* dependent)	Km[r] (Tn*903*)	*A.pp.* 50 *E.coli* 50	Tascon et al. 1993
pLS88	endogenous *H.ducrei* plasmid, used as shuttle plasmid for various *Pasteurellaceae*	Sm[r] Su[r] Km[r] all from pLS88	*A.pp.* 50, 190, 30 *H.duc.* 50, 190, 30 *H.inf.* 50, 190, 30 *E.coli* 50, 190, 30	Willson *et al.*, 1989
pVE6063	Thermosensitive replicon in *E.coli* and other bacteria, but does not replicate in *A.pp.*. Creates double recombination events. Useful for targeted insertional mutagenesis.	dependent on insert resistance used for creation of knockout mutations	1)	Maguin *et al.*, 1992 Jansen et al, 1995
pYG53	Shuttle vector with MCS based on *A.pp.* plasmid pYG10	Cm[r] (pYG10)	*A.pp.* 10 *E.coli* 10	Lalonde and O'Hanley, 1989
pYG54	Mobilizable shuttle vector derived from pYG53, containing *oriT* form RK2	Cm[r] (pYG10)	*A.pp.* 10 *E.coli* 10	Lalonde and O'Hanley, 1989

Tatum et al. (1994) were also able to do gene replacement in *P. haemolytica*. A cloned *aroA* gene was inactivating with a fragment containing a *P. haemolytica* ampicillin resistance gene. The plasmid carrying the disrupted fragment was methylated using cloned *Pha*I and introduced into *P. haemolytica* by electroporation. Only 1 Ap^r colony/µg DNA was obtained. Mutants in *aroA* (i.e. with the disrupted *aroA* gene integrated into the chromosome) were obtained after repeated passage. (Editor's note - see also abstract T27, Murphy *et al.*, which describes insertional inactivation of membrane lipoprotein expression).

GENE TRANSFER SYSTEMS

Electroporation and conjugation have been used with some success to transfer plasmids to species of the family *Pasteurellaceae*. Other transformation systems are inefficient except in a few *Haemophilus* species which are naturally competent (Albritton *et al.*, 1984; Gromkova et al 1989). Generally, electroporation gives rather low transfer frequencies and the conditions for electroporation seem to vary widely between different species and even between strains from the same species. Optimal conditions need to be established empirically for each strain used. Transformation frequencies appear to be considerably higher when the plasmid used is propagated in the same host, indicating that most species of *Pasteurellaceae* contain potent restriction barriers (Lalonde and O' Hanley, 1989; Frey, 1992; West *et al.*, 1995).

Recently, the restriction endonuclease and methylase genes of *P. haemolytica* serotype A1 have been cloned by Briggs et al. (1994). The *P. haemolytica* restriction endonuclease *Pha*I, is an isoschizomer of *Sfa*NI. Plasmids isolated from *E. coli* carrying the cloned *Pha*I-methylase genes could be introduced to *P. haemolytica* by electroporation whereas those that were not methylated could not. Interestingly, the electroporation efficiency of plasmids propagated in *P. haemolytica* (1×10^5 CFU/µg of DNA) still was 10^2 times higher than those grown in *E. coli* with the *Pha*I-methylase gene (1×10^3 CFU/µg of DNA). These result may reflect the presence of more than one restriction system (Briggs *et al.*, 1994).

Conjugative mobilization of plasmids generally gives considerably higher transfer rates compared to electroporation (Lalonde and O'Hanley, 1989; Azad *et al.*, 1994). Conjugation, however, has the disadvantage that it is necessary to modify the recipient strain so that the donor (usually *E. coli*) can counterselected. This is usually done by screening for spontaneous nalidixic, rifampicin, or streptomycin resistant mutants or by introduction of an antibiotic resistance gene on a compatible plasmid to the recipient strain (Lalonde and O'Hanley, 1989; Azad *et al.*, 1994). Alternatively, virulent *E. coli* bacteriophages such as T4 and T7 can be used for counter-selection of the donor strains. This avoids the introduction of additional antibiotic resistance genes to the recipient, which might be undesirable if the strain is used for the development of a live vaccine (J. Frey, Workshop).

CONCLUSIONS

Although not exhaustive, we hope that this review highlights, and expands on, some of the exciting new developments that were discussed at the "HAP Plasmids: Future Vectors?" workshop. It is clear, however, that despite considerable progress in recent years, a lot more work has to be done. Several plasmids have been described that will replicate in *E. coli* and in selected species of the HAP bacteria but in most cases we know little about their maintenance, copy number, host range, or incompatibility group.In addition, we know little about the expression of foreign genes in these vectors. Improved knowledge of restriction systems and development of restrictionless strains may also improve our ability to create and work with HAP shuttle vectors.

The gene replacement systems described above, have only been tested only in a narrow range of strains and most often, only in one species.The efficiency of these system is poor. Given our increased knowledge regarding the expression of various antibiotic resistance markers, electroporation conditions, and other gene transfer systems, however, improved systems are likely to emerge in the future. The ability to generate isogenic strains will be an enormous help in understanding the pathogenesis of HAP bacteria.

One of the emerging themes at the Workshop was that several promoters, which are efficiently expresed in *E. coli*, do not work well in many HAP bacteria. This fact that might be a responsible for the apparent lack of expression of various antibiotic resistance genes and transposases. In addition to antibiotic resistance genes found on naturally occurring plasmids, only the Km^r gene from Tn*903* and the Cm^r gene from *S. aureus* have shown much promise in organisms such as *A. pleuropneumoniae* and *P. haemolytica*. Other resistance genes (e.g. Ap^r from Tn*3*) may be expressed when provided with different promoters. Nevertheless, we are now in a position where we can begin to develop efficient shuttle vectors, create mutants, and complement them.

ACKNOWLEDGMENTS

We are grateful to Lynette Corbeil, Cathy Wright, and Christopher Prideaux for providing us with their notes of the Workshop and Susan West for comments and sharing unpublished results. We thank Janine Bosse for compiling the data in Table 1.

REFERENCES

Albritton, W.L., Setlow, J.K., Thomas, M., Sottnck, F. and Steigerwalt, A.G., 1984, Heterospecific transformation in the genus *Haemophilus*, *Mol. Gen. Genet.* 193:358-363.

Azad, A.K., Coote, J.G. and Parton, R., 1992, Distinct plasmid profiles of *Pasteurella haemolytica* serotypes and the characterization and amplification in *Escherichia coli* of ampicillin-resistance plasmids encoding ROB-1 β-lactamase, *J. Gen. Microbiol.* 138:1185-1196.

Azad, A.K., Coote, J.G. and Parton, R., 1994, Construction of conjugative shuttle and suicide vectors for *Pasteurella haemolytica* and *P. multocida*, *Gene* 145:81-85.

Beck, E., Ludwig, G., Auerswald, E.A., Reiss, B. and Schaller, H., 1982, Nucleotide sequence and exact localization of the neomycin phosphotransferase gene from transposon Tn*5*, *Gene* 19:327-336.

Briggs, R.E., Tatum, F.M., Casey, T.A., Frank, G.H., 1994. Characterization of a restriction endonuclease, *Pha*I, from *Pasteurella haemolytica* serotype A1 and protection of heterologous DNA by a cloned *Pha*I methyltransferase gene. *Appl. Environ.Microbiol.* 60:2006-2010.

Chang, Y-F., Ma, D.P., Bai, H.Q., Young, R., Struck, D.K., Shin, S.J. and Lein, D.H., 1992, Characterization of plasmids with antimicrobial resistant genes in *Pasteurella haemolytica* A1, *DNA Seq.* 3:89-97.

Chang, Y-F., Shi, J., Shin, S.J., and Lein, D.H., 1992, Sequence analysis of the ROB-1 β-lactamase gene from *Actinobacillus pleuropneumoniae*, *Vet. Microbiol.* 32:319-325.

Craig, F.F., Coote, J.G., Parton, R., Freer, J.H. and Gilmour, N.J., 1989, A plasmid which can be transferred between *Escherichia coli* and *Pasteurella haemolytica* by electroporation and conjugation, *J. Gen. Microbiol.* 135:2885-2890.

Frey, J., 1992, Construction of a broad host range shuttle vector for gene cloning and expression in *Actinobacillus pleuropneumoniae* and other *Pasteurellaceae*, *Res. Microbiol.* 143:263-269.

Gawron-Burke, C., and Clewell, D.B., 1982, A transposon in *Streptococcus faecalis* with fertility properties, *Nature* 300:281-284.

Gromkova, R.C., Rowji, P.B., and Koornhof, H.J., 1989, Induction of competence in nonencapsulated and encapsulated strains of *Haemophilus influenzae*, *Curr. Microbiol.* 19:241-245.

Hedges, R.W. and Datta, N., 1971, fi^- R factors giving chloramphenicol resistance, *Nature* 234:220-222.

Herrero, M., de Lorenzo, V. and Timmis, K.N., 1990, Transposon vectors containing non-antibiotic resistance selection markers for cloning and stable chromosomal insertion of foreign genes in gram-negative bacteria, *J. Bacteriol.* 172:6557-6567.

Hirsch, P.R., Wang, C.L. and Woodward, M.J., 1986, Construction of a Tn5 derivative determining resistance to gentamycin and spectinomycin using a fragment cloned from R-1033, *Gene* 48:203-209.

Holland, J., Towner, K. J., and Williams, P., 1992, Tn*916* insertion mutagenesis in *Escherichia coli* and *Haemophilus influenzae* type b following conjugative transfer *J. Gen. Microbiol.* 138:509-515.

Jansen, R., Briaire, J., Smith, H.E., Dom, P., Haesebrouck, F., Kamp, E.M., Gielkens, A.L.J. and Smits, M.A., 1995, Knockout mutants of *Actinobacillus pleuropneumoniae* serotype 1 that are devoid of RTX toxins do not activate or kill porcine neutrophils, *Infect. Immun.* 63:27-37.

Lalonde, G. and O'Hanley, P.D., 1989, Development of a shuttle vector and a conjugative transfer system for *Actinobacillus pleuropneumoniae*, *Gene* 85:243-246.

Livrelli, V.O., Darfeuille-Richaud, A., Rich, C.D., Joly, B.H. and Martel, J.-L., 1988, Genetic determinant of the ROB-1 β-lactamase in bovine and porcine *Pasteurella* strains, *Antimicrob. Agents Chemother.* 32:1282-1284.

Maguin, E., Duwat, P., Hege, T., Ehrlich, D. and Gruss, A., 1992, New thermosensitive plasmid for gram positive bacteria, *J. Bacteriol.* 174:5633-5638.

Oka, A., Sugisaki, H. and Takanami, M., 1981, Nucleotide sequence of the kanamycin resistance transposon Tn*903*, *J. Mol. Biol.* 147:217-226.

Remer, D., Frey, J., Jansen, R. and Inzana, T.J., 1993, Molecular investigation of the role of ApxI and ApxII in the virulence of *Actinobacillus pleuropneumoniae* serotype 5, *74th Annual Meeting of the Conference of Research Workers in Animal Disease, Chicago*, p. 44. Abs. #225.

Sato, S., Takamatsu, N., Okahashi, N., Matsunoshita, N., Inoue, M., Takehare, T., and Koga, T., 1992, Construction of mutants of *Actinobacillus acinetomycetemcomitans* defective in serotype b-specific polysaccharide antigen by insertion of transposon Tn*916*. *J. Gen. Microbiol.*, 138:1203-1209.

Scholz, P., Haring, V., Wittmann-Liebold, B., Ashman, K., Bagdasarian, M. and Scherzinger, E., 1989, Complete nucleotide sequence and gene organization of the broad-host-range plasmid RSF1010, *Gene* 75:271-288.

Tascon, R.I., Rodriguez-Ferri, E.F., Gutierrez-Martin, C.B., Rodriguez-Barbosa, I., Berche, P. and Vazquez-Boland, J.A., 1993, Transposon mutagenesis in *Actinobacillus pleuropneumoniae* with a Tn*10* derivative, *J. Bacteriol.* 175:5717-5722.

Tascon, R.I., Vazquez-Boland, J.A., Gutierrez-Martin, C.B., Rodriguez-Barbosa, I., and Rodriguez-Ferri, E.F., 1994, The RTX haemolysins ApxI and ApxII are major virulence factors of the swine pathogen *Actinobacillus pleuropneumoniae*: evidence from mutational analysis, *Mol. Microbiol.* 14:207-216.

Tatum, F.M., Briggs, R.E., and Halling, S.M., 1994, Molecular gene cloning and nucleotide sequencing and construction of an *aroA* mutant of *Pasteurella haemolytica* serotype A1, *Appl. Environ. Microbiol.* 60:2011-2016.

West, S.E.H., Romero, M-J.M., Regassa, L.B., Zielinski, N.A. and Welch, R.A., 1995.Construction of *Actinobacillus pleuropneumoniae - Escherichia coli* shuttle vectors: expression of antibiotic resistance genes. *Gene* (in press).

Willson, P.J., Albritton, W.L., Slaney, L. and Setlow, J.K., 1989, Characterization of a multiple antibiotic resistance plasmid from *Haemophilus ducreyi*, *Antimicrob. Agents Chemother.* 33:1627-1630.

Willson, P.J., 1990, *Haemophilus, Actinobacillus, Pasteurella*: mechanisms of resistance and antibiotic therapy, *Can. J. Vet. Res.* 54 Suppl:S73-S77.

Offered Talks and Poster Section

The conference was structured around invited keynote lectures supported by offered papers in related areas from delegates. These offered papers were allocated to talk or poster presentations and submitted abstracts collated for publication in two sections:

1) Talks (prefaced by the letter T) - Speakers are denoted by an asterisk.
2) Posters (prefaced by the letter P).

T1. Taxonomy of the *Pasteurellaceae:* A Difficult Problem Even When the Phylogeny is Known.

F.E. Dewhirst*, B.J. Paster, G.J. Fraser and I.Olsen, Department of Molecular Genetics, Forsyth Dental Center, Boston, MA 02115, USA and Department of Microbiology, Dental Faculty, University of Oslo, Oslo, Norway.

We have determined full 16S rRNA sequences for over 90 strains representing 80 species in the family *Pasteurellaceae*. Since our most recent report on the phylogeny of this group (Zbl. Bakt. 279:35-44, 1993), we have obtained sequences for 16 additional strains: Bisgaard taxa 10, 15, 16, 17, 18, and 20, *Pasteurella haemolytica* CCUG 28148, "A/P-like" CCUG 28030, H. *influenzae-murium* CCUG 6515, P. *haemolytica*- like CCUG 28018, *Haemophilus*-like CCUG 19808, *H. paragallinarum*-like CCUG 18783, P. *multocida* ss *septica* CCUG 17977, and three strains of *A. actinomycetemcomitans*. Sequences were determined by direct Sanger sequencing of purified 16S rRNA. Sequences were aligned and a homology matrix was determined by using only those positions at which 90% of strains possessed sequence information. A phylogenetic tree was constructed using the neighbour joining method. The new strains are all members of the family *Pasteurellaceae* and fall within clusters described previously. If the 16S rRNA based analysis accurately reflects the phylogeny of the *Pasteurellaceae,* our results present a dilemma for the taxonomy of the family. Branching of clusters within the family are deep (bush like) and complex. It would seem that the family needs to be divided to over 20 genera, or reduced to a single genera.

T2. Deoxyribonucleic Acid Relatedness Among Porcine V Factor-Dependent *Pasteurellaceae.*

K. Mϕller*, National Veterinary Laboratory, Bülowsvej 27, DK-1790, Copenhagen; M. Kilian, Institute of Medical Microbiology, University of Aarhus, Denmark; F. Grimont and P.A.D. Grimont, Service des Enterobacteries, Institut Pasteur, F-75724 Paris Cedex 15, France.

Deoxyribonucleic acid relatedness studies (S1 nuclease method) of selected V factor-dependent *Pasteurellacea* from the porcine respiratory tract showed that taxon Minor group and the taxa previously designated C, D/E, and F form distinct groups only distantly related to each other (1 to 26% homology), and to other selected members of the family *Pasteurellaceae* (3 to 39% homology). Three homology groups were identified in taxon Minor group and two in taxa D/E and F indicating some heterogeneity within the individual taxa. The study confirms that the taxa Minor group, C, D/E, and F warrant specific recognition. However, the exact genus affiliation can not be concluded on the basis of available data.

T3. Population Genetics of Non-Capsulate *Haemophilus influenzae.*

K.J. Forbes* and T.H. Pennington, Medical Microbiology, Aberdeen University, Aberdeen, U.K.

Haemophilus influenzae is an important human pathogen with the preponderance of invasive disease isolates being capsulate, usually type b, and non-invasive disease isolates typically being non-capsulate. Several studies have demonstrated that the genetic structure of these two groups are completely different. Linkage disequilibrium of genetic markers is found amongst the capsulate strains with extensive genetic similarity being shared between members of each clone. This is in stark contrast to non-capsulate strains where genetic similarities are minimal and groups of indistinguishable strains (clones) are often small in size and indeed frequently contain only single representatives. This genetic heterogeneity of non-capsulate strains has been analysed using genotypic methods: genomic DNA fingerprinting, Southern hybridisation, and gene RFLP's and sequencing. Each of these methods can be used to establish genetic relationships between strains but the conclusions that can be drawn from these are strongly influenced by the method used. These comparative analyses can be used to test the hypothesis that if the predominant mechanism of genomic evolution in non-capsulate strains is solely vertical transmission of genetic material, then strain phylogenies should all be identical irrespective of the method. If this is not the case, then horizontal transmission of genetic material must have disrupted the original clonal framework of the strains.

T4. Use of Biogrouping and DNA Fingerprinting to Monitor Transmission of *Pasteurella* Between Bighorn and Domestic Sheep.

A.C.S. Ward,* M.D. Jaworski and D.L.Hunter, University of Idaho, Caine Veterinary Teaching and Research Center, Caldwell, Idaho USA.

Deaths resulting from respiratory infections associated with *Pasteurella* species occur in both domestic and wild sheep populations. Exposure of animals to strains of *Pasteurella* new to their population may result in colonization of nasal mucosa, tonsillar crypts, and in some instances disease. Tests were conducted on over 300 *Pasteurella* isolates from bighorn and domestic sheep which had been in close proximity to monitor transmission. Biochemical tests (biogrouping) provided a basis for initial screening of the isolates. *Pasteurella* isolates having the same biogroup were subjected to DNA fingerprinting as a final step to determine if transmission had occurred between sheep groups. Transmission from domestic to bighorn sheep was conclusively demonstrated in contact pen exposure with disease being evident in both sheep species. Organisms with identical fingerprints were also detected in one range situation without any evidence of disease in either sheep populations. The route of transmission was not demonstrable in the latter case.

T5. Characterization of *Actinobacillus seminis*.

J.C. Low[1], I. Bromley[1], W. Donachie[2], D. Somerville[1], [1]SACVS Edinburgh, Bush Estate, Penicuik, Midlothian, [2]Moredun Research Institute, Edinburgh.

The identification of Gram-negative pleomorphic bacilli isolated from ovine semen is difficult as many of the bacterial species associated with ram infertility are biochemically inactive and published descriptions are often ambiguous and misleading. In this study, isolates of *Actinobacillus seminis* from clinical specimens, semen and vaginal swabs and other members of the *Pasteurellaceae* family have been compared on the basis of cultural characteristics, biochemical reactions, APIZYM profiles and by examining their outer membrane protein profiles in SDS-PAGE. This study was funded by SOAFD.

T6. Protective Capacity of the Antigenic 40-Kilodalton LppB Lipoprotein of *Haemophilus somnus* Against Experimental Infection in Cattle.

C.R. Rioux*[1],R.J. Harland,[1] M. Theisen[1], N.A. Rawlyk[1] and A.A. Potter[1&2], Veterinary Infectious Disease Organization[1] and Canadian Bacterial Diseases Network,[2] Saskatoon, Saskatchewan, Canada.

We have previously reported the cloning and characterization of the gene *lppB* coding for a 40-kDa outer membrane lipoprotein of *H. somnus*. *Escherichia coli* transformants producing LppB adsorbed Congo red, a property associated with virulence in pathogenic bacteria, and reacted strongly with bovine hyperimmune serum raised against *H. somnus*. The deduced amino acid sequence of LppB shows similarities to that of new lipoprotein D (NlpD) of *E. coli*. Recombinant LppB was found at low levels in *E. coli* as seroreactive polypeptides, in the 25- to 40-kDa range, which could be partially purified by aggregate preparation. To evaluate the protective capacity of LppB against experimental *H. somnus* challenge, cattle were vaccinated twice with recombinant LppB prior to intravenous infection with virulent *H. somnus*. Compared to placebo animals, cattle vaccinated with LppB showed less severe clinical signs such as fever, weight loss, and lameness. The survival rates were 0 and 88% for the placebo and vaccinated calves, respectively. LppB elicited a strong, specific antibody response in cattle. Therefore, LppB is a potential antigen for use in vaccines to prevent haemophilosis.

T7. Protection of Ruminants with *Pasteurella haemolytica* A1 Capsular Polysaccharide (CP) Vaccines Containing MDP or MDP-GDP.

K. Brogden,* B. DeBey, F. Audibert, H. Lehmkuhl and L. Chedid. USDA-ARS-NADC Ames, IA; Univ. California, Tulare, CA; and VACSYN S. A. Paris, France.

To improve CP immunogenicity, vaccines were prepared containing 1.0 mg CP with and without MDP (range 0.2 - 1.0 mg) or MDP-GDP (range 0.1 - 1.0 mg). After immunisation, vaccines with MDP or MDP-GDP induced significantly higher IgM, IgG1, and IgG2 titres. After challenge, lambs vaccinated with CP+0.05 mg MDP or CP+MDP-GDP had significantly less pulmonary consolidation and concentrations of *P. haemolytica* in lesions compared to CP vaccinated or nonvaccinated lambs. After challenge, calves vaccinated with CP containing 0.2 mg MDP, 0.5 mg MDP, or 1.0 mg MDP-GDP also had significantly less pulmonary consolidation and concentrations of *P. haemolytica* in lung lesions compared to CP vaccinated or nonvaccinated calves. Vaccines containing CP+0.5 mg MDP and CP+1.0 mg MDP-GDP induced bactericidal antibodies by 7 days and were more efficacious than two commercial vaccines.

T8. Induction of RTX Neutralizing Antibody by Exposure of Pigs to a Unique Strain of *Actinobacillus suis* and Resultant Protection Against *Actinobacillus pleuropneumoniae.*

B. Fenwick, Department of Pathology and Microbiology, Kansas State University.

Eight *Actinobacillus pleuropneumoniae* (App) free pigs from a newly derived high health status herd were inoculated intranasally with a live strain of *A. suis*. Very mild clinical signs developed approximately 8 hours after the inoculations and soon resolved. Within three weeks significant App hemolysin neutralising titers developed, yet all pigs remained negative to App by complement fixation assay and ELISA. Eight additional age matched 'control' pigs from the same herd were mixed with the *A. suis* exposed pigs and both groups challenged intranasally with the type strain of App serotype 1. Two days later all pigs were killed and post-mortem examinations conducted. Prior exposure to *A. suis* and the resultant induction of cross neutralising titers provided significant ($P<0.04$) protection against App disease. These findings have implications in the understanding of the factors that may influence the occurrence of disease by those organisms that produce related RTX proteins as well as suggesting novel methods by which to provide protection.

T9. Seroepidemiology of Undifferentiated Fever in Feedlot Calves.

P.T. Guichon*, R. Harland, C.W. Booker, and G.K Jim, Feedlot Health Management Services, 7 - 87 Elizabeth Street, Postal Bag Service #5, Okotoks, Alberta

A seroepidemiological study was conducted in western Canada to investigate the relationships between *Pasteurella haemolytica*, *Haemophilus somnus*, undifferentiated fever, and mortality in 1,219 feedlot calves. Blood samples were obtained from each animal upon arrival at the feedlot and again at approximately 33 days of the feeding period. In addition, interim samples were obtained from animals which required treatment for undifferentiated fever (Fever) and a corresponding number of healthy control animals (Control). The overall, haemophilosis, and BRD or haemophilosis mortality rates were significantly ($p<0.05$) higher in the Fever group than in the Control group. The Fever animals had significantly ($p<0.05$) lower *Pasteurella haemolytica* anti-leukotoxin (PHAL) titers at arrival than the Control animals. In addition, the factorial change in PHAL titers of the Fever animals from arrival to the convalescent sampling (4.7X) was significantly ($p<0.05$) higher than the factorial change in PHAL titers over the same interval for the control animals (3.5X). There were no significant ($p \geq 0.05$) differences in the *Haemophilus somnus* (HS) arrival titers, interim titres, or the factorial changes in titers from arrival to the interim sampling between the Fever and Control animals. However, the convalescent HS titres of the Fever animals were significantly ($p<0.05$) lower than the Control animals. The HS arrival titers of animals that died of BRD were significantly ($p<0.05$) lower than the Control animals. Moreover, the factorial change in HS titers of the Fever animals from arrival to the convalescent sampling (2.2X) was significantly ($p<0.05$) lower than the factorial change in HS titers over the same interval for the Control animals (3.2X). The HS arrival titers of animals that died of BRD were significantly ($p<0.05$) lower than the HS arrival titers of animals which lived. In summary, the results of this study show that both *Pasteurella haemolytica* and *Hemophilus somnus* are significant pathoges of the early feeding period. Moreover, animals which develop undifferentiated fever and/or die during the feeding period are less likely to respond serologically to one or both of these pathogens. Additional studies are necessary to investigate these relationships further.

T10. Large-Scale Use of PCR for the Detection of Toxigenic *Pasteurella multocida* in Nasal and Tonsillar Swabs of Pigs.

E. Kamp[1*], G. Bokken[1], T. Vermeulen[1], M. de Jong[2], H. Buys[1], F. Reek and M. Smits[1], [1]Institute for Animal Science and Health, P.O. Box 65, 8200 AB, Lelystad, [2]Animal Health Service Centre, P.O. Box 9, 7400 AA Deventer, The Netherlands.

We developed a PCR for the detection of toxigenic *Pasteurella multocida* by using two primer sets in the gene that encodes the dermonecrotic toxin. Target DNA in clinical samples was isolated with GuSCN and diatom earth as described by Boom *et al.* (J. Clin. Microbiol. (1990) 28: 495). To enable large scale use of the PCR we adapted this method to a microtiter plate format. In this way we could process 96 clinical samples simultaneously. To test the sensitivity of the PCR we examined nasal or tonsillar swabs from 345 pigs of 7 herds known to be infected with the toxigenic *P. multocida* by conventional bacteriological methods and by PCR. Toxigenic *P. multocida* were isolated from 23 samples, forty-two samples were found positive in PCR. All samples positive for toxigenic *P. multocida* were also positive in PCR. To test the specificity of the PCR we examined swabs of 211 pigs of 5 herds which have a certificate to be free from toxigenic *P. multocida*. Toxigenic *P. multocida* were not isolated from these samples. Two samples from pigs of one herd were positive in PCR. We concluded that the PCR is more sensitive than bacteriological examination and that the PCR has a high sensitivity.

T11. Field Evaluation of *Pasteurella haemolytica* Immunogen in Lambs.

F. Morales A.; L.Jaramillo M.; J. Tortora P. and F.Trigo T.*
CENID-MICROBIOLOGIA/FMVZ/UNAM Cd. Universitaria, Mexico D.F. 04510

The purpose of this work was to evaluate the efficacy and inocuity of a live *Pasteurella haemolytica* (PH) immunogen in sheep in the field. The study was conducted in five farms located in the Ajusco area, near Mexico City. Ninety six lambs were subcutaneously immunized with a live culture of PH. Eighty one lambs were sham inoculated and remained as the control group. Before and after immunization anti-capsule and anti-cytotoxin antibodies to PH were evaluated. Clinico pathological follow up was also conducted with the lambs until weaning. Anti-capsule and anticytotoxin antibody titres showed seroconversion to the immunization ($p < 0.05$). Five animals developed pneumonia, 4 in the control group (4.9%) and only one (1%) in the vaccinated group. Sera samples were taken for serological analysis when the animals presented clinical pneumonia and 15 days later. Results revealed that pneumonic animals were seronegative to anti-cytotoxin antibodies and that 15 days later presented a titre higher than in the vaccinated animals. These results suggest that the infection was caused by PH and that its cytotoxin played an important role in the disease. It can also be considered that the protection induced by the PH immunization was capable of reducing pneumonia in lambs. Lastly, the inocuity of the immunogen was demonstrated by the absence of post-vaccination reactions.

T12. Synergistic Effect of a P2-Homologous Outer Membrane Protein and ApxI Toxin in Protection Against *Actinobacillus pleuropneumoniae.*

J.F. van den Bosch*, A.N.B. Pubben, F.G.A. van Vugt, and R.P.A.M. Segers. Intervet International, P.O.Box 31, NL-5830 AA Boxmeer, The Netherlands.

A 42kDa outer membrane protein (OMP), expressed by all *A. pleuropneumoniae* (App) serotypes, was found to be homologous to the P2-OMP of *H. influenzae* type b. The strongly cytotoxic and haemolytic RTX toxin ApxI (formerly HlyI) is expressed by App serotypes 1, 5a, 5b, 9, 10 and 11. Efficacy experiments in pigs revealed that vaccines containing either OMP or ApxI did protect against mortality but not against the development of lung lesions after challenge with App serotype 1 or 5a, whereas vaccines containing both OMP and ApxI induced full protection. Experiments in mice showed a synergistic effect of OMP and ApxI antigens with regard to protection against App serotype 1 challenge. It was concluded that App subunit vaccines should contain the common P2-homologous OMP in addition to Apx toxins.

T13. A Novel *Pasteurella haemolytica* A1 Vaccine Killed by Ultraviolet Light was Efficacious in Goats Against a Transthoracic Challenge.

C.W. Purdy*, USDA-ARS, Bovine Respiratory Disease Unit, Bushland, TX; D. C. Straus, Texas Tech University, HSC, Lubbock, TX, U.S.A.

A 10 hour old pool of virulent *Pasteurella haemolytica* A1 (PhA1) containing 10^{10} CFU/ml was killed by ultraviolet light (315 nm) exposure for 1 hour. The PhA1 bacterin was tested repeatedly for viability: none was found. On days 0 and 21, 25 goats each received lml of PhA1 (10^{10}CFU/ml) subcutaneously (SC) with various carrier systems which determined the bacterin treatment groups: agar beads (AG, n=6); polyacrylamide beads (PA, n=7); saline suspension (SA, n=6); and oil adjuvant (OA, n=6). Positive controls (PC, n=9) were injected transthoracicly into the left lung with 1 ml of 7.25 X 10^5 live PhA1 in PA (1st vaccination) and 4.4 X 10^6 live PhA1 in PA (2nd vaccination). Negative controls were injected SC with only PA on days O and 21. All goats were challenged in the right lung with 2.07 X 10^8 live PhA1on day 35 and necropsied 4 days later. Vaccine efficacy was demonstrated by smaller ($P < 0.0001$) (cm^3) consolidated pneumonic tissue lesions in the vaccinated groups at the challenge site: NC, 160; PC, 1; PA,6; AG, 7; SA, 8; OA, 2.

T14. Comparison of DNA from Organisms Classified with the 3rd Taxon of *Pasteurella haemolytica* or *P. haemolytica, sensu stricto.*

Ø. Angen[1]*, J.E. Olsen[1], W. Frederiksen[2] and M. Bisgaard[1], [1]Dep. of Veterinary Microbiology, The Royal Veterinary and Agricultural University, Copenhagen, [2]Dep. of Clinical Microbiology, Statens Serum Institut, Copenhagen.

Organisms which could not be classified with *P. haemolytica* biotypes A or T were originally classified as a 3rd taxon of *P. haemolytica* by Frederiksen. Previous investigations have linked some members of this group with *P. haemolytica* biovar 1 (*P. haemolytica* sensu stricto). The results of the present study, which included ribotyping of 32 and 28 strains respectively of 3rd taxon and biovar 1, demonstrated 37 ribotypes after digestion with HindIII. Computation using Taxan4 showed the existence of 6 major clusters within the 3rd taxon. With the exception of one strain all strains of biovar 1 clustered together. Using different coefficients of similarity or clustering methods did not change the overall clustering pattern. Within the 3rd taxon a correlation was demonstrated between the observed clusters and fermentation of L(+) arabinose and decarboxylation of ornithine. Three strains representing the 3rd taxon, which according to previously published

DNA:DNA hybridization represent separate species, were allocated to 3 different clusters in the present study. However, strain P 733 of 3rd taxon which according to DNA:DNA hybridizations showed 85% similarity with the type strain of *P. haemolytica* did not cluster with biovar 1. Further investigations are needed before final conclusions can be drawn as to the taxonomical position of the 3rd taxon of *P. haemolytica*.

T15. *In vivo* Association of *Actinobacillus pleuropneumoniae* with the Respiratory Epithelium of Pigs.

P. Dom*, F. Haesebrouck, R. Ducatelle and G. Charlier[1], [1]Laboratory of Veterinary Bacteriology, University of Gent and NIDO, Brussels, Belgium.

The ability of *Actinobacillus pleuropneumoniae* serotype 2, 5, 9 and 10 strains to associate *in vivo* with the epithelium of the porcine respiratory tract was investigated after intranasal inoculation of hysterectomy-derived and colostrum-deprived pigs. At 90 minutes post-inoculation multiple focal early inflammatory pulmonary lesions were observed in which histologically different, more or less concentric zones could be distinguished. In the centre of these pneumonic areas bacteria were associated with infiltrated cells and exudate. In the surrounding zones, bacteria were intimately associated with the epithelium of the alveoli and the cilia of the terminal bronchioles, as observed by light- and electron microscopy. Bacteria were only sporadically associated with the epithelium of the upper respiratory tract. In view of these findings, we propose the hypothesis that adherence of *A. pleuropneumoniae* to lower respiratory epithelial cells, possibly leading to a local high concentration of RTX toxins at the surface of eukaryotic cells, constitutes an important initial step in pathogenesis. (This study was supported by IWONL, Brussels, Belgium).

T16. Virulence of Different Serovars of *Haemophilus parasuis* for Cesarean-Derived, Colostrum-Deprived Pigs.

V.J. Rapp-Gabrielson*, G.K. Kocur, J.T. Clark and S.K. Muir, Solvay Animal Health, Inc., 1201 Northland Dr., Mendota Heights, MN, 55120, U.S.A.

Strains representing *H. parasuis* serovars 2, 4, 5, 12, 13, 14, and a nontypeable strain, were evaluated for virulence using an intratracheal inoculation model in CDCD pigs. Inoculation of only 100 CFUs of some strains resulted in systemic infection and death within a few days post-challenge. Based on the number of CFUs required to produce death, all strains representing serovars 2, 5, 12, 13, and the nontypeable strain, were highly virulent, while two strains representing serovar 4 were of moderate virulence. Two serovar 14 strains, which differed in outer membrane protein (OMP) profile, were also examined. One serovar 14 strain caused death in all pigs inoculated with as few as 3×10^3 CFU, while the other strain caused death in only 1 of 3 pigs inoculated with as many as 1×10^8 CFU. These data concur with previous reports indicating that both serovar and OMP profile may be predictors of virulence for *H. parasuis*.

T17. Ciliostasis and Colonization of Toxigenic *Pasteurella multocida*.

J.P. Nielson[1,2]* and S. Rosendall[1], [1]University of Guelph, Ontario, Canada, [2]National Veterinary Laboratory, Copenhagen, Denmark.

The relation between ciliostasis and predisposing factors to nasal colonization with toxigenic *Pasteurella multocida* was studied. Ciliostasis was induced within 4 hours in cultured explants of nasal mucosa by acetic acid (0.01%), hydrogen peroxide (0.05%), and *Bordetella bronchiseptica* (5×10^8 CFU/ml), while PMT was ciliostatic at a concentration of 50 µg/ml only. Toxigenic *P. multocida* in concentrations from 10^8 CFU/ml to 10^{10} CFU/ml did not affect ciliary beating. Unilateral intranasal installation of 1% acetic acid or 10 % hydrogen peroxide, caused a unilateral impairment of the mucociliary transport but intraperitoneal injection of 300 ng PMT/ kg body weight did not. Pigs treated unilaterally intranasally for two days with 5 % hydrogen peroxide developed unilateral short-lived colonization with toxigenic *P. multocida*, and a significant unilateral turbinate atrophy.

These findings indicate that ciliary arrest might be a mechanism common to predisposing factors for nasal infection with toxigenic *P. multocida*.

T18. Determination of the Hyaluronic Acid Presence in the Capsular Material of *Pasteurella Multocida* Types A and D Isolated from Porcine Lungs.

J.A. Gutierrez-Pabello[1], J.E. Smith[2] and F. Suarez-Guemes[1]*, [1]Departamento de Microbiologia e Immunologia, FMVZ, UNAM Mexico D.F. 04510, [2]University of Surrey, Guildford, Surrey, UK.

A total of 20 strains of *Pasteurella multocida* was isolated from porcine lungs. Seven smooth colony strains belonged to serotype D, whereas 12 highly mucoid and one smooth strains were classified as type A by the indirect haemagglutination test. A salt gradient paper chromatography was used for the detection of hyaluronic acid. All *P. multocida* strains tested, developed a blue spot at the same level as the purified hyaluronic acid did. Serotype A strains showed a more distinct and bigger spot than serotype D cultures. Findings suggest that the capsular material of serotype D strains may contain hyaluronic acid as serotype A does, or may have a different substance that migrates to the same level.

T19. *Actinobacillus pleuropneumoniae* Encodes a Periplasmic Copper Zinc Superoxide Dismutase.

P.R. Langford* and J.S. Kroll, Molecular Infectious Diseases Group, Department of Paediatrics, St Mary's Hospital Medical School, London, W2 lPG.

Bacterial superoxide dismutases (SODs) are metallo-enzymes which form important cytosolic defences against oxygen toxicity. We have recently discovered that *A. pleuropneumoniae* has both a conventional MnSOD (SodA) and a second enzyme, CuZnSOD (SodC) - a rare bacterial example of this protein. The deduced amino acid sequence contains an N terminal domain characteristic of a signal peptide, suggesting that the enzyme is exported. Attempts to isolate periplasmic contents of *A. pleuropneumoniae* were unsuccessful and we therefore used an indirect strategy to examine CuZnSOD location. We expressed a fusion of *5'-sodC* to a promoterless gene encoding leader-peptide-deficient TEM1-β-lactamase in *E. coli*. As β-lactamase activity in the fusion is only detectable if the protein is exported to the periplasm, this allows identification of leader peptides. Clones growing on ampicillin-containing medium were found to have in-frame fusions downstream of the putative leader peptide. A clone with an in-frame fusion within this region did not have β-lactamase activity. These results strongly suggest that *A. pleuropneumoniae* CuZnSOD is periplasmic. Superoxide can cross the bacterial outer membrane but not the cytoplasmic membrane, suggesting that CuZnSOD may protect against an exogenous source of superoxide, produced for example during host defence activity.

This work was supported by the Biotechnology and Biological Sciences Reseach Council.

T20. *Pasteurella multocida* Toxin as a Mitogen.

A.J. Lax*, M.G. Smyth, P.B. Mullan and P.N. Ward. Institute for Animal Health, Compton, Newbury, Berks, RG16 0NN, U K.

We have previously shown that the *Pasteurella multocida* toxin (PMT) is a potent mitogen for cultured Swiss 3T3 fibroblast cells where it stimulates the initiation of DNA synthesis and an increase in cell number in quiescent cells. Its sequence shares a motif found in several toxins which ADP ribosylate a target[2] , but mutations within this motif do not affect its activity[3]. PMT has limited homology to the *Escherichia coli* toxin CNF1[4], but its significance is unknown. We will report here on the sensitivity of PMT to a variety of proteases, and the effects of denaturants on the toxin. The effects of site directed mutagenesis on toxic activity will also be described. We have shown that PMT is also a potent mitogen for primary derived osteoblasts and chondrocytes.

Rozengurt *et al.* (1990) Proc Nat Acad Sci USA 87, 123 - 127: 2. Lax *et al.*(1990) FEBS Letters 277, 59 - 64: 3. Ward *et al.* (1994). FEBS Letters 342, 81 - 84: 4. Falbo *et al.* (1993). Infect. Immun. 61, 4909 - 4914.

T21. Species Specificity of *Actinobacillus actinomycetemcomitans (Aa)* Leukotoxin (Ltx)

E.T. Lally and I.R. Kieba, University of Pennsylvania, Philadelphia, PA, USA 19104-6002.

Aa Ltx belongs to the RTX (repeats in toxin) family of bacterial cytolysins. While extensive amino acid homology exists amongst the various RTX, the cellular and species specificities remain unique for individual toxins. For example, *Aa* Ltx will only kill human monomyelocytes while the *P. haemolytica (Ph)* leukotoxin *(lkt)* kills bovine lymphoid cells. The present studies were undertaken to determine the domain of Ltx that is responsible for species specificity. PCR-generated fragments of the *lktA* gene from *Ph* were spliced into naturally occurring restriction sites of the *ltxA* gene. After expression in tandem with the *ltxC* gene from the Pλ promoter, cells were sonicated and the ability of the respective lysates to kill either HL-60 (human) or BL3 (bovine) cells was assessed by trypan blue exclusion. Over 75% of the *ltxA* gene product can be replaced with *lktA* with no effect on its ability to kill the human cells. The critical area required for the chimeric toxins to recognize human target cells is a 246 aa fragment (residues 694-940) that contains the GGXGXDXUX glycine-rich repeats. Initial attempts to produce mutants with splice sites within this region resulted in loss of toxicity. Recently, the three-dimensional structure of a related protein with glycine-rich repeats has been solved. Utilization of atomic coordinates from these data should permit construction of a model of the repeat region of the *A. actinomycetemcomitans* leukotoxin and subsequent development of toxin chimeras that further define the fine structure of species specificity of RTX.

T22. Identification and Structural Characterisation of an Iron- Regulated Haemopexin Receptor in *Haemophilus influenzae* Type B.

J. Holland[1]*, J.C.Y. Wong[1], P.W. Whitby[1], A. Smith[2] and P. Williams[1], [1]Department of Pharmaceutical Sciences, University of Nottingham, U.K. and [2]School of Biological Sciences, University of Missouri-Kansas City, U.S.A.

Haemophilus influenzae type b (Hib) uses haem as a source of essential porphyrin and iron and is capable of acquiring haem bound to the serum haem-binding protein haemopexin. We therefore examined the mechanism by which this pathogen acquired haem from haemopexin. Haemopexin binding to whole cells of Hib after growth in an iron-restricted medium was a specific, time-dependent, saturable process. From competitive binding assays with ^{125}I labelled haem-haemopexin, the affinity (K_2) was estimated to be 0.3 µM with approximately 3600 receptors per cell. Haemopexin affinity chromatography of solubilised membranes yielded three proteins of 57, 38 and 29 kDa. Trypsinisation of whole cells resulted in the loss of haemopexin-binding and correlated with the degradation of only the 57 kDa protein indicating that this protein, which can be renatured to bind haemopexin after SDS-PAGE and Western blotting, is the major surface exposed component of the haemopexin receptor. N-terminal amino-acid sequence analysis indicated that the 38 kDa protein was in fact the outer membrane porin protein P2. These data establish the existence of a haemopexin receptor and indicate a possible role for the P2 porin in the acquisition of haem from haemopexin.

T23. Cloning of a DNA Sequence Specific to Type B Isolates of *Pasteurella multocida* that Cause Haemorrhagic Septicaemia (HS).

K.M. Townsend*, H.J.S. Dawkins, J.M. Papadimitrou, Molecular Oncology Unit, Department of Pathology, Q.E.II Medical Centre, Nedlands WA 6009, Australia.

Haemorrhagic septicaemia (HS) is a peracute disease of cattle and buffalo that is caused by specific serotypes of the bacterium *Pasteurella multocida*. During the last fifty years, five main serotyping systems have been developed to classify the *P. multocida* species. Despite the attempts at classification, there has been no system developed to date that will definitively identify HS-causing strains of *P. multocida*. A modified subtractive hybridisation method has allowed the isolation of a DNA sequence that specifically hybridises to type B isolates that cause HS. This technique involves competitive hybridisation of *Sau* 3A l-digested DNA from the isolate 0131 (Izatnager 25) and biotin labelled sonicated DNA from 0130 (Izatnager 52). Double stranded digested DNA was enriched by paramagnetic separation and subsequently cloned into a plasmid vector. Recombinant clones were screened against a range of *P.multocida* isolates by slot blot hybridisation. Analysis by this method has identified a candidate clone for identification of Carter type B HS-causing isolates of *P.multocida*.

T24. *Haemophilus felis* as a Cause of Disease in the Cat.

D.J. Taylor* and N. Cameron, Department of Veterinary Pathology, University of Glasgow Veterinary School, Bearsden, Glasgow, G61 lQH.

21 isolates of *Haemophilus* sp were obtained from cats and six isolates were characterised biochemically and found to resemble *H. felis*. The organisms were isolated from healthy SPF cats and from cases of disease such as pleurisy, conjunctivitis, rhinitis, pharyngitis, pneumonia, bronchitis and from the vagina in infertility.Six isolates were obtained in pure culture suggesting that the organism might be a cause of disease. Further evidence for this supposition was provided by the demonstration by immunoblotting of antibody to an isolate obtained from an SPF cat with respiratory disease following successful treatment of the disease.

T25. Cloning of a DNA Region Putatively Involved in the Encapsulation of *Actinobacillus pleuropneumoniae*.

C.K. Ward and T.J. Inzana*, Virginia Tech, Blacksburg, VA.

The capsular polysaccharide (CP) is a required virulence factor of *Actinobacillus pleuropneumoniae* (Ap). Our aim is to clone and sequence the serotype 5 capsule *sfc* genes to better define the role of capsule in virulence and immunoprotection. A 5.3 kb *xbaI* fragment of strain J45 chromosomal DNA that hybridized with two DNA probes specific for the *Haemophilus influenzae* type b (Hib) Region 1 (capsule export) genes was cloned and restriction mapped. Portions of this cloned Ap DNA hybridized with genomic DNA from Ap serotypes 1, 2, 7, 9, indicating conservation among serotypes. Chromosomal walking was used to clone an adjacent 5.8 kb region of Ap DNA that would presumably encode structural genes required for CP biosynthesis. A portion of this adjacent DNA hybridized with DNA from other Ap serotype 5 strains, but not with DNA from Ap serotypes 2, 7, or 9, indicating serotype-specificity of this region. In addition, our cloned export genes could complement and partially restore the mutated *kps* export genes of *E. coli* Kl. These results indicate that the Ap capsule belongs to the conserved group II family of capsules in Gram-negative bacteria.

T26. Cloning and Characterization of a Bacteriophage Lysozyme Gene from *Haemophilus somnus*.

R.A. Pontarollo[1]*, C. R. Rioux[1], and A. A. Potter[1,2], [1]Veterinary Infectious Disease Organization and [2]Canadian Bacterial Diseases Network[2], Saskatoon, Saskatchewan, Canada.

Haemophilus somnus is an opportunistic pathogen associated with several bovine disease syndromes. The ability to scavenge the essential nutrient iron from the host is well established for bacterial pathogens. An *in vitro* plate bioassay was used to determine that *H. somnus* was capable of using bovine transferrin, haemoglobin, or haemin as a sole source of iron. A genomic cosmid library was screened for clones able to bind haemin (Hmb+). An Hmb+ clone was isolated and the phenotype was localized to a 1.6 kb fragment. DNA sequencing and Tn*pho*A mutagenesis indicated that an open reading frame *(hmb)* coding for a secreted 21.4 kDa polypeptide was responsible for the Hmb+ phenotype. Anti-Hmb antiserum detected a recombinant 22,000 M_r protein in Western blots. The predicted amino acid sequence of Hmb was homologous to the primary sequence of lysozymes (murein hydrolase; EC 3.2.1.17) from several bacteriophages. Furthermore, *hmb* could complement amber mutations in the "lysozyme" genes of bacteriophages lambda (*R* gene) and P2 (*K* gene), enabling these mutant bacteriophages to lyse a non-permissive *E. coli* host. Whole cell lysates of *E. coli* expressing Hmb could lyse EDTA-killed *E. coli* cells. These data suggest that *hmb* encodes a lysozyme of bacteriophage origin. The presence of bacteriophages in *H. somnus* is now being investigated.

T27. Construction and Analysis of a *Pasteurella haemolytica* Mutant Unable to Synthesize Three Membrane Lipoproteins.

G. L. Murphy* and L. C. Whitworth, Oklahoma State University, Dept. of Veterinary Pathology, Stillwater, OK 74078, USA.

A single genetic locus from *Pasteurella haemolytica* serotype A1, consisting of three tandemly arranged genes encoding 28-30 kDa membrane lipoproteins, was replaced with a mutated locus which carries the β-lactamase-encoding gene from a 4.2 kbp *P. haemolytica* plasmid. The inactivated locus was introduced into *P. haemolytica* by electroporation of a plasmid which carries the mutated locus but is incapable of replicating in *P. haemolytica*. Southern and western blot analyses indicate that the wild-type locus was replaced by the mutated locus through a double crossover recombination event and that the three membrane lipoproteins were no longer produced by the mutant strain. The effects of this mutation on the growth of *P. haemolytica in vitro* and on the production of other *P. haemolytica* membrane proteins, capsule, and leukotoxin were examined. These methods for mutagenesis should be useful in constructing mutant loci which can be used to analyze the roles for various *P. haemolytica* proteins in the pathogenesis of bovine pneumonic pasteurellosis.

T28. Preliminary Characterization of the Latent Haemolysin Induced in *Escherichia coli* by the *Actinobacillus pleuropneumoniae* HLYX Protein.

J.I. MacInnes[1]* and B.L. Wanner[2], [1]University of Guelph, Guelph, Ontario and [2]Purdue University, West Lafayette, Indiana.

The *hlyX* gene of the swine pathogen *A. pleuropneumoniae* encodes a protein which is very similar to the global regulatory protein, FNR of *E. coli* and like FNR, can regulate the expression of the *frdA, ndhII,* and *nar* genes. In addition, HlyX is able to induce the synthesis of a latent haemolysin. Haemolysin production is highest during late log phase and can be detected only in cells grown under anoxic conditions. Non haemolytic mutants of *E. coli*/pT51 *(hlyX+)* were created using Tn*phoA*/Tn*phoA'* elements carried on lambda phage. From P1 transduction studies, all of these mutations appeared to map to the same region. The sequence of the regions flanking one of the transposon insertions was determined, but no homology could be detected with any known *E. coli* gene. Haemolytic recombinants were also isolated in *E. coli (hlyX-)* carrying a wild type *E. coli* mini Mu library. These cells grew poorly and were only weakly haemolytic. When *hlyX was* introduced, growth rates and haemolytic activity increased.

T29. Plasmid Mediated NAD-Independent *Haemophilus Paragallinarum* and Their Use in Poultry Vaccine Manufacture.

R.R. Bragg*, L. Coetzee and J.A. Verschoor, University of Pretoria, South Africa.

Since 1990, NAD-independent isolates of *H. paragallinarum* have been isolated from chickens showing typical symptoms of infectious coryza in South Africa. These NAD-independent isolates were tested with a panel of three locally produced monoclonal antibodies (Mab) against *H. paragallinarum*. Mab patterns obtained for the NAD independent isolates were similar to those obtained with the normal, NAD dependent strains. Plasmid mediated NAD independence has been reported in *H. Parainfluenzae*. In the light of this work, it was decided to attempt plasmid transformation, with the NAD-independent isolates as plasmid donors and well characterized *H. paragallinarum* isolates as plasmid recipients. This was successfully carried out. Strain 0083, which is widely used in the manufacture of poultry vaccines against infectious coryza, was successfully transformed into a NAD-independent strain. It was established, using chicken raised antibodies against strain 0083 that the hemagglutinin of the transformed isolates was not affected during transformation. It is thus theoretically possible to use transformed NAD-independent strains of *H. paragallinarum* for the manufacture of poultry vaccines.

T30. Global Distribution of DNA Sequences Associated with Site Specific Integration of Large Antibiotic Resistance Plasmids in *Haemophilus influenzae*.

T.J. Falla, I. Dimopoulou and D.W.M. Crook, Oxford Public Health Laboratory, U.K.

A large conjugative *Haemophilus influenzae* resistance plasmid has been shown to excise from a specific location in the tRNA leucine gene of a *H. influenzae* strain isolated in the U.K. (strain 1056). The geographical extent of this phenomenon is unknown. Antibiotic resistant isolates of *H. influenzae* from North America, South America, Greece, Asia and the U.K. were analyzed by PCR and DNA sequencing to identify similarity between these disparate strains and strain 1056. PCR primers were designed to amplify the *att* P site of free plasmid and the *att* L and *att R* sites (junction fragments) of integrated plasmid. Strains from each geographical location amplified product with *att* P, *att* L and *att* R primers. PCR products of *att* P and *att* R from strains representative of each geographical location were sequenced and found to have high homology. This data shows that plasmid from a world-wide distribution share common sequences associated with recombination and excision of plasmid. These plasmids all integrate at the same site in the tRNA leucine gene of *H. influenzae,* thus suggesting a common mode of site specific recombination.

T31. Development of *Actinobacillus pleuropneumoniae* as a Bacterial Vaccine Vector.

C.T. Prideaux*, L. Pearce, C.L. Wright and A.L.M. Hodgson, CSIRO Division of Animal Health, PMBI, Parkville, Australia, 3052.

Actinobacillus pleuropneumoniae (APP) is the causative agent of porcine pleuropneumonia, a severe respiratory disease of pigs. We are currently developing APP for use as a bacterial vaccine vector system capable of delivering and expressing foreign genes in pigs. The establishment of such a system requires the identification of a number of key elements such as promoters to regulate foreign gene expression, secretory signals to allow export of foreign proteins from the bacteria, and a broad host range plasmid shuttle system suitable for the introduction, and stable expression of foreign genes in APP. Currently we are evaluating two plasmids for use in vector construction, pIG1 and pIG3, obtained from Australian pig isolates of *Pasteurella multocida* and APP respectively. These two plasmids have been evaluated for their ability to transform APP isolates, their rate of loss from APP during passage without the presence of selectable pressure, and the ability of foreign genes inserted onto these plasmids to be authentically expressed utilising various promoters.

T32. Construction and Characterization of a *Pasteurella- Actinobaclllus-Escherichia coli* Shuttle Vector.

C.L.Wright[1,2]*, R.A. Strugnell[2] and A.L.M. Hodgson[1], [1]CSIRO Division of Animal Health, Parkville, Victoria 3052, Australia, [2]Department of Microbiology, Melbourne University, Parkville, Victoria 3052, Australia.

The Gram-negative organism, serotype D *Pasteurella multocida* (Pm), is a causative agent of swine atrophic rhinitis (AR), a chronic disease of young pigs that results in the atrophy of nasal turbinates, facial deformities and reduced growth rates. A plasmid vector system for Pm will be required to enable the development of a new, live recombinant AR vaccine. A 5.4 kb streptomycin and sulphonamide resistance plasmid (pIG1) was isolated from an Australian strain of Pm and was not only found to replicate in *E.coli*, but also *Actinobacillus pleuropneumoniae* and *Pasteurella haemolytica*. Nucleotide sequence analysis of pIG1 revealed that approximately half of the plasmid consisted of the drug resistance genes and the remainder encoded replication and putative mobilisation functions. A series of vectors have been constructed that contain the replication region from pIG1, various antibiotic resistance genes and a multiple cloning site. These plasmids should facilitate the genetic manipulation of Pm and other members of the HAP group.

T33. Outer Membrane Protein and Lipopolysaccharide Analyses as Taxonomic and Epidemiological Tools for *Pasteurella haemolytica*.

R.L. Davies, Department of Microbiology, University of Glasgow, Scotland, U.K.

The outer membrane protein (OMP) and lipopolysaccharide (LPS) profiles of over 200 isolates of *P. haemolytica* were examined by SDS-PAGE and Western-blotting analysis. The isolates analyzed included representatives of all the recognized A and T serotypes as well as various untypeable isolates, and originated from both Europe and North America. Analysis of OMP and LPS profiles using these techniques was capable of differentiating between isolates having the same serotype, of demonstrating similarities between isolates of different serotypes, and of demonstrating both similarities and significant differences between untypable and typable isolates. The usefulness and applications of these techniques in taxonomic and epidemiological studies of *P. haemolytica* will be discussed in further detail.

T34. Invasive-Disease-Causing and Nasopharyngeal Isolates of *Haemophilus influenzae* Type B are Different.

N.I. Leaves*, M.. Barbour, J.Z. Jordens, T.E.A. Peto and D.W.M. Crook, PHLS Haemophilus Reference Laboratory, Oxford, U.K.

Haemophilus influenzae type b (Hib) is a commensal of the human upper respiratory tract, from where carriage can result in invasive disease. As part of an investigation into the possible bacterial factors associated with the progression from carriage to invasive disease, fifty-six naso-pharyngeal isolates of Hib (NP-Hib) from healthy subjects were compared with fifty-nine invasive isolates of Hib (I-Hib) from the same region and time period. The isolates were characterised using outer membrane proteins (OMPs), ribotyping and capsular genotyping. They were also examined for the presence of large plasmids associated with ampicillin resistance using whole plasmid probing and a PCR-based technique. The I-Hib population structure, when assayed using ribotyping and OMPs, was similar to that described previously. However, the NP-Hib population possessed a smaller proportion of the major clone (P=0.06) and a more varied selection of the more unusual subtypes; there was no difference in capsular genotype distribution. Plasmid DNA was found in a significantly higher proportion of NP-Hib than I-Hib (P = 0.006) and was not always associated with antibiotic resistance. These results suggest that the population structure of I-Hib is different from NP-Hib and that I-Hib is a subset of NP-Hib. Further, the preponderance of plasmid in NP-Hib suggests that strains containing plasmid are less likely to cause invasive Hib disease.

T35. Characterisation of Porcine Isolates of *Pasteurella multocida,* Including the use of Multilocus Enzyme Electrophoresis (MEE).

M.P. White[1]*, G. Chew[2], and D.J. Hampson[2], [1]University of Sydney, NSW, Australia, [2]Murdoch University School of Veterinary Studies, WA, Australia.

P. multocida is associated with 2 important clinical syndromes in pigs: pneumonia and atrophic rhinitis. A total of 115 predominantly porcine respiratory tract isolates which had been presumptively identified as *P. multocida* was subjected to capsular typing, somatic serotyping, toxigenicity testing, biochemical activity profiling, subspeciation and MEE. Whole cell protein profiles and outer membrane protein profiles on SDS-PAGE were produced from 25 isolates. Several significant correlations were observed between the MEE type of isolates and their capsule type, somatic serotype, biochemical activity and subspecies classification.

(This work was funded by the Australian Pig Research and Development Corporation).

T36. Multivariate Analysis of Fatty Acid Contents of Outer Membrane Vesicles from *Actinobacillus, Haemophilus* and *Pasteurella* spp.

I. Olsen*, V. Myhrvold and I. Brondz, Department of Oral Biology, University of Oslo, Oslo, Norway.

Outer membrane vesicles are released from Gram-negative cell walls during their growth. The vesicles may either penetrate tissue locally or are spread with the blood to more peripheral sites. Vesicles are probably important virulence factors due to their contents of enzymes and toxins. The present study used gas chromatography to analyse the fatty acid contents of isolated outer membrane vesicles in spp. of *Actinobacillus*, *Haemophilus* and *Pasteurella*. Both reference and clinical strains were used. Data were treated with multivariate analysis which distinguished between *Actinobacillus actinomycetemcomitans* (class 1) and *Haemophilus influenzae* (class 2) (F-test, 95% confidence limits). *Haemophilus aphrophilus*, *Haemophilus paraphrophilus*, and *Pasteurella multocida* were distinct from both these classes, except for *H. aphrophilus* ATCC 33389^T which fell in class 1 (border line). C14-3OH was a major fatty acid in all species while C10-3OH, C12-3OH, and C16-3OH were minor components. *H. aphrophilus* exhibited simpler chromatograms than *A. actinomycetemcomitans*: it did not contain C10-3OH or C16-3OH fatty acids. The abundance of 3OH fatty acids suggested that outer membrane vesicles from the examined species are rich in lipopolysaccharide.

T37. Heterogeneity of *Actinobacillus actinomycetemcomitans* Demonstrated with Multivariate Analysis of Fatty Acid Data from Outer Membrane Vesicles.

I. Brondz*, V. Myhrvold and I. Olsen, Department of Oral Biology, University of Oslo, Oslo, Norway.

We have previously used a variety of chemical characters and multivariate chemosystematics to demonstrate that *Actinobacillus actinomycetemcomitans* is a heterogeneous species (I. Brondz and I. Olsen, Oral Microbiol Immunol 1993;8:129-133). In the present study, the fatty acid contents of isolated outer membrane vesicles were used to see if this heterogeneity could be confirmed. A difference in the fatty acid contents of outer membrane vesicles from *A. actinomycetemcomitans* strains may be of importance to the variation in their virulence. Fatty acid data from *A. actinomycetemcomitans*, which were provided after gas chromatographic analysis, were treated with SIMCA analysis and compared with those of similar data from vesicles isolated from *H. aphrophilus*, *H. paraphrophilus*, *H. influenzae* and *P. multocida*. *A. actinomycetemcomitans* could be divided into two strain groups which were distinguished at 95% confidence limits (F-test). One group contained ATCC 33384^T, ATCC 29522, SUNY 366, HK 435, and Q 1247; the other comprised ATCC 29524, ATCC 29523, FDC 2112, FDC 511, and SUNY 489. This assignment of strains into groups supported our previous findings. Both groups were distinct from *H. aphrophilus*, *H. paraphrophilus*, *H. influenzae*, and *P. multocida*. Presence of OH acids suggested that vesicles of the examined species are abundant with lipopolysaccharide.

T38. Effects of Antibiotic on Host TNFα and PGE_2 Responses in Sheep after Subcutaneous Infection with *Pasteurella haemolytica* Biotype T Serotype 15.

J.C. Hodgson*, G.M. Moon, M. Quirie and W. Donachie, Moredun Research Institute, 408 Gilmerton Road, Edinburgh EH17 7JH

Pasteurella haemolytica biotype T produces systemic disease in lambs similar in speed and clinical signs to endotoxic shock. Endotoxic shock comprises a cascade of reactions involving endotoxin and inflammatory mediators, particularly TNFα and PGE_2. Preliminary results in 3 lambs showed that antibiotic (oxytetracycline) may inhibit the release of TNFα and help prolong survival. In the present work sixteen specific pathogen-free lambs aged 10 weeks were dosed subcutaneously with 3 x 10^8cfu *P. haemolytica* T15. Seven lambs served as controls, 9 were treated with oxytetracycline intramuscularly 2-3 hours post-infection (p.i.). Blood samples taken before and at intervals after infection were analysed for bacteria, TNFα and PGE_2 and selected metabolites. Bacteraemia, detected at 2 hours p.i. in all control and 7 treated lambs, was controlled by antibiotic. One lamb died immediately after treatment. All lambs became febrile (rectal temperature ≥ 40.5°C) by 2.5 ± 0.2 hours p.i. and clinical and biochemical signs for lambs which died were typical of endotoxic shock. Mean (± SEM) age at death was 8 ± 3 hours (n=7) and 20 ± 4 hours (n=7) for control and treated lambs, respectively. PGE_2 concentrations increased in 14 lambs (peak concentrations ranging from 16-307pg/ml plasma) but significant TNFα responses (peak range 4-55ng/ml plasma) occurred only in control lambs.

T39. Inflammatory Cytokines and Lung Injury in Bovine Pneumonic Pasteurellosis.

S.K. Maheswaran*, H.S. Yoo, T.R. Ames, G. Lin, M.P. Murtaugh and E.L. Townsend, College of Veterinary Medicine, University of Minnesota, St. Paul, MN 55108, U.S.A.

We hypothesise that acute lung injury seen in bovine pneumonic pasteurellosis is caused by interactions between various cellular constituents in the lung and inflammatory mediators, stemming from the initial interaction of *Pasteurella haemolytica* LPS and leukotoxin with alveolar macrophages (AMs). A large body of experimental data suggests that AM-derived tumor necrosis factor α (TNFα) and interleukin-1β(IL-1β) indeed are responsible in the pathogenesis of lung injury seen in animal models of sepsis-related human adult respiratory distress syndrome (ARDS). These findings are especially relevant because the pathogenesis of lung injury seen in bovine pneumonic pasteurellosis resembles human ARDS lung injury. In this study, we examined the kinetics of gene expression and release of TNFα and IL-1β by bovine AMs stimulated with LPS or purified leukotoxin (LPS-free) obtained from *P. haemolytica* A1. Results from Northern blot analysis indicate that there was a time-dependent expression of TNFα and IL-1β mRNA by AMs stimulated with LPS. Peak TNFα and IL-1β mRNA expression was seen at 1 hr, declined significantly by 16 hr and was undetectable at 24 hr after stimulation with 1 μg/ml of LPS. Production of bioactive

TNFα peaked at 4 hr while bioactive IL-1β production peaked at 24 hr. There was a dose dependent increase in concentration of bioactive TNFα and IL-1β secreted by bovine AMs stimulated in the presence of 0.1-10 μg/ml of LPS. The expression of cytokines mRNA and secretion by LPS-stimulated AMs were inhibited by polymyxin B. AMs stimulated with purifed leukotoxin at subcytolytic concentrations also expressed TNFα and IL-1β mRNA with peak steady state mRNA levels observed with 1 leukotoxin unit/ml. The kinetics of mRNA expression observed with leukotoxin-stimulated AMs differed from that of LPS-stimulated AMs and was not inhibited by polymyxin B. Using *in situ* hybridization, we have demonstrated TNFα and IL-1β expression associated with pneumonic lung lesions from *P. haemolytica* induced experimental pneumonic pasteurellosis. These data suggest that AM-derived inflammatory cytokines may contribute to the pathogenesis of lung injury in bovine pneumonic pasteurellosis.

T40. Passive Immune Cross-Protection in Mice with Antisera to Haemorrhagic Septicaemia and Fowl Cholera Strains of *Pasteurella multocida.*

R.B. Rimler. USDA, ARS, National Animal Disease Center, P.O. Box 70, Ames, Iowa, 50010, USA.

Mice were passively immunised with antisera made against *P. multocida* that cause haemorrhagic septicaemia (HS; serotypes B:l, B:2, B:3,4, B:4, E:2) and fowl cholera (FC; serotypes A:3, A:5). They were then challenged with homologous and heterologous serotypes. All serogroup B antisera protected against all serogroup B strains, regardless of somatic serotype. Reciprocal cross-protection did not occur among serogroup A strains. Some antisera produced cross-protection which did not correlate to specific somatic or capsule antigens. Cross-protection also occurred between certain strains that cause HS and those that cause FC. This cross-protection was not related to antigens that react in the capsule indirect haemagglutination test, H proteins, or a and b soluble antigens. Immunoblots of the various HS and FC serotypes that reacted with cross-protective antisera showed at least 20 bands of common mobility. The most intense common reactions occurred with antigens in the 59-87 kDa range.

T41. Immune Responses of Calves to Different *Pasteurella haemolytica* A1 Vaccines.

T.R. Ames[1]*, S. Srinand[1], R.E. Werdin[1], G. F. Jones[2] and S. K. Maheswaran[1], [1]College of Vet. Med., Univ. of Minnesota, St. Paul, MN 55108 and [2]BioCor Inc., Omaha, Nebraska, U.S.A.

This study was designed to evaluate the immune responses of three experimental streptomycin dependent (Sm^D) live *Pasteurella haemolytica-Pasteurella multocida* (BioCor, Omaha, NE) vaccines grown in different media (A, B, and C), two commercial bacterin/toxoid vaccines available in the United States, and two experimental subunit vaccines derived from *P. haemolytica* 12296, one consisting of leukotoxin (LKT), capsular polysaccharides (CP), and outer membrane proteins, and the other containing iron regulated outer membrane proteins (IROMPs) and CP. Four ELISAs were optimized to evaluate the ability of the vaccines to elicit antibody responses against the LKT, CP, surface antigens (SA), and IROMPs of *P. haemolytica* A1. The ideal cut-off for each ELISA was determined using a receiver operating characteristic curve (ROC) on sera from 30 field vaccinated and 30 prevaccinated cattle. At these cut-off points the specificity and sensitivity of the ELISAs against SA, IROMPs, LKT, and CP were 86% and 90%, 86% and 100%, 72% and 100%, and 88% and 100% respectively. Sm^D vaccine trial was performed in conventionally raised calves. Sm^D vaccinates showed significant increases in antibody levels against CP and SA, while no increases were detected against LKT and IROMPs. Analysis of magnitude of antibody responses in Sm^D vaccinates indicated a better efficacy of the vaccine produced in media C. Studies evaluating serologic responses to LKT, SA, CP, and IROMPs generated by commercial and subunit vaccines are underway. Preliminary results with commercial vaccines indicate poor antibody responses against LKT. In addition, challenge studies will be used to determine the vaccine efficacy in terms of lung lesion and clinical scores and their correlation to the magnitude of antibody responses to LKT, CP, SA, and IROMPs.

T42. Nad (V-Factor) - Independent and Typical *Haemophilus paragallinarum* Infection in Commercial Chickens: a Five Year Field Study.

R.F. Horner[1]*, G.C. Bishop[1], C. Jarvis[1], T. Coetzer[2], [1]Allerton Laboratory, Private Bag X2, Cascades 3202, South Africa, [2]Department of Biochemistry, University of Natal, Pietermaritzburg, South Africa.

An unusual bacterium causing respiratory disease in chickens emerged in South Africa in February 1989. The disease clinically resembled infectious coryza but the organism differed from typical *Haemophilus paragallinarum* especially in that it did not require V-factor for growth. The bacterium has been termed an NAD-independent *H. paragallinarum*. A study of avian Haemophili isolated from diseased chickens in Natal over the past five years showed three types to be present; typical *H. paragallinarum*, NAD-independent *H. paragallinarum* and *H. avium* (now transferred to the genus *Pasteurella*). Before the end of 1989 the NAD-independent *H. paragallinarum* had become the predominant isolate and thereafter was isolated from commercial chickens in other regions of South Africa. The disease affected all strains of chickens in an overall age range of 14 days to 66 weeks and was responsible for upper respiratory disease of layers and upper and lower respiratory disease in broilers. It was commonly isolated from diseased adult birds previously vaccinated against typical *H. paragallinarum*. Broilers were most commonly infected from 3 weeks of age and layers within the placement to peak production period.

T43. Antigens Involved in Cross-Protection Between *Actinobacillus pleuropneumoniae* Biotypes-Serotypes in Pigs.

F. Haesebrouck* and P. Dom, Laboratory of Veterinary Bacteriology, University of Gent, Casinoplein 24, B-9000 Gent, Belgium.

Four groups of 5 gnotobiotic pigs were intranasally inoculated with an *Actinobacillus pleuropneumoniae* biotype 1-serotype 2 (*App* 1-2) strain (producing RTX toxins ApxII and ApxIII), an *App* 1-10 strain (producing ApxI), an *App* 2-2 strain (producing ApxII) or saline (controls), respectively. Six weeks later, all pigs were exposed to *App* 1-2 endobronchial challenge. Severe clinical signs were only observed in all control pigs, one pig immunized with *App* 1-10 and two pigs immunized with *App* 2-2. These pigs died within 36 hours post inoculation. In the other pigs, clinical signs were mild or absent, and none of these pigs died. At the time of challenge neutralizing antibodies against ApxI only, Apx II only and both Apx II and III were present in sera of pigs immunized with *App* 1-10, *App* 2-2 and *App* 1-2, respectively. Opsonizing antibodies against *App* 1-2 were only detected in sera of 8 of the 12 pigs that survived. SDS-PAGE and Western Blot analysis revealed that supernatants of 12-hour-old cultures of the *App* strains contained 3 common antigens. These results indicate that immune mechanisms other than Apx neutralizing antibodies were involved in cross-protection of pigs immunized against *App* 1-10 and challenged with *App* 1-2. (This study was supported by IWONL, Brussels, Belgium).

T44. Lesions of *Actinobacillus suis* Infection in Swine.

D.M. Middleton*, Dept. of Veterinary Pathology and J.M. Chirino-Trejo, Department of Veterinary Microbiology, Western College of Veterinary Medicine, University of Saskatchewan, Saskatoon, SK, Canada.

Gross, histological and microbiological findings from a retrospective study of 36 confirmed cases of *Actinobacillus suis* infection in swine necropsied since 1988 will be presented and discussed. In the majority of cases, *A. suis* was isolated alone; in the remainder usually with one other of a variety of pathogens, including *Streptococcus suis* and *Pasteurella* species, with one case of concurrent cytomegalovirus infection. Age of affected animals ranged from 4 days to young adult, with the majority of cases occurring from 2 to 16 weeks. The multiple manifestations of the disease include septicaemia; polyserositis; arthritis and periarthritis; endocarditis; embolic pneumonia, hepatitis and nephritis. Pleuro-pneumonia has been seen also, and with increasing frequency in the last two years. An association with minimal disease herd status has been noted. Pathogenesis will be discussed. Is this an emerging disease?

T45. Efficacy And Safety Of A *Pasteurella haemolytica* A1 Bacterin Toxoid (One Shot^{TM}) Under Experimental Conditions.

R. Newsham*, T. Kaufman and K.I. Dayalu, SmithKline Beecham Animal Health, Lincoln, Nebraska, U. S. A.

A *P. haemolytica* Bacterin Toxoid vaccine (One Shot^{TM}) has been developed for the protection of cattle against pneumonic pasteurellosis. Beef calves challenged at two weeks post-vaccination showed an 81% (subcutaneous [SC] vaccination) to 94% (intramuscular [IM] vaccination) reduction in lung lesions compared to controls. One SC dose offered little protection to animals challenged with *P. haemolytica* at 3 days post-vaccination, but provided a 52% reduction in lung involvement, compared to controls, when challenged at 7 days post-vaccination. An 87% reduction in lung lesions, compared to controls, occurred when animals maintained under field conditions were challenged at 4.25 months post-SC vaccination. *P. haemolytica* was consistently re-isolated from lung lesions. Serological status of animals did not necessarily correlate with protection. Vaccination of cattle at twice the recommended antigen level or with five vaccinations over a four month period caused no significant systemic reactions. A significant rise in body temperature occurred at 8 hours post-vaccination, but returned to normal levels by 24 hours post-vaccination. Localized swellings at the injection site in some animals were encountered by either route of administration, but decreased in size over time. The vaccine is considered safe and efficacious by either route of administration.

P1. The Efficacy of a *Pasteurella haemolytica* A1 Vaccine Given as a Single Dose Against Experimental Pneumonic Challenge In Cattle.

J.A. Rice Conlon*, G.F. Gallo, P.E. Shewen and C. Adlam, Langford/Cyanamid, 131 Malcom Road, Guelph, Ontario, N1K 1A8, Canada.

The efficacy of single dose administration of a commercially available *Pasteurella haemolytica* A1 culture supernatant vaccine (PRESPONSE) was tested in an experimental challenge model in cattle. One dose of vaccine was as effective in protecting cattle from development of pneumonia as was administration of the recommended 2 doses. Also, leukotoxin neutralizing and bacterial agglutinating titres were similar in both groups on the day of challenge and in both cases, the titres were correlated with protection. It is hypothesized that calves are naturally primed to *P. haemolytica* through commensal colonization of the upper respiratory tract, and therefore, respond in an anamnestic nature to delivery of a single dose of the vaccine. Both serologic and protection results presented here support this theory of natural priming.

P2. Experimental Evaluation of *Pasteurella haemolytica* Immunogen in Lambs.

F. Morales A.; L. Jaramillo M.; J. Tortora P. and F. Trigo T.*, CENID -MICROBIOLOGIA/FMVZ-UNAM Cd. Universitaria, Mexico D.F. 04510.

This study was conducted to evaluate the protective effect of live *Pasteurella haemolytica* (PH) and its leukotoxin on lambs upon challenge. Forty lambs were used and randomly distributed in four groups. Group one was the control group, receiving subcutaneously only growth media. Group two was subcutaneously immunized with a live culture of PH Al. Group three was intradermically immunized with a live culture of PH. Group four received PH leukotoxin, which was subcutaneously administered with complete Freund's adjuvant. All animals were exposed with PI-3 virus and then challenged with a field strain of PH Al, and subsequently slaughtered to evaluate pulmonary lesions. Control lambs presented fever after challenge. Anticapsule and anti-leukotoxin antibody titres were higher in all immunized groups compared with the control group ($p< 0.05$). Pulmonary lesions were more severe in the control group (<0.05), but there was not significant difference among the immunized groups. These results show that the protection conferred with live PH vaccine or its leukotoxin is satisfactory to protect lambs against pulmonary lesions produced experimentally by PH challenge. Furthermore, inocuity of the immunogens used in the present study due to the absence of post-vacunal reactions, was demonstrated.

P3. Vaccination of Lambs Against Pneumonic Pasteurellosis in Experimental Conditions.

F. Aguilar R.; L. Jaramillo M.; F. Morales A.; F. Trigo T. and F. Suarez G.*, CENID-MICROBIOLOGIA/FMVZ-UNAM Cd.Universitaria Mexico D.F. 04510.

The aim was to evaluate the effects of vaccination with different antigens of *Pasteurella haemolytica* (PH), on the control of pneumonic pasteurellosis in lambs. Forty two lambs were divided into six groups, with seven animals each. Five groups were inoculated subcutaneously with different immunogens of PH: group A, was vaccinated with 0.5 ml of live PH harvested in the log phase, at concentration of $1x10^9$ CFU/ml; group B, was the control group; group C, with a commercial bacterin; group D, 2 ml of leukotoxin; group E, 2ml of surface antigens extract (SAE) plus leukotoxin; group F, SAE plus leukotoxin plus adjuvant. On day 35, all lambs were exposed to PI-3 virus; on day 42 were transthoracically challenged with 2 ml of a suspension of live PH (1×10^9 CFU/ml). All lambs were euthanized seven days after challenge. Sera samples were obtained on days 0,7,14,21,35,42. Each serum was tested for antibodies to PH by indirect hemagglutination test, and with the simple visual assay for determination of leukotoxin neutralizing antibody titres. Mean antibody titres were compared using the variance analysis, and the Tukey test, which showed a significant difference ($P < 0.05$) in those lambs immunized with leukotoxin, live PH, and commercial bacterin. Pneumonic lesions in lambs from groups A, C, and D were smaller than in groups B, E, and F. Vaccination of lambs with live PH and leukotoxin enhanced lung resistance as shown by reduction in lesions and increase in antibody titers.

P4. Serotypes of *Pasteurella haemolytica* and *Pasteurella multocida* Isolated From Pneumonic Domestic Ruminants in Mexico.

F. V.; F.Trigo T.; L. Jaramillo M.; F.Aguilar R.; G.Tapia P. and F. Suarez G.*, CENID-MICROBIOLOGIA/FMVZ-UNAM Cd. Universitaria , Mexico D.F. 04510.

This study was conducted to isolate and determine the capsular and somatic serotypes of *Pasteurella multocida* (PM) and the capsular serotypes of *Pasteurella haemolytica* (PH) from bovine, ovine and caprine lungs with inflammatory lesions. A total of 232 lungs with inflammatory lesions were collected and these yielded 117 isolates of Pasteurella sp. Forty two strains were identified as PH and 75 as PM. All PH were biotype A. The serotypes, which were determined by indirect haemagglutination, were as follows: in sheep there were 28 strains (39% were A8, 32% A2, 11% A1, 7% A6, 4% A5, and 7% untyped); in cattle there were 12 strains (58% were A1, 17% were A2 and A6, and 8% untyped); in goats there were 2 strains (A5 and A8). For PM, 60% were capsular type A and 27% capsular type D, according with the acriflavine and hyaluronidase techniques. Somatic serotypes were determined by gel immunodiffusion and represented 72% in sheep for serotype 3; 9% for serotype 15; 4% for serotype 7; 2% for serotypes 4,10,12, and 14; and 7% untyped. In cattle 77% were serotype 3, 8% serotype 4, 4% serotypes 7 and 12, and 8% were untyped; whereas in goats, with 3 strains one was serotype 3, one serotype 15 and one untyped. The distribution of serotypes is compared and contrasted with other related studies.

P5. Comparative Evaluation of Two Commercial Atrophic Rhinitis Vaccines

D. Coyle*, A.G. Ostle, C. Frank, B. Kregness, J. Rehder, M. Welter, Ambico, Inc., 902 Sugar Grove Ave., Dallas Center, IA 50063 USA.

A dual-adjuvant atrophic rhinitis (AR)/pneumonia vaccine (Ambico B*P*M) containing *Bordetella bronchiseptica, Pasteurella multocida* type A, *Pasteurella multocida* type D (PmD) toxoid whole cells and *Mycoplasma hyopneumoniae* was compared to a commercial oil-adjuvanted AR vaccine containing *B. bronchiseptica* and PmD toxoid in a passive immunity pig model. Groups of four pregnant gilts were vaccinated according to label directions with either of the products. Four additional gilts remained unvaccinated as controls. The litters were not vaccinated. Reproduction of AR was achieved by infecting half of each litter at 4 days post-farrow with 10^9 cfu *Bordetella bronchiseptica* and at 8 days post-farrow with 10^9 cfu toxigenic PmD. The litters were euthanized six weeks post-infection. The B*P*M vaccine reduced nasal turbinate atrophy by 85% and the oil-adjuvanted product reduced turbinate atrophy by 74% in comparison to controls. The oil-adjuvanted product induced greater serum and colostrum antibody titres vs. *B. bronchiseptica* and PmD toxin, but this product did not confer greater protection to pigs nursing vaccinated gilts. It is possible that other immunogens from PmD are also effective in protecting pigs in addition to the PmD-toxoid.

P6. Modulation of Local Defence Mechanisms by Aerogenous Immunizations Against Porcine *Pasteurella multocida* Pneumonias.

H. Kohler*, A. Berndt, G. Muller, Federal Health Office, Institute of Veterinary Medicine, Jena Branch, 07722 Jena, Germany.

In investigations into pathogenesis and immunoprophylaxis of porcine *Pasteurella multocida* (*P.m.*)-induced pneumonias, only aerogenous immunizations with *P.m.* bacterins via aerosol caused an increase in the pulmonary clearance associated with a decrease in the severity of experimentally induced *P.m.*-pneumonias. In addition, specific IgA antibodies in the bronchoalveolar lavage fluid were elevated, whereas specific IgG antibodies showed no reaction. To determine cellular effector mechanisms involved in the defence of *P.m.* infections, the phagocytosis of fluoresceinated *P.m.*-whole cells by alveolar macrophages (AM) in comparison to blood monocytes (BM) was measured using flow cytometry. Furthermore, the accessory function of AM and BM was examined by detecting their ability to stimulate the proliferation of blood T-cells. In aerogenously immunized pigs, the percentage of blood monocytes phagocytosing *P.m.* bacteria (%P) was significantly higher than in control animals, associated with an increased mean fluorescence intensity (MFI). AM of immunized pigs showed also an increased %P, but a decreased MFI, possibly reflecting a faster degradation of ingested bacteria. The accessory function of AM but not BM was enhanced after aerogenous immunization. These data indicate an important role of specific local IgA antibodies and activated AM in the defence of *P.m.* pneumonias in pigs.

P7. Capsule Depolymerization of *Pasteurella multocida*.

T. Magyar*, and R. B. Rimler, Vet. Med. Res. Inst., P.O.Box 18. Budapest, Hungary, Nat. Anim. Des. Cntr., P.O.Box 70. Ames, Iowa, USA.

Capsules of serogroups A, D and F *P. multocida* were depolymerized by treating with chondroitinase AC. After treatment, antigens were absorbed onto red blood cells and tested in indirect haemagglutination (IHA) tests with anti-capsule sera. Treatment of serogroup D and F strains resulted in increased IHA titres. Whereas, IHA titres to serogroup A were diminished. Enzyme treated bacteria were agglutinable in antisera, but non-treated bacteria were not. Chondroitinase AC treatment of capsulated *P.multocida* enhanced phagocytosis by swine neutrophils. However, the degree of phagocytosis was less than that of naturally-occurring, non-capsulated variants derived from parent capsulated strains.

P8. Immune Response of Colostrum Fed Dairy Calves to Capsular Polysaccharide of *Pasteurella haemolytica* A1.

D.C. Hodgins*, P.E. Shewen, Dept. of Vet. Micro. & Immunol., Univ. of Guelph, Guelph Ont., Canada, NIG 2W1.

The ability of young colostrum fed dairy calves to respond to the capsular polysaccharide of *Pasteurella haemolytic* A1 was investigated. Calves were vaccinated, at 2 and 4 weeks of age with a culture supernatant vaccine (Presponse, Langford, Inc.); blood samples were collected at weekly intervals. Antibodies binding the capsular polysaccharide were determined using isotype specific (IgG1, IgG2, and IgM) enzyme immunoassays. A fourfold increase in IgM antibodies was evident in 18% of the calves after one dose of vaccine, with an additional 23% seroconverting after the second dose. Increases in IgG1 and IgG2 antibody titres of this magnitude were present in 9% and 23% of the calves respectively, after the second dose of vaccine. There was a negative correlation between pre-vaccination IgG1 antibody titre and IgG1 IgG2 and IgM responses to vaccination (r= -0.78, -0.19, and -0.75, respectively). Thus vaccination can induce an active immune response to capsular polysaccharide in calves with low passive (IgG1) antibody titres.

P9. Epidemiology of Human Infections by *Pasteurella, Actinobacillus* and Related Species (or Group) in France.

F. Escande*, Laboratoire des *Pasteurella*, Institut Pasteur, Paris, and the *Pasteurella* and Related Bacteria Working Group: J.L. Avril, J. Beytout, H. Chardon, J. Croize, H. Dabernat, P.Y. Donnio, J. Frottier, B. Joly, G. Laurans, J. Lemozy, C. Lion, A. Marmonier, G. Paul, J.M. Scheftel, M. Simonet - France.

A retrospective study of *Pasteurella, Actinobacillus* and related species infections was performed by the *Pasteurella* and Related Bacteria Working Group from 1992 to 1993. Among the 379 cases recorded, wound infections (bites, scratches and punctures) were the most common forms of pasteurellosis (68.6%). In human infections unrelated to animal bites, respiratory tract diseases and septicemia were the predominant infections with respectively 17 and 9%. Next in importance were urogenital (2%), central nervous system (1.3%), abdominal (0.5%), and various (1.6%) infections. The infections were principally caused by *P. multocida* (61.4%), *P. canis* (9.6%), *P. stomatis* (4%), *P. dagmatis* (2.3%), and related species (EF-4, *Neisseria weaveri, Weeksella zoohelcum* and *Capnocytophaga canimorsus:* 16.4%). In a few cases, *Pasteurella (pneumotropica, bettyae* and SP group: 3%), and *Actinobacillus (ureae* and *lignieresii:* 2.3%) were isolated. The majority of animal wound infections were treated with β-lactamins (penicillins) or cyclines; with other forms, β-lactamins (penicillins and cephalosporins) were more likely.

P10. *In vitro* Activity of Selected Antimicrobial Agents Against HAP Organisms Isolated from BRD, SRD, and ORD.

J.L. Watts[1]*, S.A. Salmon[1], R.J. Yancey, Jr[1], L.J. Hoffman[2] and H.C. Wegener[3], [1]The Upjohn Company, Kalamazoo, MI, [2]Iowa State University, Ames, IA, and [3]National Veterinary Laboratory, Copenhagen, Denmark.

The minimum inhibitory concentrations (MIC) for eleven antimicrobial agents commonly used in veterinary therapy were determined with 1047 strains of HAP organisms isolated from bovine (BRD), swine (SRD), and ovine (ORD) respiratory disease. MIC determinations were conducted in accordance with NCCLS guidelines. Ceftiofur was the most active compound with an MIC_{90} of ≤0.06 μg/ml for all isolates tested. Enrofloxacin was also very active with MIC_{90} of ≤0.06 μg/ml for the BRD and SRD strains and 0.13 μ g/ml for the ORD strains. In comparison, ampicillin was less active with an MIC_{90} of 32.0 μ g/ml for *P. haemolytica* and 16.0 μg/ml for *A. pleuropneumoniae.* Similarly, tetracycline was much less active with MIC_{90} ranging from 16.0 to 32.0 μg/ml for BRD and SRD isolates. Sulfamethazine demonstrated poor activity against the isolates tested with MIC_{90} of ≥256.0 μg/ml for the BRD and SRD strains tested.

P11. Serious Infection of the Forefinger Due to *Pasteurella* "Sp" Group Following a Guinea Pig Bite.

C. Lion*, M.L. Dupuy, M.C. Conroy, Laboratoire de Bactériologie, CHU Nancy; F. Escande, C.I.P., Institut Pasteur, Paris, France.

Case report: a previously healthy 31-year-old woman was bitten on April 23rd, 1993, at the left forefinger by her guinea pig. After cleansing of the wound, she was empirically treated with oxacillin and anti-inflammatory. Two days later, she was admitted to the hospital because a cellulitis of the hand: the forefinger looked inflammatory with purulent discharge, lymphangytis of the forearm and axillary adenopathy. After surgical treatment, she received pristinamycin for 5 days then cloxacillin. On May 4th, because a severe arthritis of the interphalangeal joint with necrosis of the tendon, antibiotherapy was changed for amoxycillin-clavulanate. On May 5th, the gravity of the sepsis necessitated an amputation at the second phalanx of the forefinger, and the patient improved with amoxicillin-clavulanate for 10 days. Bacterial culture of the joint obtained on May 4th evidenced *Pasteurella* "SP" group susceptible *in vivo* to amoxycillin-clavulanate. *Pasteurella* "SP" group is a rare organism usually isolated from guinea pig but seldom from humans, likely because it is not widely spread and probably because of its low virulence: actually it is almost always found in patients with predisposing conditions. This report also shows it can be responsible for very severe infection even in individuals without any particular risk factor.

P12. Bacteriophage Genes From *Haemophilus somnus*: Homology to Genes From Bacteriophages P2 and Hp1.

R.A. Pontarollo[1]*, C.R. Rioux[1], and A.A. Potter[1,2], [1]Veterinary Infectious Disease Organization and [2]Canadian Bacterial Diseases Network, Saskatoon, Saskatchewan, Canada.

Haemophilus somnus is a bovine pathogen and etiological agent of "hemophilosis". While investigating iron uptake in *H. somnus*, a 7.8 kb fragment of genomic DNA coding for a hemin-binding (Hmb+) phenotype was cloned. DNA sequencing (7808 bp) indicated a series of unidirectional open reading frames (orfs). Database analysis of *H. somnus* predicted polypeptides showed homology to proteins in other bacteriophages. The greatest homology occurred between 7 non-consecutive *H. somnus* polypeptides and the DNA packaging, capsid, and tail proteins of bacteriophage P2. Interrupting the P2 homology were 4 polypeptides that were similar to the predicted lysis proteins of the *H. influenzae* temperate bacteriophage HP1. A prominent feature was the identical gene order within the regions of homology. The *H. somnus* phage genes are organized like the P2 genes, however, HP1 lysis genes have replaced the P2 lysis genes. This type of recombination between bacteriophages is consistent with Botstein's modular evolution theory. The highly conserved gene order in bacteriophages, particularly in the area known as the "lysis cassette", increases the likelihood of successful recombination between distantly related bacteriophages. Polyclonal antiserum against *H. somnus* orf2 (P2 major capsid protein homologue) will help determine whether this DNA is part of a functional bacteriophage.

P13. Molecular Investigation of the Role of ApxI and ApxII in the Virulence of *Actinobacillus pleuropneumoniae* Serotype 5.

D. Reilmer[1]*, J. Frey[2], R. Jansen[3], H. Veit[1], E. Kamp[3], T. Inzana[1]*, [1]Virginia Tech, Blacksburg, VA, [2]Univ. Berne, Berne, Switzerland, [3]Central Vet. Inst., Lelystad, The Netherlands.

The *apxI* gene locus of a non-haemolytic mutant of *Actinobacillus pleuropneumoniae* (Ap) serotype 5 contained a major deletion that spanned most of *apxICABD*. Use of monoclonal antibodies confirmed that ApxI was not synthesized, and that ApxII was synthesized but not exported. The *apxICABD* genes and *apxIBD* genes were cloned into a broad host-range vector and electroporated into the mutant to obtain strains JF1064 and JF1060, respectively. JF1064 exported ApxI and ApxII, and produced haemolytic activity exceeding that of parent strain J45; JF1060 produced only weak haemolytic activity. In mouse virulence studies JF1064 was significantly more virulent than J45, while JF1060 was significantly less virulent than J45 or JF1064, but more virulent than the nonhaemolytic strain. JF1064 and J45 were similar in the capability to cause pleuropneumonia in pigs. JF1060 maintained the ability to induce pleuropneumonia in pigs, although higher doses were required to induce similar lesions. This study shows a direct correlation between Apx toxin production and virulence in Ap.

P14. Characterization of a Restriction Endonuclease From *Pasteurella haemolytica* Serotype Al and Construction of a Gene-Replacement *Aroa* Mutant.

R.E. Briggs*, F.M. Tatum, National Animal Disease Center, USDA, ARS, Ames, IA 50010. USA.

A new restriction endonuclease, *PhaI*, was isolated from *P. haemolytica* serotype 1, strain NADC D60, obtained from pneumonic bovine lung. *PhaI* recognizes the 5 base non-palindromic sequences 5'-GCATC-3' and 5'-GATGC-3'. Cleavage occurs 5 bases 3' from the former recognition site and 9 bases 5' from the latter. A gene encoding for a methyltransferase which protects against *PhaI* cleavage was cloned from *P. haemolytica* into *E. coli*. Whereas unmethylated plasmid DNA containing a *P. haemolytica* origin of replication was unable to transform *P. haemolytica* when introduced by electroporation, the same plasmid DNA obtained from *E. coli* which contained cloned *PhaI* methyltransferase could do so. The *aroA* gene of *P. haemolytica* serotype A1 was cloned and sequenced. A *P. haemolytica* ampicillin-resistance fragment was cloned into the unique *NdeI* site of *aroA*. A hybrid plasmid was constructed by joining the *aroA* replacement plasmid with the 4.2 kb *P. haemolytica* plasmid which encodes streptomycin resistance. Following *PhaI* methylation, the hybrid plasmid was introduced into P. *haemolytica* by electroporation. Allelic exchange between the replacement plasmid and chromosome of P. *haemolytica* gave rise to an ampicillin-resistant mutant which was unable to grow on medium deficient in tryptophan. Although transformation efficiency with methylated hybrid plasmid was $< 10^3/\mu g$, the hybrid was capable of unstable replication in *P. haemolytica,* so this system may be suitable for construction of additional gene-replacement mutants

P15. Restriction Barriers in *Pasteurella multocida* Type D Strains From Pigs.

I.C. Hoskins* and A.J. Lax, Institute for Animal Health, Compton, Berks. RG16 ONN.

A system for transfer of foreign DNA into type D *P. multocida* is being developed. A natural streptomycin/sulphonamide resistance plasmid, pPM1, which could be transferred into *Escherichia coli* DH5 alpha was used. It was found that this plasmid could not be transferred between certain *P. multocida* strains or between *E. coli* and some *P. multocida* strains by electroporation (1). This evidence suggests that there are several restriction barriers present in these *P. multocida* strains. To overcome the restriction barrier(s) a toxigenic strain LFB3 was mutated using ethyl-methanesulphonate (2). The mutation resulted in a strain which could take up pPM1. Plasmid pPM1 isolated from this strain could be transferred to wild type LFB3.

(1) Conditions for transformation of *Pasteurella multocida* by electroporation. Jablonski L., Sriranganathan N., Boyle S.M., and Carter G.R. 1992. Microbial Pathogenesis 12 63-68.

(2) Experiments in Molecular Genetics. J.H. Miller. 1973. Cold Spring Harbor.

P16. A Native Plasmid Of *Pasteurella haemolytica* Serotype A1: DNA Sequence Analysis and Investigation of its Potential as a Vector.

A.R. Wood*, F.A. Lainson and W. Donachie. Moredun Research Institute, 408 Gilmerton Road, Edinburgh, EH17 7JH, Scotland, UK.

The complete nucleotide sequence was determined for a 4.3 kilobase pair plasmid, pAB2, isolated from a bovine strain of *Pasteurella haemolytica* serotype A1 . The plasmid encodes a Rob-1 type beta-lactamase and contains a region with significant DNA homology to elements of the mobilization (mob) region of the *E. coli*, ColE1. pAB2 can be transferred to *P. haemolytica* by transformation and electroporation and can also be mobilized for conjugative transfer using *E. coli* helper functions. An insertion mutant of pAB2 (pTC2/81) carrying a copy of Tn5 was transferred to *E. coli* K12 by conjugation. This could be transferred to *E. coli* HB101, but not to *P. haemolytica* serotypes A1 or A2 by transformation . However, a derivative of pTC2/81 containing only a fragment of Tn5 was able to transform *P. haemolytica.* Two further constructs containing the pAB2 presumptive origin of replication and the ampicillin resistance gene and cloned kanamycin resistance gene or a fragment of the *P. haemolytica* A1 leukotoxin A gene, were similarly able to transform *E. coli* but not *P. haemolytica.* This has demonstrated that the pAB2 is capable of acting as an *E. coli* / *P. haemolytica* shuttle vector. However, the nature of the cloned DNA sequences are important for transformation and stable plasmid replication. This suggests the presence of a restriction system or DNA modification system in *P. haemolytica.* To overcome this attempts have been made to use chemical mutagenesis to select mutants which are susceptible to transformation.

P17. Characterization Of *Actinobacillus Pleuropneumoniae Strains* By Ribosomal Intergenic Sequences.

V. Fussing[1]*, F.E. Dewhirst[2] and B.J. Paster[2], [1]Department of Microbiology, National Veterinary Laboratory, Copenhagen, Denmark, [2]Department of Molecular Genetics, Forsyth Dental Centre, Boston, Massachusetts 02115.

Actinobacillus pleuropneumoniae is the aetiologic agent of contagious porcine pleuropneumoniae. The disease is widespread among pigs, and results in mortality, chronic lung lesions, and pleurisy. The aim of this study, was to characterize the species further and determine the intraspecies relationship, based on the ribosomal intergenic regions. Eight strains, each representing 1 of the 8 serotypes found in Denmark, 13 international reference strains and 3 strains of the closely related species *Actinobacillus lignieressi,* were included in the study. The intergenic regions were amplified by PCR, using primers matching the conserved regions in the surrounding ribosomal genes. This resulted in amplification of 2 intergenic regions of approximately 500 and 600-650 bases, from all the strains. The intergenic regions were sequenced with the Taquence cycle sequencing kit. From sequence similarity, a phylogenetic tree was constructed using the Neighbor-Joining method, and analyzed by bootstrapping. The intergenic sequences of these strains allowed a clear discrimination between the 2 closely related species, *Actinobacillus pleuropneumoniae* and *Actinobacillus lignieressi.*

P18. The *Haemophilus Influenzae Sxy-1* Mutation is in a Newly-identified Gene Essential for Competence.

P.M. Williams, C. Ma. and R.J. Redfield*, Department of Zoology, University of British Columbia, Vancouver, Canada.

DNA is abundant in mucosal environments and may be a significant source of nucleotides for those mucosal pathogens able to take it up. In *Haemophilus influenzae*, competence for DNA uptake is increased 100-fold to 1000-fold by the sxy-1 mutation, which we have mapped to an ORF adjacent to the rec-1 locus. Insertional inactivation of sxy entirely prevents both DNA uptake and transformation, indicating that it encodes a function essential for competence. In contrast, a multicopy plasmid containing the wildtype sxy gene confers constitutive competence on wildtype cells, suggesting that Sxy may be a positive regulator of competence whose activity is increased by the sxy-1 mutation. However, sxy-1 is a point mutation causing only a Val-Ile substitution in the Sxy protein. Three other point mutations in sxy cause similar levels of hypercompetence but do not change the specified protein at all. However, each of these four point mutations is predicted to destabilize a region of potential secondary structure at the 5' end of the sxy mRNA. In wildtype cells formation of this stem may depend on the rate of transcription, thus permitting translation of sxy mRNA and induction of competence only when nucleotide pools are depleted.

P19. Evaluation of a PCR Detection System of *Actinobacillus pleuropneumoniae* in Mixed Cultures from the Respiratory Tract of Pigs.

T. Gram*, M.J. Jacobsen, P. Ahrens and J.P. Nielsen, National Veterinary Laboratory, Bülowsvej 27, 1790 Copenhagen V, Denmark.

A PCR for detection of *A. pleuropneumoniae* was evaluated for its suitability as a diagnostic tool. All of the 102 tested Danish isolates of *A. pleuropneumoniae* reacted in the PCR and a 985-bp product was amplified. Mixed cultures from 70 lungs with clinical signs of pleuropneumonia were tested. The results of the PCR assay were compared with those of culture, and all isolation positive lung cultures tested positive in the PCR. Approximately one third of the isolation negative lungs were, however, positive in the PCR test, suggesting a superior sensitivity of the PCR test to that of culture. Cultured samples from tonsils of 101 pigs from 9 different herds were examined by PCR for the presence of the

985-bp target sequence and 65% reacted positive compared to 23% positive by culture. A significantly higher detection level was achieved, when the tonsils were precultured on selective media as compared to non-selective media. The results show that a combination of culture and PCR is a valuable method for the identification of *A. pleuropneumoniae* in pigs.

P20. Localization of an Epitope Involved in Neutralizing the Leukotoxin from an A1 Serotype of *Pasteurella haemolytica*.

F.A. Lainson*, K. Aitchison and W. Donachie, Moredun Research Institute 408 Gilmerton Road Edinburgh EH17 7JH.

Fragments of the Lkt A gene from A1 serotype of *P. haemolytica* were generated by restriction digests or by PCR. These were cloned into a plasmid vector for expression as a fusion protein with β-galactosidase. The resulting fusion proteins were characterized by immunoblotting for their reaction with a monoclonal antibody which has a high toxin neutralizing titre. The epitope was localized to a 20 AA region located at the C terminus of the LktA molecule and the epitope of a non-neutralizing antibody was located a further 20 AA downstream. Although the minimal recombinant peptides were not antigenic, a protein encompassing a 200 AA region of the C terminus was immunogenic and evoked production of neutralizing antibodies in SPF lambs.

Neither of the monoclonals used reacted with T10 serotype LktA on blots or neutralized T10 toxin. Comparison of the deduced AA sequence in the region of these epitopes showed considerable sequence divergence between A1 and T10 toxins.

P21. Serological Profile of a Porcine Herd Using a Hemolysin Neutralization Assay to Detect *Actinobacillus pleuropneumoniae* Antibodies.

J.A. Gutierrez-Pabello[1], J.M. Doporto[2], M.A. Monrroy[2] and F. Suarez-Guemes[1], [1]Departamento de Microbiologia e Immunologia, FMVZ UNAM, Mexico D.F. 04510, [2]Grupo Roussel, Ave. Miguel A. de Quevedo, esq. Ave. Universidad, Mexico D.F.

The purpose of this study was to determine the serological status of a pig herd in relation to *Actinobacillus pleuropneumoniae* infection. Two gilts, four sows and their litters were bled six times at 1, 5, 9, 13, 19, and 25 weeks of age. One hundred per cent of the females, 68% of the piglets from the gilts, and 93% of the piglets from the sows were classified as positive on week one, by the hemolysin neutralization assay. By week 5 none of the piglets showed positive titres, they remained seronegative until week 13. Seroconversion arose when the animals were moved to the finishing pen. When the study was finished, at week 25, 20% of the sampled population was positive, and 26% was classified as suspicious. Clinical disease did not appear during the study, and pleuropneumonic lesions were not found at the slaughter house. According to these results, herd immunity is a key in the infected farms to diminish the clinical disease.

P22. The Occurrence of *Pasteurella haemolytica* Serotypes Amongst Cattle, Sheep and Goats in Southern Africa.

M.W. Odendaal* and M.M. Henton, Onderstepoort Veterinary Institute,
Onderstepoort 0110 South Africa.

Pasteurella haemolytica is a frequent cause of disease amongst cattle and sheep in southern Africa, causing millions of rands in losses. In cattle it is usually seen amongst animals in feedlots, whilst it occurs in sheep under intensive as well as extensive conditions. Over an eight year period from September 1986 to March 1994, a total of 497 specimens were received from sheep and goats for the isolation of *Pasteurella* organisms, from seven different geographical areas in southern Africa. In all these regions *P. haemolytica* serotype 6 was the most frequent, followed closely by types 9 and 15. One hundred and sixty eight isolates could not be typed (33.8 %). Four disease syndromes were associated with these serotypes, pneumonia being the largest with 51.3 % cases, blue udder in ewes (28.14 %), septicaemia in 8.66 % and other miscellaneous causes 11.9 %. From cattle, 96 specimens were submitted for the isolation and typing of *Pasteurella* organisms, 41 % belonged to *P. haemolytica* type 1 and originated from pneumonia specimens whilst 25 % were not typeable. Serotype 1 was isolated from 56.71 % of all cases presented for examination, from the whole region.

P23. Serological and Molecular Characterisation of V-Factor (NAD)- Independent *Haemophilus paragallinarum*.

J.K. Miflin[1], R..F. Horner[2]*, P.J. Blackall[1], X. Chen[1], G.C. Bishop[2], C.J Morrow[3], T.Yamaguchi[4], Y.Iritani[4], [1]Animal Research Institute, Yeerongpilly 4105, Australia, [2]Allerton Veterinary Laboratory, Private Bag X2, Cascades 3202, Natal, South Africa, [3]Victorian Institute of Animal Sciences, Attwood 3049, Australia, [4]Shionogi & Co. Ltd., Shiga 520-34, Japan.

Fifteen isolates of V-factor (NAD)-independent strains of *H.paragallinarum* isolated from poultry in South Africa were characterised by biochemical typing, serotyping, restriction endonuclease analysis (REA) and ribotyping. The isolates were deliberately selected to be as diverse as possible. All the isolates were shown to be typical of the species in biochemical properties and all were serovar A. REA was performed with three enzymes. All the isolates gave identical REA profiles. Ribotyping was performed using a probe which consisted of the plasmid pUC19 into which the 16S rRNA operon of *H. paragallinarum* had been inserted. All the isolates gave the same ribotyping profile. Our results strongly suggest that these NAD-independent isolates are clonal in nature.

P24. A Co-agglutination Test for Serotyping *Pasteurella haemolytica.*

L. Fodor*, Z. Pénzes and J. Varga, Department of Microbiology and Infectious Diseases, Univ. Vet.Sci., Budapest.

A co-agglutination test was adopted for detection of *Pasteurella haemolytica* type specific antigens in lung lesions even if no viable bacteria could be cultured The co-agglutinating reagents were prepared by coating protein-A producing *Staphylococcus aureus* cells with hyperimmune sera raised against *P. haemolytica* type strains. Homologous reactions of bacterial suspension, saline extract and boiled saline extract antigens were good, but some one-way cross-reactions appeared. Ninety-three per cent of 64 field strains examined could be serotyped using the coagglutination test. The coagglutination test detected *P. haemolytica* type specific antigens in the lung specimens of 4 calves, 5 sheep which had succumbed due to naturally occurring pneumonia, 20 calves experimentally infected with *P. haemolytica* Al, and in 36% of the lung specimens of slaughtered field sheep with chronic lung lesions. This test is recommended as an additional method for fast and reliable serotyping of *P. haemolytica*.

P25. A Soluble 70 kilodalton Protein Present in Cells of *Pasteurella haemolytica* A2

R.C. Davies and W. Donachie, Moredun Research Institute, 408 Gilmerton Road, Edinburgh EH17 7JH.

A monoclonal antibody (Mab) was raised against whole live A2 *P. haemolytica* cells in mice primed with outer membrane proteins of iron-deficient A2 cells. It recognises a cytoplasmic / periplasmic 70kDa protein found in A2 iron- deficient cells as did serum from either a convalescent sheep or an animal vaccinated with a sodium salicylate extract of A2 iron-deficient cells. The Mab recognises the antigen in A2 cells grown in either nutrient broth culture or αα'dipyridyl-treated culture broths, but not in cells of several other serotypes eg A1, 6, 7, 9, T3, 4, 10, 15 whether iron-replete or iron-deficient. The soluble protein was purified by affinity chromatography on a column of mouse IgG - Sepharose and after SDS/PAGE and transfer to PVDF had an N-terminal sequence of 18 amino acids which was similar to that of the DnaK heat shock "chaperone" protein of *E. coli.* However, the protein did not seem to possess the "consensus" sequence belonging to the general family of heat shock proteins. Its location in the cell could be modified by the presence of iron in the growth medium. Thus the protein could be differentiated from the other immunologically active iron- regulated OMPs of *P. haemolytica* A2.

P26. Rapid Preparation of Competent Cells by Filtration.

P.M. Williams, W. Hung, and R.J. Redfield*, Departments of Zoology and Biochemistry, University of British Columbia, Vancouver, Canada.

Collection and washing of cells by centrifugation is a time consuming and inconvenient component of many microbiological protocols. We have found that filtration onto disposable membrane filters can efficiently replace the centrifugation steps in many procedures. To demonstrate the utility of filtration, we have substituted it for centrifugation in standard competence protocols for *Escherichia coli, Haemophilus influenzae*, and *Saccharomyces cerevisiae.* With all three protocols filtered cells showed transformation efficiencies comparable to or higher than those of cells prepared by centrifugation. Furthermore, the filtered cells were competent sooner, both because filtration is faster than centrifugation and, in the case of *E. coli*, because competence was highest immediately after filtration. The procedure can be easily adapted to collect cells for many other purposes, and should be especially useful in metabolic studies and in preparing cells for electroporation, as it facilitates both washing and resuspension of cells.

P27. Variations of the Chromosomal Structure Between Serotypes 1-12 and Biovars 1 and 2 of *Actinobacillus pleuropneumoniae.*

J. Frey[1]*, M.A. Holloway[1], K. Tarasjuk[2] and J. Nicolet[1], [1]Institute for Veterinary Bacteriology, University of Berne, Switzerland and [2]National Veterinary Research Institute, Pulawy Poland.

Entire chromosomal DNA of all 12 *Actinobacillus pleuropneumoniae* serotype reference strains and of field strains from some selected serotypes and also from biovar 2 strains has been analyzed using infrequently-cutting restriction enzymes and field inversion gel electrophoresis (FIGE). The size of the chromosome of the *A. pleuropneumoniae* type and serotype 1 reference strain 4074^T was calculated to be 1850 +/- 100 kilobasepairs (kbp). The other serotype reference strains have a very similar chromosome size. The restriction fragment patterns obtained from the different serotype reference strains show that their chromosomal structures differ. However, the serologically related serotypes 1, 9,11, serotypes 5a and 5b and also serotypes 6 and 8 have very similar patterns. Filed strains with the same serotype show the same patterns, even when they originated from different geographical areas. Analysis of PCR-amplified 16S rRNA genes with frequently-cutting restriction enzymes does not distinguish between serotypes and biovars.

P28. Clinical Isolates of *Actinobacillus actinomycetemcomitans* From Three European Cities Carry Highly Related Prophages.

K. Willi, H. Sandmeier, J. Meyer*, Department of Preventive Dentistry and Oral Microbiology, Dental Institute, CH-4051 Basel, Switzerland.

Actinobacillus actinomycetemcomitans is a putative periodontal pathogen. Others had suggested that bacteriophages of *A. actinomycetemcomitans* might be associated with disease activity. We characterized five lysogenic *A. actinomycetemcomitans* strains isolated from different patients from three geographic locations (Basel, Freiburg i. Br., Amsterdam). They belonged to two genotypes as revealed by RFLP analysis. One of the strains showed a DNA restriction modification system. The temperate bacteriophages induced from these hosts were related: The phages have the same virion structure. The phage DNA genomes were investigated by restriction fragment pattern analysis and Southern hybridizations. All five phages had a genome size of about 43 kb and showed identical or similar restriction fragment patterns. Hybridization experiments indicated extensive DNA homologies. However, the phages formed three immunity groups. These results may suggest, but do not prove, a common genetic trait providing a selective advantage to lysogenic *A. actinomycetemcomitans*.

P29. High-Molecular-Mass LPS are Involved in *Actinobacillus pleuropneumoniae* Adherence to Porcine Respiratory Tract Cells.

S.-E. Paradis, D. Dubreuil, S. Rioux, M. Gottschalk and M. Jacques*, Faculte de Medecine Veterinaire, Universite de Montreal, C.P. 5000, St-Hyacinthe, Quebec, Canada J2S 7C6.

Actinobacillus pleuropneumoniae is the causative agent of porcine pleuropneumonia. We have identified the lipopolysaccharides (LPS) as the major adhesin of *A. pleuropneumoniae*. Using immunoelectron microscopy and flow cytometry, we showed in the present study that LPS were well exposed at the surface of this encapsulated micro-organism. Inhibition of adherence of *A. pleuropneumoniae* with extracted LPS was performed on porcine trachea frozen sections. Acid hydrolysis of LPS revealed that the active component of LPS was not lipid A but the polysaccharides. LPS from *A. pleuropneumoniae* serotypes 1 and 2 were separated by chromatography on Sephacryl S-300 SF according to their molecular mass. The adherence inhibitory activity was found in the high-molecular-mass fractions. These fractions contained KDO and neutral sugars, and were recognized by a monoclonal antibody directed against *A. pleuropneumonia-* O-antigen but not recognized by a monoclonal antibody against capsular antigen.

P30. Serotyping of *Actinobacillus pleuropneumoniae* Serotype 5 Strains Using a Monoclonal Antibody Against Lipopolysaccharides O-Chain in a Latex Agglutination Test.

J. Daniel Dubreuil[1]*, N. Mayott[1], E. Stenbaek[2] and M. Gottschalk[1], [1]Department of Pathology and Microbiology, Faculty of Veterinary Medicine, Montreal University, Quebec, Canada, [2]National Veterinary Institute, Copenhagen, Denmark.

A latex agglutination test has been developed for serotyping *Actinobacillus pleuropneumoniae* serotype 5 strains. Latex particles were sensitized with a murine monoclonal antibody recognizing an epitope on the serotype 5 LPS-O chain as shown by SDS-PAGE and Western blotting. A total of 144 *A. pleuropneumoniae* and 14 common bacterial species associated with swine were tested by mixing 50µl of latex reagent with the same volume of a dense suspension (MacFarland #5) of bacterial cells grown for 18 h on PPLO agar plates. All *A. pleuropneumoniae* strains had been previously serotyped using a standard procedure. The latex agglutination test was rapid (few minutes), easy to perform and overall a good correlation was found with the standard techniques. In addition, the sensitized latex particles were stable for at least two months.

P31. Purification and Characterisation of Bovine Lipopolysaccharide Binding Protein.

N.U. Horadagoda[1]*, P.D. Eckersall[1], L. Andrew[2], P. Gallay[3], D. Heumann[3] and H. Alison Gibbs[1], [1]Department of Veterinary Medicine and LRF Virus Centre, [2]Department of Veterinary Pathology, Glasgow University Veterinary School, Glasgow G61 lQH, U.K and [3]Departmentof Internal Medicine, CHUV-101 1 Lausanne, Switzerland.

Lipopolysaccharide binding protein (LBP) is a recently described acute phase protein which is implicated in modulating the host responses to lipopolysaccharide (LPS) from Gram-negative bacteria. LBP-like activity has been identified in bovine serum and in this study LBP was purified from acute phase serum using ion exchange chromatography over Bio-Rex 70 and Econo-Pac Q. On SDS-PAGE, bovine LBP demonstrated a single band with a molecular radius of 58 kDa. Functionally, bovine LBP resembled LBPs from other species in that it enhanced binding of LPS to monocytes and increased the sensitivity of monocytes to LPS by at least a 100-fold. However, immunoglobulins to rabbit LBP or recombinant human LBP did not cross- react with bovine LBP. Studies on LBP activity during the acute phase response demonstrated a four-fold increase 36 hours after a single intratracheal inoculation of *Pasteurella haemolytica* A1. Isolation of LBP will allow further investigations on LBP-mediated responses to LPS in cattle.

P32. Identification of a Conserved, Phase Variable Epitope in the Lipooligosaccharide of *Haemophilus somnus*.

T. Inzana[1]*, A. Campagnari[2], A. Lesse[2], M. Apicella[3], [1]Virginia Tech, Blacksburg, VA, [2]SUNY Buffalo, Buffalo, NY, [3]Univ. Iowa, Iowa City, IA.

*Haemophilus somnus (*Hs) is a multisystemic, pathogenic bacterium of bovines. As with *Haemophilus influenzae* (Hi) and *Neisseria gonorrhoeae* (Ng), the lipooligosaccharide (LOS) of disease isolates of Hs undergoes phenotypic phase variation *in vivo* and *in vitro*, whereas preputial isolates do not. Monoclonal antibodies (MAb) to the LOS of Hi, Hi biogroup *aegyptius*, and Ng were examined for reactivity with the LOS of Hs strain 738 and LOSs from 13 phase variants of strain 738. Hs LOS bands of M_r 4200 to 4300 reacted with 1 MAb to Ng LOS, 3 MAb to Hi biogroup *aegyptius* and 1 MAb to Hi LOS. Hs LOS bands <4000 M_r did not react with any MAb. All phase variants of strain 738 tested reacted with the same MAb as strain 738. Hs preputial isolate IP produced a single LOS band of M_r 3300, did not react with any MAb, and did not undergo LOS phase variation. A $(CAAT)_5$ probe reacted with Hs 738 and Hi DNA of 4-5 kb, but did not react with Hs DNA from non-disease isolate IP. Thus, at least 1 phase variable epitope seems to be conserved in Hs disease isolates, and a common mechanism for LOS phase variation may occur in mucosal Gram-negative bacteria.

P33. The Influence of *Haemophilus somnus* Lipooligosaccharide on Acrosomal Status of Rabbit Sperms.

A. Chelmonska-Soyta*, J. Mazur[1], C. Lugowski[2], S. Stafaniak[1], M. Nikolajczuk[1], [1]Faculty of Veterinary Medicine, Agriculture University of Wroclaw, Institute of Immunology and [2]Experimental Therapy, Pol.Ac.Sci., Wroclaw, Poland.

The ejaculated rabbit sperms were incubated with different doses of *H.somnus* lipooligosaccharide (LOS).The number of viable cells, their motility and acrosomal status were examined at different time intervals and capacitation conditions. The percentage of sperms with partially or completely damaged acrosomes was dependent on the LOS concentration and on the time of incubation. The viability and motility of rabbit spermatozoa were not affected during the incubation in comparison to control samples. It seems that endotoxin of *H.somnus* may induce the acrosomal reaction of rabbit sperms. This phenomenon may contribute to poor reproductive performances of infected females.

P34. Comparison of Bovine and Ovine Isolates of *Haemophilus somnus.*

A.C.S. Ward[1]*, J.M. Eddow[2], M.D. Laworski[1] and L.B. Corbeil[2], [1]University of Idaho, Caine Veterinary Teaching and Research Center, Caldwell, Idaho and [2]University of California, San Diego, California USA.

Recent studies indicate that *Haemophilus agni* and *Histophilus ovis* from sheep and *Haemophilus somnus* from cattle are biochemically and culturally very similar. These organisms produce varied and frequently severe infections in the two animal species. Isolates from different types of infections in the two animal species were evaluated to detect differences at the biochemical, molecular, and antigenic levels. Six different biogroups were detected by an expanded panel of biochemical tests. Restriction enzyme analysis (REA) revealed marked differences between bovine and ovine isolates, between bovine disease and preputial isolates and even among ovine carrier isolates. All isolates had the major 78, 60, and 40kDa antigens on Western blots. However, minor differences were detectable in the low molecular weight antigens. Although marked differences were detected in biochemical tests and in REA profiles, similarities of antigens of ovine and bovine isolates suggest that vaccines containing these major antigens may provide protection against disease in both sheep and cattle.

P35. Development of Disrupted Whole-cell *Actinobacillus pleuropneumoniae* (App) Vaccines.

B.J. Thacker, M.H. Mulks*, Michigan State University, East Lansing, Michigan and F. Udeze and R.R. Simonson, Oxford Laboratories, Worthington, Minnesota, USA.

We previously reported the development of an effective App whole cell vaccine utilizing growth conditions that optimize antigen production, disruption/ inactivation of cells by sonication to improve antigen presentation, and a light oil adjuvant (Emulsigen^{TM}). In the present study, we evaluated other adjuvants and disruption/inactivation methods to improve commercial feasibility. In preliminary experiments, based on serum antibody responses, we found light oil adjuvants(Emulsigen and Imugen^{TM}) were superior to vegetable oil and aluminum hydroxide. Heat inactivation (HI) or disruption via a "French press" (FP) were similar to sonication, and the optimal organism concentration was approximately 10^9 per dose. The following vaccines (Imugen adjuvant) were evaluated against homologous experimental challenge with serotype 5: A-- HI/$10^{9.5}$, B-- FP/10^{9}·5, C-- FP/$10^{8.9}$, CON-- control. The percent pneumonia and mortality by vaccine were: A-- 10.6, 1/7; B-- 4.4, 0/7; C-- 12.1, 1/7; and CON-- 37.5, 7/7. The modifications did not alter the efficacy of our original sonicated-whole-cell vaccine.

P36. Specificity and Sensitivity of an ELISA for *Actinobacillus pleuropneumoniae* (App).

M.H. Mulks* and B.J. Thacker, Michigan State University, East Lansing, Michigan, USA.

We previously reported the development of an ELISA for measuring App antibodies using an outer membrane antigen prepared by sucrose density gradient centrifugation. This ELISA has been useful for evaluating serum immune responses within controlled experiments. However under field conditions, pigs will often develop low to moderate antibody levels even though they appear to be App-free. Recently, reference sera from pigs experimentally infected with gram-negative bacteria were made available. The sera were tested blind and a key was provided later by the reference laboratory. The number of positive sera/total tested, by organism for each coating antigen serotype (1 and 5, respectively) was: App serotype 1-- 23/24, 20/24; App serotype 5-- 28/28, 26/28; App serotype 7-- 11/22, 19/22; *Actinobacillus suis*-- 14/16, 15/16; *Haemophilus parasuis*-- 0/16, 10/16; *Pasteurella multocida*-- 2/18, 3/18; and *E. coli*-- 0/10, 1/10. Immunoblot analysis revealed several obvious cross-reactive proteins, and suggested that capsular polysaccharide and lipopolysaccharide also cross-react. Development of App specific serotests can be hindered by cross-reactive antigens from other Gram-negative bacteria.

P37. Cloning And Expression Of *Actinobacillus pleuropneumoniae* (App) Riboflavin Synthesis Genes.

T.E. Fuller* and M.H. Mulks, Michigan State University, East Lansing, MI, USA 48824.

Riboflavin synthesis genes from APP serotype 5 were cloned and expressed in E. *coli DH5S*-α. APP serotype 5 chromosomal DNA was isolated and restricted with *HindIII*. Fragments ranging from 4 to 7 kb in length were ligated into the cloning vector pUC 19. Transformation of *E. coli* with the ligation mixture yielded one clone with a 5.0 kb insert that produced a water-soluble extracellular yellow compound. The absorption spectra of the compound were similar to that of riboflavin with absorption peaks at 444.5 and 373 nm. Shifts in absorption due to acidification and reduction with sodium dithionate also compared well with riboflavin. Mass spectroscopy with positive and negative ion fast atom bombardment confirmed that the extracellular product was riboflavin (Vitamin B-2). Riboflavin is produced at 40 µg/mL in a 24 hour culture grown in minimal salts media with N-Z amine. Production is not dependent on IPTG induction indicating presence of the endogenous APP promoter. Partial double-strand sequencing data indicates the presence of at least two genes with good homology at the amino acid level to *rib* A and *rib* H from *Bacillus subtilis* . It is possible that an entire operon with homology to the *Bacillus rib* GBAHT operon may be present in APP. Further sequencing will determine whether homologues to other *rib* genes exist in APP.

P38. Evaluation of a Polyclonal Blocking ELISA Detecting Antibodies to *Actinobacillus pleuropneumoniae* Serotype 2.

V. Sørensen[1]*, K. Barfod[1], R. Nielsen[2], N.C. Feld[2], J.P. Nielsen[2], J. Christensen[1], [1]The Federation of Danish Pig Producers and Slaughterhouses, DK-4000 Roskilde, [2] National Veterinary Laboratory, DK-1790 Copenhagen V, Denmark.

A polyclonal blocking ELISA detecting antibodies to *A.pleuropneumoniae* serotype 2 was compared to the conventionally used Complement-Fixation (CF) assay in three different samples: 1) test samples from SPF-breeding and multiplying herds. 2) consecutive serum samples from SPF sows. 3) aerosol-inoculated SPF pigs. 1) The ELISA was shown to have a higher herd-sensitivity, but a lower herd-specificity than the CF-assay. 2) The SPF sows were serological negative in three consecutive samplings, but sows with blocking percentages in the 'cut-off' area, tends to remain there in all three samplings. 3) However, some of the aerosol-inoculated pigs showed specific serological reactions in the 'cut-off' area of the blocking percentages and thereby excluded the possibility of adjusting the 'cut-off' level in order to solve the herd specificity problem. Serological reactions in the 'cut-off' area of this ELISA need to be verified by other signs of *A.pleuropneumoniae* serotype 2 infections in herd health surveillance programs.

P39. Screening of *Actlnobacillus pleuropneumoniae* Field Isolates on Expression of ApxI, ApxII, ApxIII Toxins and 42kDa-OMP.

J.F. van den Bosch[1]*, R. Nielsen[2], J. Nicolet[3] and J. Frey[3], [1]Intervet International, NL-5830 AA Boxmeer, [2]The Netherlands; National Veterinary Lab., DK-1503 Copenhagen, [3]Denmark; University of Berne, CH-3012 Berne, Switzerland.

A total of 135 *A. pleuropneumoniae* (App) strains, approx. 10 representatives of each of the presently known 13 serotypes, was isolated from diseased pigs in various countries all over the world. The strains were tested for expression of the RTX toxins ApxI, ApxII and ApxIII, and for expression of a common 42kDa P2-homologous outer membrane protein (OMP), by Western blotting using specific monoclonal (toxins) and polyclonal (OMP) antibodies. The vast majority of 96% of the strains had the same antigen expression profile as the corresponding App serotype reference strain. Combined with efficacy testing in pigs against various App serotypes, these results provide strong evidence that an App subunit vaccine containing ApxI, ApxII, ApxIII and OMP antigens will induce protection against all encountered App strains and serotypes.

P40. An *Actinobacillus pleuropneumoniae* Subunit Vaccine Inducing Protection in Pigs Against all Serotypes.

J.F. van den Bosch*, A.N.B. Pubben, I.M.C.A. Jongenelen, and R.P.A.M. Segers, Intervet International, P.O.Box 31, NL-5830 AA Boxmeer, The Netherlands.

At present 13 different *A. pleuropneumoniae* (App) serotypes are known, which all can cause disease in fattener pigs. The different App serotypes express one or two of three different RTX toxins (ApxI-III), resulting in four different toxin profiles among serotypes. In addition, all serotypes express a P2-homologous 42kDa outer membrane protein (OMP). A subunit vaccine containing purified ApxI, ApxII, ApxIII and OMP in an aqueous adjuvant was tested for efficacy in SPF pigs against challenge with various App serotypes. The vaccine induced excellent protection against challenge with App serotypes 1, 2, 5a, 7, 9 and 10. Since these serotypes are representatives of all four different toxin profiles seen among the 13 App serotypes, it is concluded that the vaccine will be protective against all presently known App serotypes.

P41. Role of Apx Toxins of *Actinobacillus pleuropneumoniae* in *vitro*.

M. Smits[1], R. Jansen[1], E. Kamp[1]*, L. van Leengoed[1], P. Dom[2] and F. Haesebrouck[2], [1]Institute for Animal Science and Health, P.O. Box 65, 8200 AB, Lelystad, [2]The Netherlands and Laboratory of Veterinary Bacteriology, University of Ghent, Casinoplein 24, B-9000 Ghent, Belgium.

The Apx toxins of *Actinobacillus pleuropneumoniae* are thought to be important virulence factors. To study the role of Apxl and Apxll, we constructed knock-out mutants of strain S 4047 (serotype 1). Phenotypes of the mutants were ApxI-/ApxII+, ApxI+/ApxII-, and ApxI-/ApxII-. The effects of the mutants on porcine PMNs were studied in a chemiluminescence assay and in a chemotaxis assay. Wild type and the ApxI-/ApxII+ and ApxI+/ApxII- mutants first activated and then killed the PMNs. This was not observed for the ApxI-/ApxII- mutant. Wild type and the ApxI+/ApxII- mutant clearly attracted PMNs but in the vicinity of the bacteria the directed migration was arrested. This was not observed for the ApxI-/ApxII+ and the ApxI-/ApxII- mutants. Our results with these mutants agree with those obtained with affinity purified toxins alone.

P42. Production and Characterization of Monoclonal Antibodies Against the 42 kDa Outer Membrane Protein of *Pasteurella haemolytica* A1.

K. Pandher, D. Styre, and G. L. Murphy*, Oklahoma State University, Dept. of Veterinary Pathology, Stillwater, OK 74078, USA.

Pasteurella haemolytica serotype A1 is the organism most frequently associated with a fibrinous pleuropneumonia of feedlot beef cattle in the United States. *P. haemolytica* A1 produces major outer membrane proteins (OMPs) of approximately 30 kDa and 42 kDa. In this study, we purified the 42 kDa OMP after preparative SDS-polyacrylamide gel electrophoresis of *P. haemolytica* A1 outer membranes. The gel-purified protein was used to immunize mice for the production of monoclonal antibodies (MAbs). Hybridomas were subsequently screened by ELISA against *P. haemolytica* A1 cell envelopes. Seven individual hybridomas, which produced MAbs that recognized the 42 kDa OMP, were isolated. Western blot analyses revealed that all seven MAbs recognize a protein of similar molecular weight (42 kDa) in *P. haemolytica* strains of serotypes A1, A2, A6, and A9 as well as in untypeable strains. Immunoelectron microscopy and whole cell ELISA assays suggest that some of the MAbs are directed against surface-exposed regions of the 42 kDa protein.

P43. *Pasteurella haemolytica* Leukotoxin and its Interaction With Target Cells.

M. Saadati, H.A. Gibbs*, G.D. Westrop, R. Parton, J.G. Coote, Departments of Microbiology and Veterinary Medicine, University of Glasgow, Glasgow G12 8QQ.

P. haemolytica leukotoxin (LKT) is one of the RTX family of toxins and is thought to be one of the major virulence factors in pneumonic pasteurellosis in cattle and sheep. Because of its apparent specificity for ruminant leukocytes, it may be a prime determinant of the host range of *P. haemolytica*. A large number of *P. haemolytica* isolates, including representatives of all serotypes and untypables, has been examined for LKT production in supernates of log phase cultures. There was some variation in the amount and molecular weight of LKT produced by different strains, as judged by SDS-PAGE and immunoblotting with monoclonal antibodies. There were also marked differences in leukotoxic activity as measured by a chemiluminescence assay with bovine and ovine neutrophils. Some strains produced apparently normal amounts of LKT protein but which had low toxicity. Data will be presented on the target cell specificity of these LKT preparations in terms of their cell-binding, leukotoxic and haemolytic activities for different mammalian species. The properties of the native LKT from *P. haemolytica* will be compared with those of recombinant LKT from *E. coli.*

P44. Evaluation of a *Pasteurella haemolytica* Recombinant Sialoglycoprotease Fusion Protein (Rgcp-F) Vaccine.

P.E.Shewen[1], A. Perets[1]*, C.W. Lee[1], R.Y.C. Lo[2], [1]Dept. of Vet Micro and Imm., [2]Dept. of Micro., University of Guelph, Guelph, Ontario, Canada.

Recombinant sialoglycoprotease fusion protein (rGcp-F) was used in a vaccination and challenge trial in cattle. Calves received two doses of one of the following vaccines: saline plus adjuvant, Presponse (culture supernate vaccine), rGcp-F alone with adjuvant, Presponse enriched with rGcp-F, Presponse enriched with rLkt (recombinant leukotoxin), Presponse enriched with rGcp-F and rLkt. Following first vaccination with rGcp-F most calves seroconverted as shown by Western Blot. As expected, vaccination with Presponse was approximately 70% efficacious. Enrichment with both rGcp-F and rLKT enhanced protection, but enrichment with either alone resulted in improvement that was not significantly better than Presponse alone. This failure of rLKT alone to enhance Presponse, which was expected from previous experiments, may reflect the unusually severe challenge in this particular trial (all controls died within 5 days). Given this, it was surprising to discover that rGcp-F alone provided protection that was significantly better than saline (no vaccine) and not statistically less than Presponse. Protective efficacy approximated 40%. This is the first demonstration of protection using a single *P. haemolytica* antigen.

P45. The Distribution of the Sialoglycoprotease Gene and Activity Among *Pasteurella haemolytica* Serotypes 1 to 16.

C.W. Lee[1]*, R.Y.C. Lo[2], P.E. Shewen[1], K.M. Abdullah[3] and A.Mellors[3], [1]Departments of Veterinary Microbiology and Immunology, [2]Microbiology, and [3]Chemistry and Biochemistry, University of Guelph, Guelph, Ontario, NlG 2W1, CANADA.

An examination of the distribution of the sialoglycoprotease (Gcp) gene and activity among *Pasteurella haemolytica* serotypes 1 to 16 was recently completed; along with other selected Pasteurellacae, including several clinical isolates of *P. haemolytica* serotype 1. Southern blot hybridization analysis and Gcp activity assay indicated that virtually all recognized *P. haemolytica* A biotypes contained the *gcp* gene and proteolytic activity. The exception was serotype 11 which revealed an alternative genetic organization and exhibited no proteolytic activity [Abdullah *et al;* Biochem. Soc. Trans. 18: 901-903]. Furthermore, all recognized T biotypes were negative for both gene and activity. This may have relevance in the differential pathogenicity and host range of A and T biotypes.

P46. Cloning and Molecular Characterization of the *GalE* Gene of *Pasteurella haemolytica* A1.

M.D. Potter* and R.Y.C. Lo, Department of Microbiology, University of Guelph, Guelph, Ontario, Canada. NlG 2Wl.

The enzyme UDP-galactose-4-epimerase (GalE) catalyzes the epimerization of UDP-galactose to UDP-glucose. This enzyme has been shown to play an important role in the biosynthesis of lipopolysaccharide (LPS) in a variety of Gram-negative bacteria such as *Escherichia coli, Salmonella typhimurium, Neisseria meningitidis,* and *Haemophilus influenzae.* The LPS of P. *haemolytica* Al has been implicated as a virulence factor in the organism' s pathogenesis. To investigate the biosynthesis of P. *haemolytica* Al LPS at the molecular level, the *galE* gene was cloned on a 5.5 kbp plasmid insert. This recombinant plasmid could complement and restore the synthesis of the entire O-antigen in a *S. typhimurium galE* mutant. Several proteins are expressed from the *galE* complementing DNA, including a protein of approximately 37 kDa which is similar in size to other GalE proteins. A 2.3 kbp sub-clone that encodes the 37 kDa protein is homologous to the *H. influenzae* type b *galE* gene.

P47. The Evaluation of Usefulness of Selected Polyclonal Antibodies for Immunocytochemical Diagnosis of *Haemophilus somnus*.

T. Stefaniak*, A. Chelmonska-Soyta, J. Molenda[1], M. Nikolajczuk, Dept. of Vet. Prevention and Immunology Agriculture University of Wroclaw, [1]Dept. of Veterinary Hygiene, Wroclaw.

The results of bacteriological examination in diagnosis of *Haemophilus somnus* infection are often difficult to interpret because of high cultivation requirements for this micro-organism and the overgrowth with saprophytic flora. We decided to work out the immunohistochemical method for detection of *H.somnus in situ*, based on using the monovalent sera against selected components of outer membrane protein (OMP) complex. The OMP of *H.somnus* was isolated according to Kania *et al.*(1992). After the separation of the OMP in SDS-PAGE, the major components: 67, 40 and 22 kDa were cut out and used for immunization of rabbit and goat, according to Boulard and Lecroisey (1982). The developed immune sera were controlled with ELISA and immunoblotting against H.somnus and other related bacteria whole antigens. Rabbit sera against 67 and 22 kDa OMP showed wide cross reactivity with *Pasteurella multocida, P.haemolytica* and Escherichia coli. Only goat immune serum against 40kDa OMP showed strong reactivity with *H.somnus* and very weak cross-reactivity with *P.multocida*. IgG class (both IgGl and IgG2 isotype) was isolated from immune serum and negatively absorbed with supernatant of *P.multocida* sonicate coupled to sepharose 4B. This antibody was successfully used in immunocytochemical detection of *H.somnus*.

P48. Protective Efficacy of a *Pasteurella haemolytica* A1 Iron-Regulated Bacterin Vaccine in Calves.

S.B. Houghton[1]*, M. Quirie[2], M. Lavery[1], R.C. Davies[2] and W. Donachie[2], [1]Hoechst Animal Health, Walton, Milton Keynes, UK, [2]Moredun Research Institute, Edinburgh, Scotland ,UK.

The protective efficacy of an iron-regulated bacterin vaccine was tested in hysterotomy-derived 8-week old male calves which were kept in isolation and were free of *Pasteurella* at the time of challenge. Cells of *P.haemolytica* A1 (strain M4/1/2) grown under iron restricted conditions using αα-dipyridyl , were inactivated with formalin and combined with aluminium hydroxide gel (alhydrogel). Five calves were vaccinated subcutaneously in the neck with a 2ml dose on two occasions, 4 weeks apart. The vaccinated calves and a control group of 6 unvaccinated calves were challenged intra-tracheally with 60 mls of a log phase culture of strain M4/1/2 .using an endobronchoscope. Surviving calves were monitored daily for clinical signs and necropsied on the fourth day after challenge. All control calves were removed due to severity of disease before the scheduled necropsy time with a mean total disease score (TDS) of 39.3 ± 1.2. All vaccinated animals survived with a significantly lower (p= 0.01) TDS of 11.8 ± 5.1 . Western blotting showed a strong response to iron-regulated protein (IRP) bands in the sera of vaccinated calves.

P49. Transferrin-binding Protein From *Actinobacillus pleuropneumoniae* Serotype 7: Purification, Characterization and Potential Serodiagnostic Use.

P. Heegaard*, J. Klausen and L. Ole Andresen, National Veterinary Laboratory, 27 Bulowsvej, DK 1790 Copenhagen V, Denmark.

Transferrin-binding protein from *Actinobacillus pleuropneumoniae* serotype 7 (WF83) was purified by a published method employing affinity-chromatography on immobilised hemin. An essentially pure protein was obtained (as judged from silver-stained SDS-polyacrylamide gel electrophoresis). The protein had an apparent molecular weight of 66 kD in contrast to the previously reported molecular weight for this protein (60 kD); also, we found no difference in its expression in an iron-restricted as compared to a non-restricted culture. The hemin- and swine transferrin-binding activities of the protein was found to be interdependent. We found a preferential reaction with the purified protein in ELISA of sera from a range of experimentally *Actinobacillus pleuropneumoniae*-infected swine as compared to sera from non-infected swine and from swine infected with other agents.

P50. Analysis of the *tbp* Loci in *Pasteurella haemolytica* Serotypes 1 to 16, *Actinobacillus suis* and *A. pleuropneumoniae*.

T.K.W. Woo, L.L. Burrows and R.Y.C. Lo*, Department of Microbiology, University of Guelph, Guelph, Ontario, Canada. NlG 2Wl.

Pasteurella and *Actinobacillus* species exhibit an iron uptake system which is capable of obtaining iron directly from host transferrin. This uptake is mediated by specific transferrin binding proteins in the bacterial outer membrane. The gene encoding for the P. *haemolytica* Al transferrin binding protein Tbpl (M.W. 100 kDa) has been cloned and sequenced. A *tbpA* specific probe was used to examine chromosomal DNA digests from the sixteen serotypes of P. *haemolytica, A. suis* and two *A. pleuropneumoniae* strains. The results showed that all 16 serotypes of *P. haemolytica* carry the *tbpA* gene, however, there are considerable differences in the restriction fragments that hybridized with the probe between the A and the T biotypes. Both *A. suis,* and *A. pleuropneumoniae* (CMS and Shoppe) examined also showed hybridization with the probe, though with less homology. A restriction map of the *tbpA* region in some of these strains was constructed to examine the restriction site polymorphism and to compare the degree of relatedness between the *tbpA* genes.

P51. Conservation and Antigenic Cross-Reactivity of the Transferrin Binding Proteins of *Haemophilus* and *Actinobacillus*.

T. Parsons*, J. Holland and P. Williams, Department of Pharmaceutical Sciences, University of Nottingham, United Kingdom.

In extracellular body fluids, *Haemophilus influenzae* acquires iron from the iron-transporting glycoprotein transferrin via a receptor mediated process. This involves two iron-regulated outer membrane proteins referred to as TBPI and TBP2 which show considerable preference for the human form of transferrin. Since the TBPs have attracted considerable attention as potential vaccine components, we used transferrin affinity chromatography of detergent solubilised membranes together with SDS-PAGE to examine their conservation amongst *H.influenzae* type b strains belonging to each of the recognised outer membrane subtypes as well as non-typable strains. Virtually all strains yielded a TBPI of 105 kDa and a TBP2 of 90 kDa which on immunoblots cross-reacted with a monospecific antibody raised against the denatured 68 kDa TBP2 of *Neisseria meningitidis* and with an antiserum raised against the *H.influenzae* native TBPI/TBP2 complex. In addition, these antibodies cross-reacted with the TBPs of the *Actinobacillus pleuropneumoniae* even though this porcine pathogen recognises pig but not human transferrin. These data demonstrate the existence of shared epitopes on the TBPs of

H.influenzae, N. meningitidis and *A.pleuropneumoniae* despite their transferrin species specificity.

P52. Intracellular survival of *Haemophilus somnus* in bovine blood monocytes: effect of cytokine and LPS treatment.

S. Gomis[1,2]*, W. Chen[1], D.L. Godson[1], H. Hughes[3], A Potter[1,4], [1]Veterinary Infectious Disease Organization, [2]Department of Veterinary Pathology, University of Saskatchewan, Canada, [3]M6 Pharmaceuticals New York, USA and [4]Canadian Bacterial Diseases Network.

The interactions between bovine blood monocytes (BBM) and *H. somnus* are known to be complex. In order to study this interaction further, a colorimetric assay using MTT was developed to assess the survival of *H. somnus* within cultured BBM. By using this system, it was found that (1) *H. somnus* was able to survive within BBM *in vitro* and the kinetics of its survival was similar to that seen in BBM isolated from experimentally infected cattle; (2) treatment of BBM with varying concentrations of rBoIFN-γ, rBoTNF-α, rBoIL-1β or rBoGM-CSF had no effect on the survival of *H. somnus;* and (3) treatment of BBM with E. *coli* LPS inhibited their survival. These results suggest that the ability of *H. somnus* to survive in both untreated and cytokine-treated BBM could be an important virulence mechanism, and that LPS is able to trigger BBM to control *H. somnus*.

P53. Inhibition of the Growth of Biotype A Strains of *Pasteurella haemolytica* by Sheep Blood

R.P. Lacroix* and R.P. Jenkins, Agriculture and Agri-Food Canada, Biologics Evaluation Laboratory, Food Production and Inspection, 3851 Fallowfield Road, Nepean, Ontario, Canada K2H 8P9.

Commercial preparations of defibrinated sheep blood inhibited the growth of biotype A strains of *Pasteurella haemolytica* but not T biotypes. This inhibitory activity was demonstrable only in the poured agar medium containing either of two blood lots and was present for up to three months at 4°C. Subsequent batches of plates prepared using the same lots of blood which had been stored an additional two weeks at 4°C, did not retain inhibitory activity. To determine the cause of inhibition, the potential role of serum antibody was examined by the use of immunoblotting and the Rapid Plate Agglutination (RPA) technique. Although inhibitory blood lots contained antibodies to 16 *P. haemolytica* serotypes (representing both biotypes) as shown in immunoblots, the serum of non-inhibitory lots yielded similar results. Serum fractions from inhibitory and non-inhibitory lots were indistinguishable by RPA. Based upon previously documented evidence that A and T biotypes of *Pasteurella haemolytica* are differentiated by their sensitivity to penicillin and observations that the inhibitory lots would not support the growth of other penicillin sensitive organisms such as *Bacillus subtilis* and *Pasteurella multocida*, we suggest that this antibiotic may have been responsible for this observed phenomenon.

P54. Synergy between an ovine isolate of *Bordetella parapertussis* and *Pasteurella haemolytica* A2 in mice.

J.F. Porter, C.S. Mason, N. Kreuger, K. Connor and W. Donachie, Moredun Research Institute, 408 Gilmerton Road, Edinburgh, EH17 7JH.

When administered intranasally to Swiss-white mice an ovine isolate of *Bordetella parapertussis* caused lung lesions of a severity which was dependent upon the infecting dose. While there were few overt clinical signs of disease, histopathological examination of the lungs revealed a catarrhal pneumonia characterized by hyperaemia, haemorrhage and oedema. *Pasteurella haemolytica* on its own was not pathogenic and was rapidly cleared from the lungs. Pre-infection of mice with 7.5 $\times 10^6$ colony-forming units (cfu) of *B. parapertussis* followed by intranasal infection with 1.4 $\times$ 10^5 cfu of *P. haemolytica* resulted in increased severity of the lung lesions. The timing between the administration of

B. parapertussis and *P. haemolytica* appeared to be important in determining the severity of the disease with the most severe lesions observed when *P. haemolytica* was administered three days after *B. parapertussis*. In addition to the Swiss-whites a second strain of mouse was used in a similar disease protocol. When *B. parapertussis* (6 x 10^7 cfu total) was administered to Porton mice 33% died within 48h and the organism was isolated from the lungs along with another bacterium. This bacterium (closest identification by API strips: *Actinobacillus ureae*) was also found in the upper respiratory tract, but not the lungs, of control uninfected mice and in lower numbers than observed in the infected groups. Therefore it appeared that the damage to the respiratory tract caused by *B. parapertussis* may have allowed the *A. ureae* to penetrate deeper and to colonise the lung, causing disease.

P55. Pneumonic Pasteurellosis in Sheep and Goats in Malaysia.

A.R. Bahaman, Faculty of Veterinary medicine and Animal Science, Universiti Pertanian Malaysia, 43400 Serdang, Selangor, Malaysia.

Pneumonic pasteurellosis is a major constraint to the development of the sheep and goat industry in Malaysia. The disease is endemic and is responsible for 39 percent of the mortalities seen in the small ruminants in the country. Imported vaccines were used to control the infection but with limited success. Subsequently, an oil adjuvant combined pasteurella vaccine incorporating local strains of *Pasteurella multocida* and *P. haemolytica* was developed and showed considerable success in controlling the disease in certain areas of the country. Bacteriological study done showed the presence of five serotypes prevailing in different parts of the country. The major serotype seen in the country is A2 (37%) followed by serotypes A7 and A1. Protein profile of the isolates on polyacrylamide gel electrophoresis irrespectively of their origin (animal host, organs and geo-location) showed no distinct difference amongst them. However, on immunostaining revealed 5 to 7 major proteins mainly in the higher molecular weight region that were immunogenic.

CONTRIBUTORS

L.A. Babiuk
Veterinary Infectious Disease Organization, 124 Veterinary Road, Saskatoon, SK, S7N 0W0

M. Bisgaard
Department of Veterinary Microbiology, The Royal Veterinary and Agricultural University, DK-1870 Frederiksberg C, Copenhagen, Denmark

M. Campos
Veterinary Infectious Disease Organization, 124 Veterinary Road, Saskatoon, SK, S7N 0W0

A.W. Confer
Department of Veterinary Pathology, College of Veterinary Medicine, Oklahoma State University, Stillwater, OK 74078 USA

L.B. Corbeil
Department of Pathology, University of California, San Diego 92103-8416

K.D. Clinkenbeard
Department of Veterinary Pathology, College of Veterinary Medicine, Oklahoma State University, Stillwater, OK 74078 USA

C.J. Czuprynski
Dept. Pathobiological Sciences, School of Veterinary Medicine, University of Wisconsin, Madison, WI 53706

M.C.L. de Alwis
Veterinary Research Institute, PO Box 28, Peradeniya, Sri Lanka

W. Donachie
Moredun Research Institute, 408 Gilmerton Road, Edinburgh EH17 7JH, Scotland.

B. Fenwick
Department of Pathology and Microbiology, College of Veterinary Medicine, Kansas State University, Manhattan, KS 66506

S.K. Field
Department of Molecular Biology, Bristol-Meyers Squibb Pharmaceutical Research Institute, Princeton, NJ 08534 USA

J. Frey
Institute for Veterinary Bacteriology, University of Berne, Laenggasstr. 122, CH-3012 Berne, Switzerland

R.P. Gogolewski
P.O. Box 135, Ingleburn, NSW 2565, Australia

R. Harland
Veterinary Infectious Disease Organization, 124 Veterinary Road, Saskatoon, SK, S7N 0W0

S. Houghton
Vaccine Research and Development, Hoechst Animal Health, Walton Manor, Walton, Milton Keynes MK7 7AJ

T.J. Inzana
The Center for Molecular Medicine and Infectious Diseases, Virginia Polytechnic Institute and State University, Blacksburg, VA 24061

S.D. Kirby
Department of Microbiology and Infectious Diseases, Faculty of Medicine, University of Calgary, Calgary, Alberta, Canada

R.Y.C. Lo
Department of Microbiology, University of Guelph, Guelph, Ontario, Canada N1G 2W1

J.I. MacInnes
Department of Veterinary Microbiology and Immunology, University of Guelph, Guelph, Ontario N1G 2W1, Canada

D.C. Morrison
Cancer Center, University of Kansas Medical Center, Kansas City, KS 66160 USA

M. Morsy
Veterinary Infectious Disease Organization, 124 Veterinary Road, Saskatoon, SK, S7N 0W0

G.L. Murphy
Department of Veterinary Pathology, College of Veterinary Medicine, Oklahoma State University, Stillwater, OK 74078 USA

J.A. Ogunnariwo
Department of Microbiology and Infectious Diseases, Faculty of Medicine, University of Calgary, Calgary, Alberta, Canada

M. Pollack
Department of Medicine, Uniformed Services University of the Health Sciences, Bethesda, MD 20814 USA

A.B. Schryvers
Department of Microbiology and Infectious Diseases, Faculty of Medicine, University of Calgary, Calgary, Alberta, Canada

P.E. Shewen
Department of Veterinary Microbiology and Immunology, Ontario Veterinary College, University of Guelph, Guelph, Canada N1G 2W1

L.R. Stephens
Victorian Institute of Animal Science, Attwood, Victoria 3049 Australia

L. van Alphen
Department of Medical Microbiology, Academic Medical Center, Room 162L, Faculty of Medicine, Meibergdreef 15, NL - 1105 AZ Amsterdam, The Netherlands

INDEX

T indicates a talk abstract
P indicates a poster abstract

GPSR Compliance
The European Union's (EU) General Product Safety Regulation (GPSR) is a set of rules that requires consumer products to be safe and our obligations to ensure this.

If you have any concerns about our products, you can contact us on

ProductSafety@springernature.com

In case Publisher is established outside the EU, the EU authorized representative is:

Springer Nature Customer Service Center GmbH
Europaplatz 3
69115 Heidelberg, Germany

www.ingramcontent.com/pod-product-compliance
Ingram Content Group UK Ltd.
Pitfield, Milton Keynes, MK11 3LW, UK
UKHW021333070726
13610UKWH00011B/56

9781489909794